CONTEMPORARY DEVELOPMENTS IN CHEMISTRY

Lectures delivered at Columbia University in the special course in Chemistry given in the Summer Session of 1926 on the occasion of the opening of the Chandler Chemical Laboratories.

NEW YORK
COLUMBIA UNIVERSITY PRESS
1927

CONTENTS

CONTENTS

SYNTHETIC ORGANIC CHEMISTRY
IN THE
STUDY OF ODOROUS COMPOUNDS

BY

MARSTON TAYLOR BOGERT

PROFESSOR OF ORGANIC CHEMISTRY
COLUMBIA UNIVERSITY

Lecture delivered at Columbia University in the special course on Contemporary Developments in Chemistry given in the Summer Session of 1926 on the occasion of the opening of the Chandler Chemical Laboratories

COLUMBIA UNIVERSITY PRESS

NEW YORK

1927

Printed from type. Published January, 1927

Printed in the United States of America

TABLE OF CONTENTS

SYNTHETIC ORGANIC CHEMISTRY IN THE STUDY OF ODOROUS COMPOUNDS

by

MARSTON TAYLOR BOGERT[1]

Phylogenetically the oldest of our five special senses, animal olfactory organs were well developed and efficient long before the human animal, as we now know him, was evolved. From the dawn of his history, therefore, man has distinguished between odors, and has derived enjoyment from those which appealed to him.

In ancient times, the then known world was ransacked for precious perfume substances, great caravans made long and perilous journeys across whole continents and galleys plied the seven seas on similar errands, for only the richest could afford such luxuries. The use of incense in religious worship, of perfumes and of perfumed unguents goes back no one knows how many thousands of years, and was an old and well established custom in the days of the ancient Persians, Babylonians, Assyrians, Egyptians and Hindus, as is attested by primitive inscriptions, hieroglyphics and tablets. Certain of the vases found in the recently opened tomb of the pharaoh, Tut-ankh-amen, it is believed served to contain such unguents.

Nearly 3700 years ago, Joseph was carried away a captive by "a company of Ishmaelites" who "came from Gilead with their camels bearing spicery and balm and myrrh, going to carry it down to Egypt."[2] Some 1700 years later, at the beginning of our own Christian Era, when the three Wise Men of the East selected gifts most worthy of a new born king, they brought gold and frankincense and myrrh,[3] two of the three being perfumes regarded as of equal or even greater value than the pure gold itself. During the active ministry of our Saviour,[4] Mary anointed

1 Lecture delivered July 6, 1926, at the Columbia University Summer Session, in the Course on Contemporary Developments in Chemistry.
2 Genesis, XXXVII, 25.
3 Matthew, II, 11.
4 Matthew, XXVI, 6-9; Mark, XIV 3-5; Luke, VII, 37-50; John XI, 2 and XII, 3-8.

his feet with oil of spikenard from an alabaster box, so that "the house was filled with the odour of the ointment," and Judas Iscariot complained of such extravagance because he said that this perfume could have been sold for more than 300 denarii. Perfumes also were conspicuous among the priceless treasures sent by the Grand Vizier, Haroun-al-Raschid, as gifts to Charlemagne.

To-day, Scientific Research, the greatest and truest democratizer, has brought these precious perfumes and hundreds of others within the reach of all. Poor indeed must the woman be who cannot afford perfume in some form—in soap, cosmetics or extracts for the handkerchief.

Although of but very recent development, the scientific study of odorous compounds is one of the most fascinating and alluring in all the vast domain of organic chemistry. Everybody loves the flowers, and the perfume has been well described as the soul of the flower. Although the composition of these aromas is one of Nature's intimate and most closely guarded secrets, the prying and insatiable curiosity of the scientific investigator has finally solved many of these riddles and the composition of most natural perfumes is now known, at least approximately, to the organic chemist. Some he has been able to reproduce exactly, others he can but imitate.

Based upon the knowledge thus accumulated, and with the resources of synthetic organic chemistry at his command, he has proceeded to build up from simpler compounds, not only the odorous components of natural perfumes themselves, but also hosts of new aromatic substances which, so far as we are aware, do not occur in nature. They are made in man's laboratory, instead of in that of the plant, sometimes by the conversion of a natural product into a more valuable one, like the citral of lemon grass into the violet perfume ionone, and sometimes by building up the desired compound step by step, from the elements themselves or from relatively simple substances obtained from industrial wastes and by-products, from petroleum or from other cheap and easily accessible materials. The transmutation of such a foul black viscous mass as coal tar into delicious perfumes is one of his miracles, akin to his production of dyes of all the hues of the rainbow and of life-restoring drugs from this same unpromising raw material. Quite recently, he has taken that deadly gas phosgene, the chief killing gas of the late war, together with the castor oil so dear to our childhood, waved his magic wand over the mixture and from it there has emerged a beautiful new perfume, methyl heptine carbonate, possessing when dilute the delicious fragrance of fresh violets.

Odor

The perception of odor arises from stimulation of the olfactory, or first cranial nerve. Since its receptive surfaces are intimately associated with those of the fifth, or trigeminus nerve, which is concerned with sensation, motion and taste, it is often a difficult experimental problem to distinguish between true smells, i.e., those which stimulate the olfactory alone, and those which affect the fifth nerve as well, like acid fumes and other irritants.

The only satisfactory definition of a perfume is "a pleasing odor," and what pleases me may not please you at all. We all recognize the fact that such differences exist between nations, classes and individuals, and this is not merely a matter of olfactory experience or education, but undoubtedly is due in part to variations in the material composing the nerve substance which result in slightly different chemical reactions with the odorous substance and different products. The character of the stimulus transmitted to the olfactory area of the cortex therefore varies accordingly. Individual distinctions in the chemical composition of the human body are matters of common knowledge and account for our idiosyncrasies in the direction of tolerance or non-tolerance of food and drugs, and there is no reason why the olfactory nerve substance should be any exception to this general observation; but discussions of the different theories concerning odor, of the mechanism of the reactions involved, of the physiology or psychology of the sense of smell, cannot be undertaken here.

Odors are due to chemical changes which occur both in the animate (animal and vegetable) and in the inanimate or mineral world. Some, like hydrogen sulfide or the *n*-heptane of petroleum and of *Pinus sabiniana,* are common to both. The majority, however, are peculiar to one kingdom or the other. The same will be seen to be true when animal or vegetable odors are compared. A few, like the trimethylamine of herring brine and of the cotton plant, or the indole of human excreta and of orange blossoms, are found in both groups. Others are characteristic of one group, one order, family, genus or species.

Of the natural products of interest to the perfumer, all are of vegetable origin, with the exception of musk, ambergris, civet and castor. These animal products are among the most ancient of all perfumes. They have been known and prized for centuries, and the first two command high and steadily mounting prices, the best gray ambergris selling for considerably more than its weight in pure gold. Yet we are still ignorant of the molecular structure of their chief odorous components and hence the synthesis of the perfume or fixative principles of ambergris and of castor, of the muskone of genuine musk, or civetone of civet, remains unaccomplished, an enticing and lucrative prize awaiting the winner.* On the other hand, the odorous constituents of a large number of our perfume products of vegetable origin are well known, their molecular constitutions have been determined and their syntheses completed, and each year similar triumphs for the organic chemist are added to an already long list.

It is worth noting that all real perfumes are carbon compounds, *i.e.*, organic, and that the odoriferous inorganic substances are nearly always unpleasant and not infrequently toxic, like ammonia, hydrogen sulfide, carbon disulfide, hydrogen selenide, hydrogen telluride, arsine and phosphine.

Description and Classification of Odor

No sense is more marvelously delicate than that of smell, and minute traces of contaminants in a perfume cause a change in odor which is obvious even to the inexpert. In fact there is good reason for asserting that the discriminatory power of the sense of smell is greater than that of our other special senses of sight, hearing, touch or taste. We teach our young people to distinguish musical sounds and shades of color, harmony in musical and in pictorial art, but the vast possibilities of artistic enjoyment peculiar to our wonderful olfactory endowment they are left to learn in the slow and casual school of practical experience, with the result that even our language lacks specific descriptive adjectives by which different odors can be characterized in the same way that we can describe different colors and shades. Even the organic chemist himself, in the preparation of new compounds, either neglects to record their odors or is forced by the poverty of our vocabulary to employ such vague and meaningless terms as "peculiar," "characteristic" or "nondescript." There are masterpieces in olfactory art, just as in pictorial art, but the individuals whose training enables them fully to understand and appreciate the former are few indeed.

Smell has been not only a "neglected sense," but it has also been scorned and more or less despised by a considerable proportion of the male sex, on the ground that its cultivation was a frivolous and effeminate pursuit. This is not the place to offer any extended refutation of this criticism, but it may be pointed

* Since the above was written, the brilliant investigations of Walbaum (*J. Prakt. Chem.*, **113**, 166–70 (1926), Ruszicka (*Helvetica Chim. Acta*, **9**, 230–48, 249–64, 339–54 (1926)), Chuit (*Helvetica Chim. Acta*, **9**, 264–78 (1926)) and their co-workers have appeared, which throw considerable light upon the probable chemical constitution of both Civetone and Muskone, and have led to the manufacture of the new synthetic musk "Exaltone," now selling at $300 an ounce.—M. T. B.

out that if we can learn the connection between odor and molecular structure we shall have approached much nearer to the solution of the most important chemical problem of all, namely the connection between molecular structure and those physical and chemical changes which determine the complex cycle we call Life.

It follows from what has just been said concerning the difference in the reaction of individuals to the same odorous compound, that the description of a perfume will vary with the describer, and that physiological, psychological or esthetic classifications of odors are unsatisfactory. Comparisons of odor based upon the statements found in the literature are at best only approximate and may be quite misleading, even when the products under examination, as has not always been the case, were of high purity.

An exact and impersonal scientific classification of odors, comparable to that available for colors, seems clearly unattainable in the light of our present knowledge. The best that can be done is a more or less superficial grouping into types or classes which to the observer seem to bear a sort of family resemblance to one another, and we therefore speak of (for example) lily type, rose type, musk type, of odor, but the boundaries of these groups are both vague and variable.

Nature never standarizes her products, whether they be medicinals, dyes or perfumes, and the same plant species yields products which change with the changing conditions of climate, season, soil, temperature and many other factors; so that biological or botanical classifications are also open to objections. In the case of such natural products, the only practical standardization is a chemical one based upon the percentage of active principle present, be it drug, dye or perfume, and in standardizations of that kind the legal minimum invariably becomes the basis for the calculation of market value, although the final criterion will always remain its effect upon or appeal to an individual living organism.

Synthetic versus Natural Products

The inevitable outcome of this inconstancy of natural products, with its attendant difficulties in duplicating exactly the mixtures and blends prepared therefrom, together with the high cost and violent price fluctuations of many of these initial materials, has been to direct increasing attention to the synthetics and their development since the war has proceeded with steadily gathering momentum.

Not so many years ago, practically all dyes in common use were natural products and the announcement, in 1856, of Perkin's epoch-marking discovery of "Mauveine," the first of the so-called "aniline dyes," attracted considerable interest. The suggestion, however, that some day such artificial products might take the place of the natural ones was received with incredulity, scorn and ridicule in many quarters. As though man-made dyes could hope ever to displace such God-given tinctorial substances as madder, indigo and cochineal. The more pious of the community even attacked such efforts as blasphemous. Undaunted, nevertheless, the chemist continued to apply his own God-given intelligence to the solution of the problem and was soon able to show conclusively that the tinctorial power of these natural products resided solely in certain of their chemical components and that the rest usually was either wholly valueless or might be distinctly injurious. He then proceeded to prepare in his own laboratory, in one hundred per cent purity, these active dye principles and to study more closely the connection between molecular structure and tinctorial properties, with the result that tens of thousands of dyes are now available at low price, of every hue of the rainbow, and which do not occur in nature, while the best of the natural dyes, like indigo and the alizarin of madder, are now made from coal tar. The finest and fastest dyes known are not obtained from the animate world, but from the inanimate, not from plants or animals but from the laboratories and factories of the organic chemist.

As the direct consequence of the discovery by Graebe and Liebermann, in 1868, of a commercially profitable method of manufacturing from coal tar the alizarin which is the chief tinctorial constituent of the madder root, and within the brief period of ten years, the output of madder fell from over 500,000 tons per annum to a total of less than 500 tons, and this too in spite of attempted protective legislation on the part of the French government, for the industry was largely a French one, decreeing that none of the bright red trousers then worn by their soldiers should be accepted unless dyed with madder. The experience of natural versus synthetic indigo has been entirely similar, and many other illustrations might be cited whose consideration should convince anyone of the futility of attempting to substitute political protection for scientific investigation. The anxiety among our own wood alcohol distillers upon the arrival from Germany of the first consignment of synthetic methanol is still fresh in our minds. Compared to the great industry of synthetic dyes, that of natural dyes already has become relatively insignificant and is destined to shrink still further.

The history of medicaments is entirely parallel to that of the dyes. Although the development of synthetic drugs has not advanced quite so far, due largely to the greater difficulty and complexity of the problem, it is slowly but surely providing a really scientific materia medica to replace the shot-gun remedies and natural products of uncertain and varying composition upon which the physician of a few decades ago was entirely dependent. Even the enzymes, vitamins, endocrine secretions, immunizing sera, insulin and other vital products, must owe their physiological activity to certain definite chemical components, and when these constituents and their molecular structure are known their synthesis will follow. Think for a moment what it means to suffering humanity to have available such products of the organic laboratory as chloroform, ether, arsphenamine (salvarsan), procaine (novocaine), acetylsalicylic acid (aspirin), phenacetine, barbital (veronal), trional, luminal, cinchophen (atophan), and the like. In its fight against disease and death, medicine has no more potent or resourceful ally than organic chemistry and this fact is being recognized increasingly.

If it seems to you that I have dwelt over long upon the subjects of dyes and drugs, and have been guilty of wandering far afield from my text, my answer must be that the development of synthetic perfumes now actively under way is advancing along entirely similar lines and is writing a similar history.

But a few years since, while the use of synthetics in soaps and cheap perfumes was tolerated as inevitable, the addition of such products to a fine natural floral blend was regarded as *infra dig.*, and in a high-class and expensive toilet perfume such an interloper was made to feel that in more senses than one he was a "rank" outsider. Day by day this prejudice is being overcome, as new compounds of delicious fragrance emerge from the laboratory of the synthetic organic chemist, and the percentage of such products in all classes of perfumes is steadily mounting. It may sound heretical and even sacrilegious to some of my good friends among the perfumers, but the day is fast approaching when the great flower fields of southern France will be devoted mainly to supplying visitors to the Riviera with cut flowers, and as raw material for the manufacture of perfumes they will give place to the laboratories and factories of the synthetic organic chemist. To anyone familiar with the history of dyes and drugs, and in touch with the trend of recent progress in the perfume field, that conclusion is certain and ineluctable, and therein lies the great hope and promise for the future of the perfume industry of our own land, where labor costs are so high in comparison with other countries, for the percentage of labor entering into the cost of the synthetic is generally far less than that involved in growing, cultivating, harvesting and preparing floral products.

The great advantages which synthetic products possess over natural ones include the following:

(1) *Purity.*—The synthetic is a homogeneous individual substance, of fixed and definite physical and chemical properties, whose purity or adulteration is easily established; whereas the natural product is usually a mixture whose inconstant composition renders its integrity difficult of proof, the evidence of its sophistication often laborious and inconclusive, and the exact duplication of blends wellnigh impossible.

(2) *Contaminants.*—These are normal constituents of genuine natural products and are frequently distinctly injurious to the fine quality of the odor. They may, and often do, lead finally to resinification, darkening and decomposition.

(3) *Harvest.*—The horticulturist, like any other farmer, is dependent for his crop upon the uncontrollable vagaries and uncertainties of the weather, and must wage a never ending battle with plant diseases, parasites and other pests. His yields hence are uncertain and subject to wide fluctuations. Not so is the case of the synthetic, where process and yield are under control and relatively constant.

(4) *Labor.*—The advantage in favor of the synthetic product has been stated above.

(5) *Price.*—Natural products tend to rise in cost with mounting labor charges, and prices fluctuate with the character and amount of the harvest, as witness the French experience with jasmine, or our own this year with oil of peppermint. The price in the case of synthetics, *per contra,* varies much less and is consistently downward, as new or improved methods are discovered, increased manufacturing economies developed, and wastes and by-products utilized.

(6) *Interpretation of changes in odor.*—Alteration of odor due to such effects as ageing, light, heat, moisture, or the changes occurring in soaps and cosmetics from alkali or other ingredients, can be interpreted intelligently only when pure chemical compounds are studied, or all the components of a mixture and their relative amounts are known accurately.

(7) *Molecular structure and odor, fixative properties, etc.*—The connection between molecular constitution and odor, or fixative properties, manifestly can be established only by the use of pure compounds and not of indefinite mixtures.

(8) *Solvents.*—The same is true in the determination of the effect of solvents upon solutes, both physically and chemically.

(9) *Future Promise.*—Compared with the rich promise of the synthetic organic laboratory, that held out by natural substances is poor indeed. There is little likelihood of the dis-

covery of new animal perfumes of importance, and while unquestionably novel plant products will be introduced from time to time, their number will be small and will gradually decrease. Contrast the vast army of new and valuable synthetic dyes and drugs with the corporal's guard of useful natural products discovered and introduced within the same period.

The possibilities for new synthetics are limitless. We have scarcely crossed the threshold of this wonderful treasure house of the organic chemist, crowded with splendid opportunities for service, through scientific investigation, to chemistry and medicine, to industry and art, to civilization and progress. Who can assess the sum total of human enjoyment resulting from the utilization of the beautiful synthetic perfumes already available?

The simple compounds which function as initial materials for the synthetic organic chemist, like carbon monoxid, methane, ethylene, acetylene, the lower acyclic alcohols, aldehydes, ketones, acids and esters, are daily growing cheaper and more easily obtainable; methods of synthesis are multiplying and becoming less difficult; and we are about entering upon a new era in the development of acyclic chemistry which may rival or surpass our recent era of coal tar chemistry, and which will mean new perfumes, new solvents and new products of all kinds.

How far we have advanced along this road, which we believe leads to the final solution of the problem, will be evident if we consider briefly the factors upon which odor appears to depend, and then trace the influence of these factors, so far as we may, in the case of one or two groups of organic compounds.

Factors upon which Odor Appears to Depend

These may be presented briefly as follows:

I. Psychological and esthetic factors.

II. Physiological effects whose physical or chemical explanation is still obscure.

III. Physical factors, such as vapor pressure, solubility and partition coefficients between water and lipoids, dilution and ionization.—To possess an odor, a compound must be sufficiently volatile to come into actual physical contact with the olfactory organs in the nasal cavity, and sufficiently soluble in both water and lipoids to be taken up by the moisture of the olfactory epithelium and to react with the nerve substance as pointed out by Backmann[5] in 1917.

The concentration of a substance, or the total weight of odorous material reaching a given area of the olfactory organs in unit

[5]Backmann, J., *Physiol. Path. General,* 1917, 17, 1. See also Kremer, J., *Chem. Soc.,* Abstr. 1917, i, 607.

time, is a factor of considerable moment, for some odors (like indole, scatole, civet or castor), which are disgusting and revolting in ordinary concentrations, become agreeable if sufficiently diluted. The pleasing violet odor of ionone appears only in high dilution.

IV. Chemical factors.—To the organic chemist, the central problem is that of the influence of molecular constitution upon odor, for he is confident that its answer will enable him to synthesize any desired odor at will; just as the determination of the connection between molecular structure and tinctorial or physiological properties has pointed the way to the synthesis of new dyes and drugs. In spite of our very imperfect and fragmentary knowledge in this field, definite connections of this kind have been established already, restricted in scope it is true, but sufficiently encouraging to assure us that we are on the right track and to beckon eagerly on. Sweeping generalizations are as yet unwarranted, and before any can be formulated intelligently much additional information must be gathered by the study of comparatively small groups of compounds closely related chemically.

(1) The architectural plan of the carbon skeleton composing the framework of the molecule, whether acyclic, mono- or polycarbocyclic.—That the nucleus is not the sole determining factor is manifest from the totally different odors of derivatives of the same nucleus, like EtBr, EtOH, EtSH and EtCOOH, or $PhNO_2$, PhOH, PhOMe, PhAc and PhCOOEt, and from the fact that derivatives of different nuclei may have similar odors, like the violet scent of the ionones, of the heptine carboxylic esters and of certain halogen substitution products of cymene compounds, or the camphoraceous odors of R_2NCH_2CN, $Me_2C(OH)CCl_3$, $MeCH(OH)CMe_3$, and of many other compounds from nuclei quite different from that of camphor itself.

The transition of an acyclic to a cyclic structure may result in an improvement to the odor, as in the conversion of the odorless pseudo-ionone to ionone; in a change in the opposite direction, as of citronellal to isopulegol; or of practically no noteworthy alteration whatever, as in the case of alpha-terpineol, ethyl-dimethyl-allyl carbinol, and dimethyl-allyl carbinol, the two acyclic carbinols having much the same odor as the cyclic terpineol.

(2) Unsaturation.—Unsaturations between C and C, whether olefinic or acetylenic, their number and location, are important. Chemically, physiologically and mentally, unsaturation is characterized by greater activity and mutability than the saturated and satisfied condition.

The influence of unsaturation will appear upon comparing compounds of similar carbon skeletons but differing in extent or location of the unsaturation, like menthane, menthene, limonene and cymene; menthol, terpineol and thymol; *p*-propyl anisole and anethole; mellilotin and coumarin; tetrahydroionone and ionone; hydrocinnamic aldehyde, cinnamic aldehyde and phenylpropiolic aldehyde; methyl heptine carbonate and methyl caprylate, or *n*-heptane with amyl acetylene.

Sometimes the change from saturation to unsaturation is not accompanied by any marked alteration in odor, as may be seen by contrasting undecylic alcohol and aldehyde with the undecylenic isologs.

How a change in the location of the unsaturation may affect the odor is well exemplified in the case of the terpenes and their derivatives, in the ionones and irones, or in the isomeric allyl and propenyl ethers like anethol and estragol, eugenol and isoeugenol, or safrol and isosafrol.

(3) Osmophores.—The presence of certain atomic groupings which have been designated, from analogy with the chromophores of the dye chemist or the pharmacophores of the drug chemist, as "osmophores," "odorophores" or "aromatophores." Those commonly associated with pleasant odors are called "euosmophores," those characteristic of bad odors "cacosmophores." To the former class belong the OH, OR, COH, COR, COOR, NO_2, CN and N_3 groups; to the latter, the SH, SR, CSH, COSR, analogous Se and Te compounds, and the NC group. Unsaturations between C and C, referred to in the foregoing paragraph, should also be included as osmophores. As in the case of unsaturations, the kind, number and location of the osmophores is material.

That the location of the osmophore has considerable bearing upon the nature of the aroma can be shown by comparing the hyacinth fragrance of *beta*-bromostyrene with the totally different odor of its lacrimating *alpha*-bromo isomer, or the pleasant odors of the *beta*-naphthyl ethers with the unpleasant *alpha*-naphthyl isomers.

Of the disubstitution products of the benzene series, agreeable odors seem to occur more frequently in the *o*- or *p*- than in the *m*-series, as may be illustrated by comparing anisic or salicylic aldehydes, salicylic or anthranilic esters, with their *m*-isomers.

Passing to the trisubstitution products in this same series, the 1:3:4 arrangement is found so often among the better known perfumes of this group that it has been termed by Cohn the "ideal" configuration. Thymol, eugenol, safrol, vanillin and heliotropin, illustrate the point.

Multiplication of osmophores may strengthen the odor, as happens in the case of the polynitro synthetic musks compared with their mononitro analogs; but in other instances it so alters the physical properties (especially the solubility and vapor pressure) that the odor is lost completely.

That the osmophore, like any of these other factors, is not the only determining influence is similarly manifested by the existence of compounds carrying the same osmophore but possessing different odors, like phenol and benzyl alcohol, and of others having similar odors but different osmophores, like nitrobenzene, benzaldehyde and benzonitrile, or vanillin, *p*-nitro- and *p*-cyano-guaiacol.

(4) Homology.—The odor of the immediately contiguous members of homologous series is generally of the same type, although that of widely separated members may be quite dissimilar. Increasing molecular weight is usually accompanied by increase in potential physiological effect, but this is offset by decrease in vapor pressure and in solubility. The effect observed is the resultant of these forces, so that the curve of activity rises at first with increase in molecular weight and then falls again as diminishing volatility and solubility make their influence felt. Comparisons between homologs should be made only when the compounds are strictly comparable structurally, and a primary should not be compared with a tertiary alcohol, or benzyl alcohol with phenol, or benzaldehyde with phenylacetic aldehyde.

Good illustrations of similarity of odor in homologs are furnished by the lower acyclic alcohols, mercaptans, ethers, aldehydes, ketones, acids, esters, amines, and the like, by nitrobenzene and its homologs, by benzaldehyde and the toluic aldehydes (but not by cumic aldehyde), by vanillin or coumarin and their nuclear methyl homologs, by the methyl and ethyl ethers of *beta*-naphthol, or the methyl and ethyl esters of salicylic acid.

(5) Substituents.—As applies in the case of other physiological effects, the odor of a compound varies with the number, location and character of the substituents present, whether these substituents happen to be osmophores or not. In the benzene series, for example, the *o*- and *p*-bi-derivatives are more apt to possess pleasant odors than the *m*-isomers, and in the tri-derivatives it is the 1:3:4 configuration which is found most often in perfumes, as has been pointed out for the osmophores themselves.

(6) Spatial configuration.—It has been observed repeatedly that stereoisomers exhibit quantitative or even qualitative variations in physiological effects; *d*-nicotine is much less toxic than the naturally occurring *l*-form, *l*-adrenaline of the suprarenal capsules is far more potent than the *d*-isomer, the *syn*-oxime of perilla aldehyde is not sweet at all although its *anti*-isomer is

3000 times as sweet as cane sugar. The existence of similar distinctions in odor between stereoisomers therefore is to be expected and is noted frequently.

Connection between Odor and Molecular Structure in Some Groups of Organic Compounds

How the synthetic organic chemist is attacking the problem will become evident as we consider a few specific illustrations.

The formulas used in the figures are skeletonized in most instances and represent only the bonds, it being understood that angles and termini are occupied by carbon atoms whose remaining affinities are saturated by hydrogen atoms. Thus, in Figure 1, the 2-methyl-3-ethyl-1,5-hexadiene and the limonene if written out in full would appear as follows:

$$CH_2 : CH \cdot CH_2 \cdot CH(C_2H_5) \cdot C(CH_3) : CH_2$$

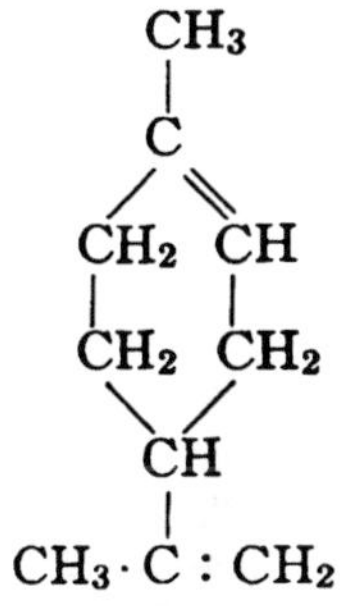

The Kekulé formula for benzene derivatives is employed for simplicity of architectural analogy with the hydrobenzenes and does not signify any preference for this particular representation of such molecules.

Hydrocarbons

Although most of the volatile hydrocarbons possess distinctive odors, and many occur in essential oils, diphenylmethane is about the only synthetic odorous hydrocarbon used to any considerable extent by the practical perfumer, and even this finds its chief service in soaps rather than in fine toilet perfumes. Yet the investigation of the hydrocarbons merits greater attention, since they may be regarded justly as the mother-substances of all other groups of organic compounds, and the nature of the carbon skeleton must be a fundamental factor in determining the physiological and chemical properties of all organic molecules.

The influence of unsaturation can be noted in the different odors of ethyl-benzene, styrene and phenylacetylene, or of menthane, carvomenthene, limonene and cymene; and the change due

FIG. 1.

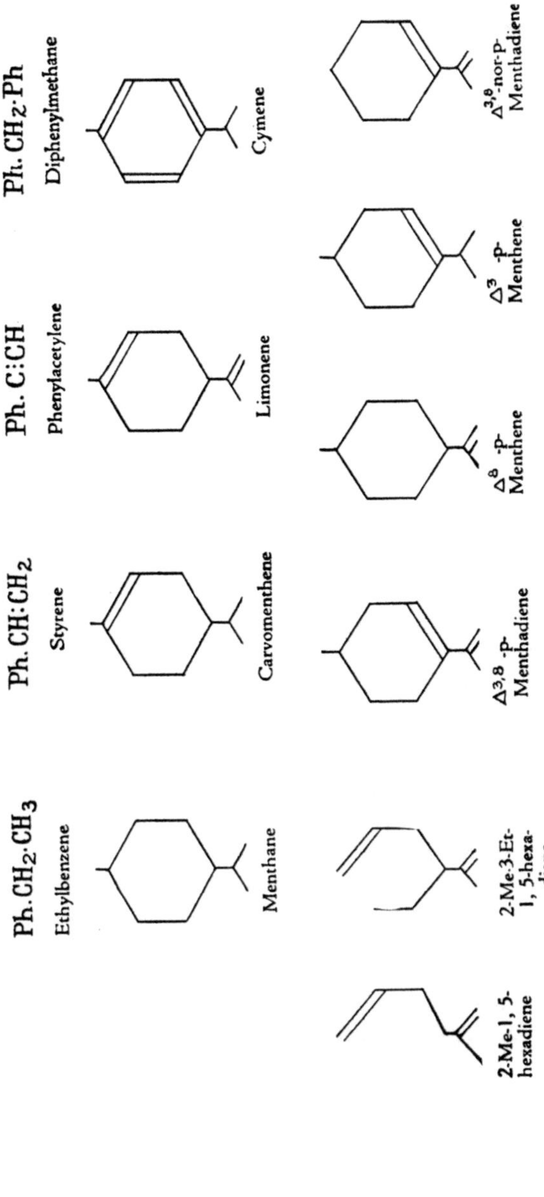

$Ph.CH_2 \cdot CH_3$
Ethylbenzene
$Ph.CH:CH_2$
Styrene
Ph. C:CH
Phenylacetylene
$Ph.CH_2 \cdot Ph$
Diphenylmethane
Menthane
Carvomenthene
Limonene
Cymene
2-Me-1, 5-hexadiene
2-Me-3-Et-1, 5-hexa-diene
$\Delta^{3,8}$-p-Menthadiene
Δ^{8}-p-Menthene
Δ^{3}-p-Menthene
$\Delta^{3,8}$-nor-p-Menthadiene

to replacement of a methyl by a phenyl group is manifest on comparing ethylbenzene with diphenylmethane.

In the menthenes and menthadienes, an excellent opportunity is presented of studying the effect of number and location of olefin bonds upon the odor of isomeric compounds with similar carbon skeletons. Determination of the exact location of these unsaturations is often difficult and the structure assigned to several of the cyclo-olefins and -diolefins therefore is still in some doubt.

The odor of 2-methyl-1,5-hexadiene is acrid and unpleasant, whereas that of its 3-ethyl derivative is pleasant and suggestive of lemon and peppermint. The odor of this latter compound may perhaps foreshadow the approach of its structure to that of limonene. Further, $\Delta^{3,8(9)}$-*p*-menthadiene has a more pronounced lemon odor than limonene (dipentene) itself, from which it would seem that olefin bonds in position 3:8(9) have more influence in this respect than when in position 1:8(9). On the other hand, the removal of one of the double bonds eliminates also the lemon odor, since $\Delta^{8(9)}$-*p*-menthene has but a faint parsley odor, and the weak aromas of Δ^{1}- and Δ^{3}-*p*-menthenes have nothing in common with lemon.

The methyl group in position 7 modifies but is not indispensable for the production of a lemon aroma, since the $\Delta^{3,8(9)}$-*nor-p*-menthadiene has a fragrance of similar type. Nor is the *para*-position a *sine qua non* for this smell, as it is encountered also in sylvestrene (carvestrene), the limonene of the *meta* series.

Alcohols, Phenols and Their Ethers

In the organic compounds containing only carbon, hydrogen and oxygen, are found the vast majority of vegetable products, and to this same class belong most of the important perfumes, natural and synthetic.

The chief groups of interest to the student of odors are the alcohols and phenols, their ethers, the aldehydes, ketones, acids and esters, characterized respectively by the osmophores —OH, —OR, —COH, —COR, —COOH and —COOR.

Compounds containing more than one free hydroxyl rarely give rise to the sensation of smell. This may be due to the fact that increase in the number of such groups is generally followed by increased solubility in water and decreased solubility in lipoids, with the result that the partition is so heavily in favor of the former that but very little of the compound dissolves in the lipoids or enters into reaction with the material of the olfactory nerve. Replacement of the hydroxyl hydrogen by a hydrocarbon radical, transforms the alcohol or acid into its ether or ester and has the opposite effect, the distribution then favoring the lipoid rather than the water, which likewise may be carried to the point where insolubility in water will preclude odor through failure of the moisture of the olfactory epithelium to take up any of the compound.

Among the numerous series of alcohols, phenols and ethers, only a few have been selected for purposes of illustration.

(1) *Citronellol Series.*—Citronellol has an especial attraction for us, because it was discovered in our own organic laboratories by F. D. Dodge[6] in 1889. It has been the subject of several researches, since it is one of the chief components of otto of rose.

Austerweil and Cochin[7] investigated the effect upon the rose odor of citronellol caused by substitutions in the methylene hydrogens of the terminal —CH_2OH group. They synthesized the mono- and di-methyl, mono- and di-ethyl, mono-propyl, mono-butyl and mono-phenyl derivatives, and reported that the mono-methyl derivative had an odor similar to that of citronellol but more pronounced, recalling tea roses; that the di-methyl was rather camphoraceous, perhaps due to its tertiary structure; that the mono-ethyl had a very fine rose odor, the di-ethyl a similar but stronger one, and the mono-phenyl a very strong one. From which they concluded that the rose odor probably resided largely in the —$CH_2 \cdot CR_2OH$ group, in which R = H, alkyl or aryl.

Quite recently, von Braun and Gossel[8] have synthesized various analogs of citronellol, involving retention of the methyl group in position 6, removing one or both methyls of the terminal —C: CMe_2, and adding methyls to the carbons corresponding to positions 5 or 7 in the original citronellol molecule.

Gildemeister tested the odors of the new products and characterized the odor of II as like hexenol, of III and IV as like mint, of V as of weak citronellol type, growing stronger in VI, still stronger in VII, but that VIII was the only one with a full citronellol aroma.

From these results it appeared that removal of one of the methyls of the —C: CMe_2 to position 5 greatly weakened the rose odor, to position 7 weakened it yet more, and its complete elimination obliterated the odor almost wholly. When both methyls of the —C: CMe_2 were removed, the rose odor disappeared entirely and could not be restored by introduction of methyls at 7, or 5 and 7. The methyls of the —C: CMe_2 group therefore are of much greater significance for the rose odor than similar groups in positions 5 and 7; from which it is argued, but

6 *J. Am. Chem. Soc.*, **11**, 456 (1889).
7 Austerweil and Cochin, *Compt. rend.*, **150**, 1693 (1910).
8 Von Braun and Gossel, *Ber.*, **57B**, 373 (1924).

Fig. 2.

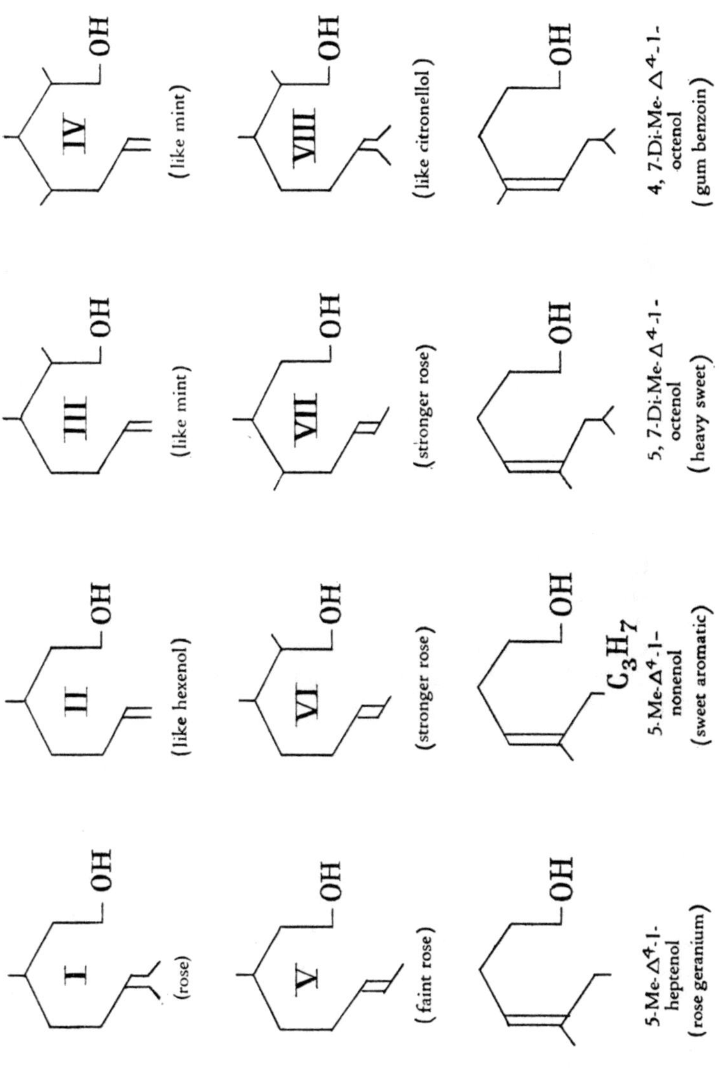
I
OH
(rose)
II
OH
(like hexenol)
III
OH
(like mint)
IV
OH
(like mint)
V
OH
(faint rose)
VI
OH
(stronger rose)
VII
OH
(stronger rose)
VIII
OH
(like citronellol)
OH
5-Me-Δ4-1-
heptenol
(rose geranium)
OH
C_3H_7
5-Me-Δ4-1-
nonenol
(sweet aromatic)
OH
5, 7-Di-Me-Δ4-1-
octenol
(heavy sweet)
OH
4, 7-Di-Me-Δ4-1-
octenol
(gum benzoin)

still remains to be proved, that the methyl group in position 6 also is probably of subordinate importance only.

The rose odor is lost by reduction to dihydro-citronellol, or by oxidation of the alcoholic OH. Hence, both the olefin bond and the alcoholic OH are essential factors.

The following odorous alcohols in this series were synthesized recently by Bogert and Slocum:[9]

5-Methyl-$\Delta^{4, \text{ or } 5}$-1-heptenol = pleasing odor of rose geranium type.

5-Methyl-$\Delta^{4, \text{ or } 5}$-1-nonenol = sweet aromatic odor.

5,7-Dimethyl-$\Delta^{4, \text{ or } 5}$-1-octenol = heavy sweet aromatic odor.

4,7-Dimethyl-$\Delta^{3, \text{ or } 4}$-1-octenol = sweet aromatic odor like gum benzoin.

Their fragrance was weak and not at all comparable in power or value with that of citronellol, indicating perhaps that the location of the olefin bond with reference to the OH is an important factor also. This assumption is supported by the fact that the aroma of 2,6-dimethyl-Δ^{5}-1-heptenol is fruity and not rose-like.[10]

Geraniol Series.—Paralleling their studies of citronellol, Austerweil and Cochin [11] conducted a similar investigation of geraniol analogs, replacing the methylene hydrogens of the terminal —CH_2OH group by various radicals, and reported the odors of these compounds to be as follows:

Geraniol = sweet, delicate and rose-like.
1-Methyl = pronounced geranium type.
1-Ethyl, 1,1-Diethyl = more like original geraniol.
1-Phenyl = geranium, strongly suggestive of roses.

From which observations they drew analogous conclusions regarding the significance of the methylene hydrogens in the —CH_2OH group. Because of the rosy shade of the mono-phenyl's fragrance, they were of the opinion that the more the influence of the radicals in position 1 predominated, the less became the influence of the adjacent olefin bond, although in general the effect of the introduction of the second unsaturation in passing from citronellol to the geraniol is to alter the fragrance from the rose to the geranium type.

Bogert and Slocum,[9] in the course of the investigation already cited, synthesized the 4,6-dimethyl-3,5-heptadien-1-ol and found that it exhaled an intense cedar-like fragrance.

9 Bogert and Slocum, *Am. Perf. Ess. Oil Review*, **18**, 621 (1924).
10 Semmler, *Die aetherischen Oele*, **I**, 249–250.
11 Austerweil and Cochin, *Compt. rend.*, **151**, 440 (1911).

FIG. 3.

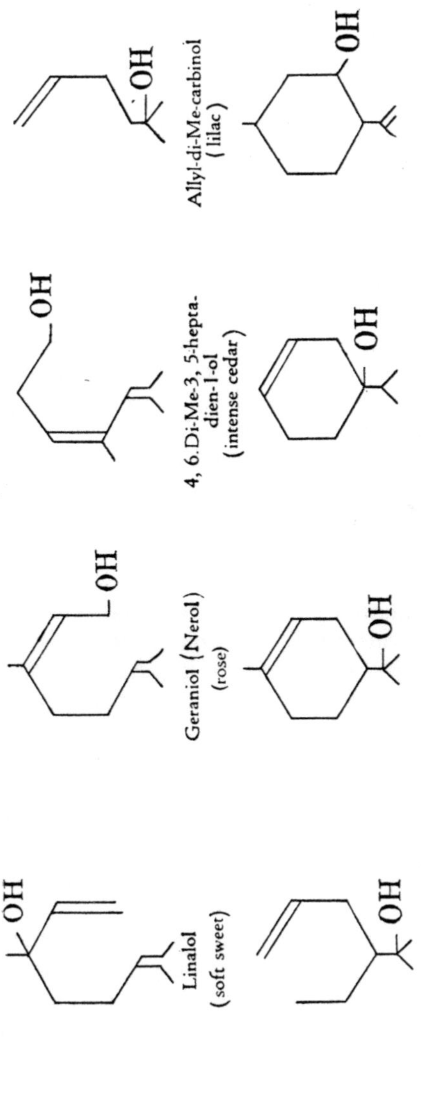

Linalol (soft sweet)

Geraniol {(Nerol) (rose)

4, 6.Di-Me-3, 5-hepta-dien-1-ol (intense cedar)

Allyl-di-Me-carbinol (lilac)

Allyl-di-Me-Et-carbinol (lilac)

α-Terpineol (lilac)

Terpinenol-4 (pine)

Isopulegol (like menthol)

Terpineol Series.—Terpineol has a fine lily-of-the-valley or lilac fragrance and is very extensively used as a basis of perfumes of this type. As ordinarily prepared, it is a mixture of various isomers in which the *alpha* form preponderates. Comparison of odor and structure of these isomers, however, must await further evidence as to their exact configurations and the aromas of the pure products.

Perkin and Pickles[12] have shown an interesting connection between structure and odor in the case of the three compounds represented in Figure 3, allyl-dimethyl-carbinol, allyl-dimethyl-ethyl-carbinol and *alpha*-terpineol, all of which have odors of the lilac type. Comparison of this series with that of 2-methyl-, and 2-methyl-3-ethyl-1,5-hexadiene and limonene (Figure 1), will show that in both cases final cyclization is not accompanied by any marked change in type of odor.

Terpinenol-4, which differs from *alpha*-terpineol in carrying its OH in position 4 instead of in 8, has a less agreeable odor.

Isopulegol, which resembles menthol structurally, with an olefin bond at 8(9), likewise resembles it in odor, but has a totally different aroma from citronellol or from citronellal, from the latter of which it can be obtained easily by cyclization. Carvomenthol and dihydro-carveol present a similar pair to menthol and isopulegol, but with the OH at 2 instead of 3 and with rather more agreeable odors.

Monohydroxy Phenols and Their Ethers.—Conversion of phenol into its methyl or ethyl ether marks the transition from the familiar carbolic smell to a much more pleasing one, and in diphenyl ether we discover an inexpensive, powerful though rather harsh geranium perfume, of remarkable stability, which has a widespread use in soaps.

p-Cresol retains much of the creosote smell of phenol, but its alkyl ethers are much more fragrant than the corresponding ethers of phenol and find frequent use in perfume blends.

The influence of the position of the OH on the odor is manifest on comparing carvomenthol with menthol, or carvacrol with thymol.

Anethole is characterized by its strong aniseed aroma, and either the terminal methyl of the propenyl group, or of the OMe group, can be replaced by ethyl without altering this odor appreciably. When the unsaturation in the side chain is moved out next to the end carbon, the odor is greatly weakened, estragol having but little of the aniseed quality of odor.

Ethers of Di- and Poly-hydroxy Phenols.—The most interesting ethers of the di-hydroxy phenols are those derived from

12 Perkin and Pickles, *J. Chem. Soc.*, **87**, 655 (1905).

FIG. 4.

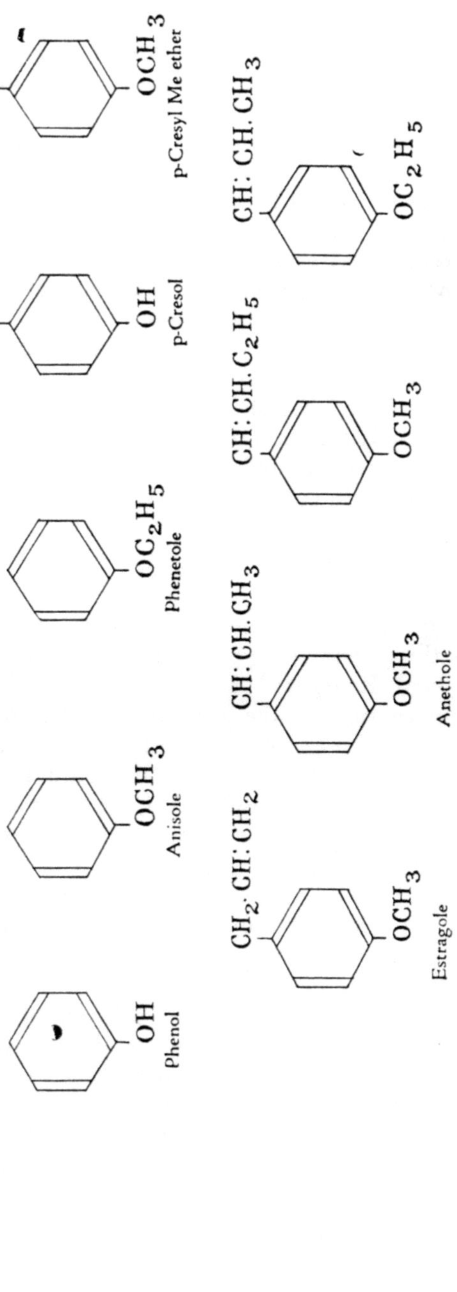
OH
Phenol
OCH 3
Anisole
OC2H5
Phenetole
CH 3
OH
p-Cresol
CH 3
OCH 3
p-Cresyl Me ether
CH2. CH: CH2
OCH 3
Estragole
CH: CH. CH3
OCH 3
Anethole
CH: CH. C2 H5
OCH 3
CH: CH. CH 3
OC2 H5

catechol. The beech tar creosote odor of guaiacol is much less in evidence in veratrol, and with suitable substituents *para* to the OH leads to valuable perfumes (iso-eugenol, vanillin, etc.). The eugenol of oil of cloves is readily rearranged to iso-eugenol, a product of carnation fragrance largely employed as a basis for perfumes of this type. This shifting of the olefin bond nearer to the nucleus is often accompanied by an improvement in the odor, as noted above and as in passing from safrol to iso-safrol. Chavibetol, the isomer of eugenol, with the OH and OMe groups reversed, has the odor of betel nut oil and not of cloves.

The striking difference in odor between methyl eugenol and the powerful sassafras aroma of safrol may have something to do with the cyclic structure of a methylene ether as over against one with two open OMe side chains.

Of the ethers of polyhydroxy phenols displayed in Figure 5, asarone is present in the essential oil of *Asarum arifolium*, and elemicin in that of Manila elemi. The occurrence of 2,3,4,5-tetramethoxy-allylbenzene, myristicin and apiol, in different parsley oils is interesting and suggestive. Dillapiol, as its name recalls, is present in certain dill oils.

Ketones

Perfumes of ketonic structure include such beautiful synthetics as the ionone violet products, due to the brilliant work of Tiemann and Krueger,[13] Mehrling and Welde,[14] and others. They are isomeric with the irone of the orris root.

As ordinarily prepared by condensing citral with acetone to pseudo-ionone, and then closing the chain to the two ionones (*alpha* and *beta*), the transition of the odorless acyclic pseudo-ionone to the fragrant cyclic ionones illustrates the effect of cyclization in this particular group.

Instead of condensing citral with acetone to pseudo-ionone, the citral may be closed up first to the cyclocitral and this in turn condensed with acetone. By selection of the appropriate cyclocitral, *alpha*- and *beta*-ionone and *alpha*- and *beta*-irone may be prepared, as shown in figure 6, all four of which have violet odors. Or, if acetone homologs are used, homologs of ionone and of irone result, of similar odor type. Further, in *alpha*-ionone, its $—CH:CH·CO·CH_3$ side chain may have its terminal $—COCH_3$ replaced by either —CHO, $—CR_2OH$ or $—COOCH_3$, or in *beta*-ionone by $—CO·CH:CMe_2$, without any change of odor type.

13 Tiemann and Krueger, *Ber.*, **26**, 2675 (1893).
14 Mehrling and Welde, *Ann.*, **366**, 119 (1909).

FIG. 5.

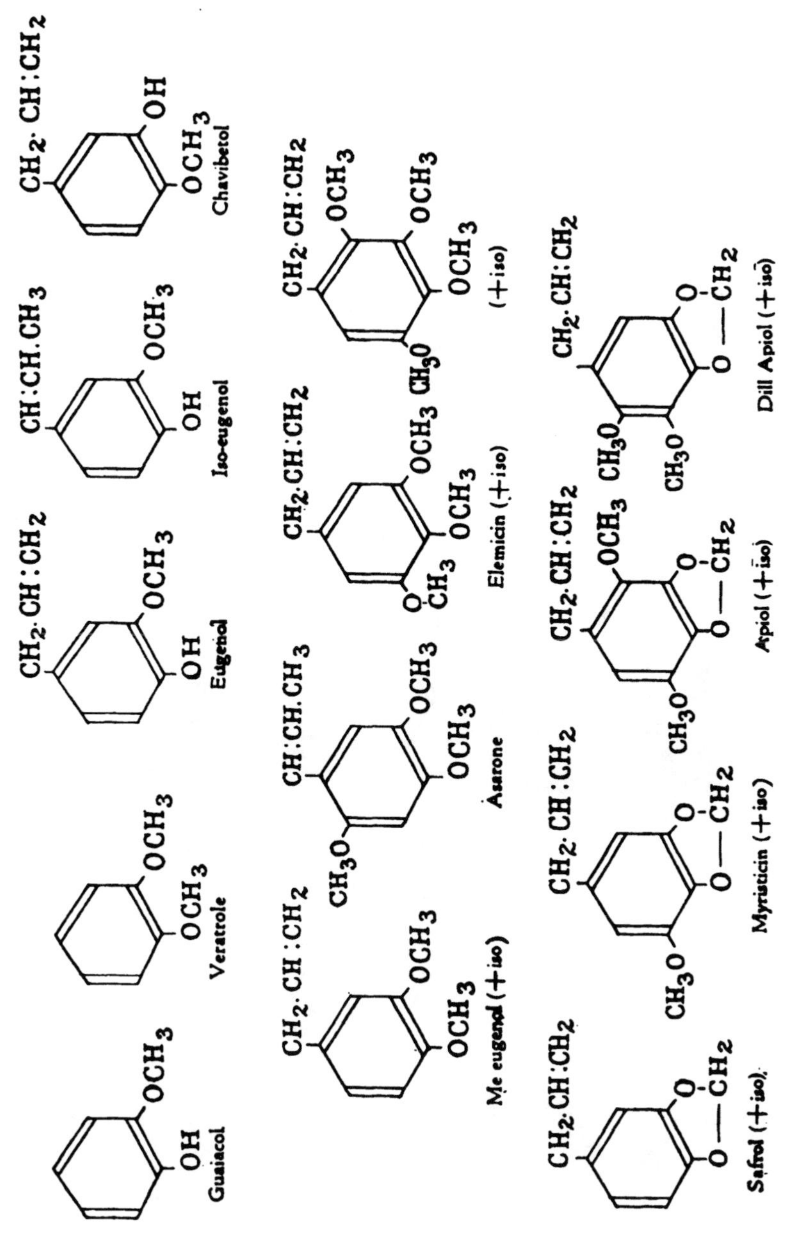
Guaiacol
Veratrole
Eugenol
Iso-eugenol
Chavibetol
Me eugenol (+iso)
Asarone
Elemicin (+iso)
(+iso)
Safrol (+iso)
Myristicin (+iso)
Apiol (+iso)
Dill Apiol (+iso)

Fig. 6.

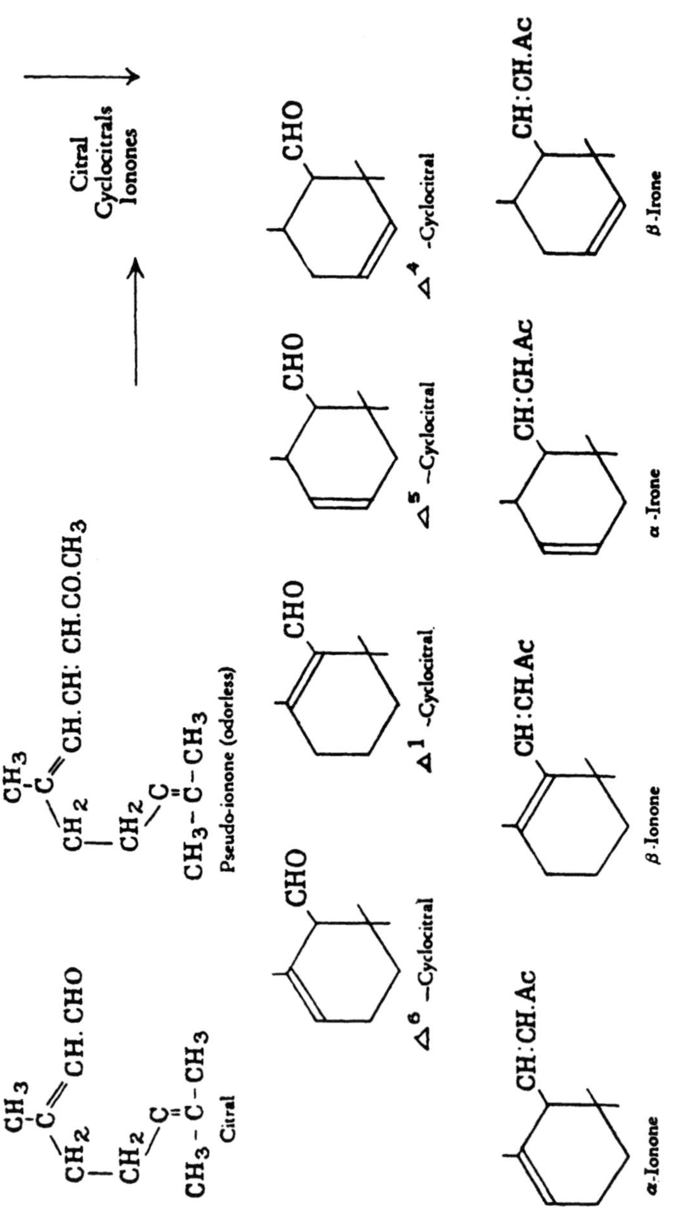
Citral
Pseudo-ionone (odorless)
Citral
Cyclocitrals
Ionones
CHO
CH:CH.Ac
Δ^6 -Cyclocitral
Δ^1 -Cyclocitral
Δ^5 -Cyclocitral
Δ^4 -Cyclocitral
α-Ionone
β-Ionone
α-Irone
β-Irone

Mehrling and Welde, in the prosecution of their skillful experiments to determine the cause of the violet odor in these ketones, found that by condensing the appropriate alhehyde with acetone, the ketones shown in Nos. IV–VII of Figure 7, were obtained, of which only the first had a violet odor, the smell of the others resembling fenchone and camphor; from which they concluded that it was essential to the production of violet perfumes in this series that the nucleus should be a 1,3-dimethyl-, or 1,3,3,-trimethyl-cyclohexene, that the aldehyde group should be adjacent to at least one of these methyls, and that the intensity of the violet odor increased with the number of methyl groups adjoining the aldehyde group.

They also condensed the corresponding aldehydes with acetone to the ketones shown in Nos. I–III of Figure 7, and discovered that the second and third of these had the violet odor, whilst the first was odorless.

To determine the part played by the $—CMe_2—\overset{|}{C}(CHO)—$ $\underset{}{\overset{|}{—CMe}}$-grouping, *i*-propyl butyl aldehyde was condensed with acetone, as depicted in Figure 7, and the product proved to have a floral, but not a violet odor.

The importance of the double bond in the side chain of ionone appears when H_2 is added to it, the dihydroionone having but a very faint cedar odor. The odor of tetrahydroionone is also of cedar type.

Nitro Derivatives

Another chapter wherein the contributions of the organic chemist have been conspicuous is that of the synthetic musks, a chapter due chiefly to the extensive and intensive investigations of Baur and his co-workers.[15]

Examination of the formulas exhibited in Figure 8 will disclose that the compounds having a good musk odor have the following in common: the tertiary butyl group at 1, a methyl at 3, and two nitros at 2,4 or 2,6, that is one or the other of the consecutive arrangements $Bu:NO_2:Me:NO_2 = 1:2:3:4$ or $1:2:3:6$, with other groups (preferably one an osmophore) occupying one or both of the remaining positions.

The architectural specifications for a good musk are therefore as follows:

Position 1 = tertiary butyl group.
Position 2 = NO_2 group.

15 Bauer et al, *Compt. rend.*, **111**, 238 (1890); *Ber.*, **24**, 2832 (1891); etc.

Fig. 7.

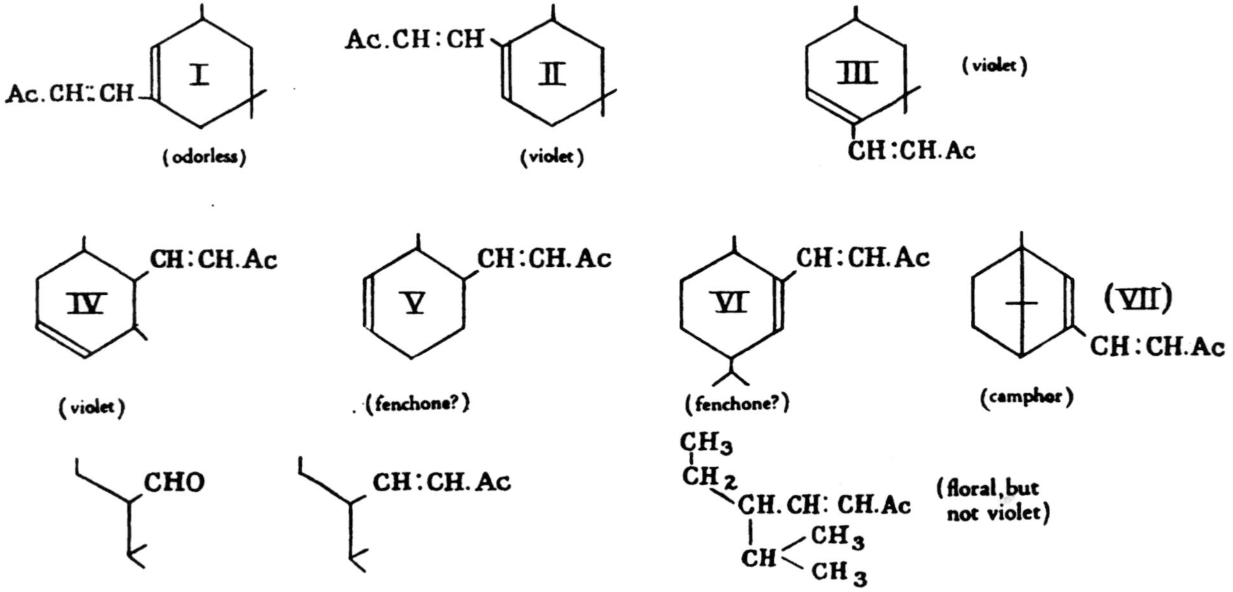

Addition of H_2 to side chain of Ionone = only very faint odor.
In Ionone, —Ac side chain replaced by —CHO, —COOMe or —CR_2OH = little change.
In Ionone, —Ac side chain replaced by —COCH:CMe_2 = little change.

FIG. 8.

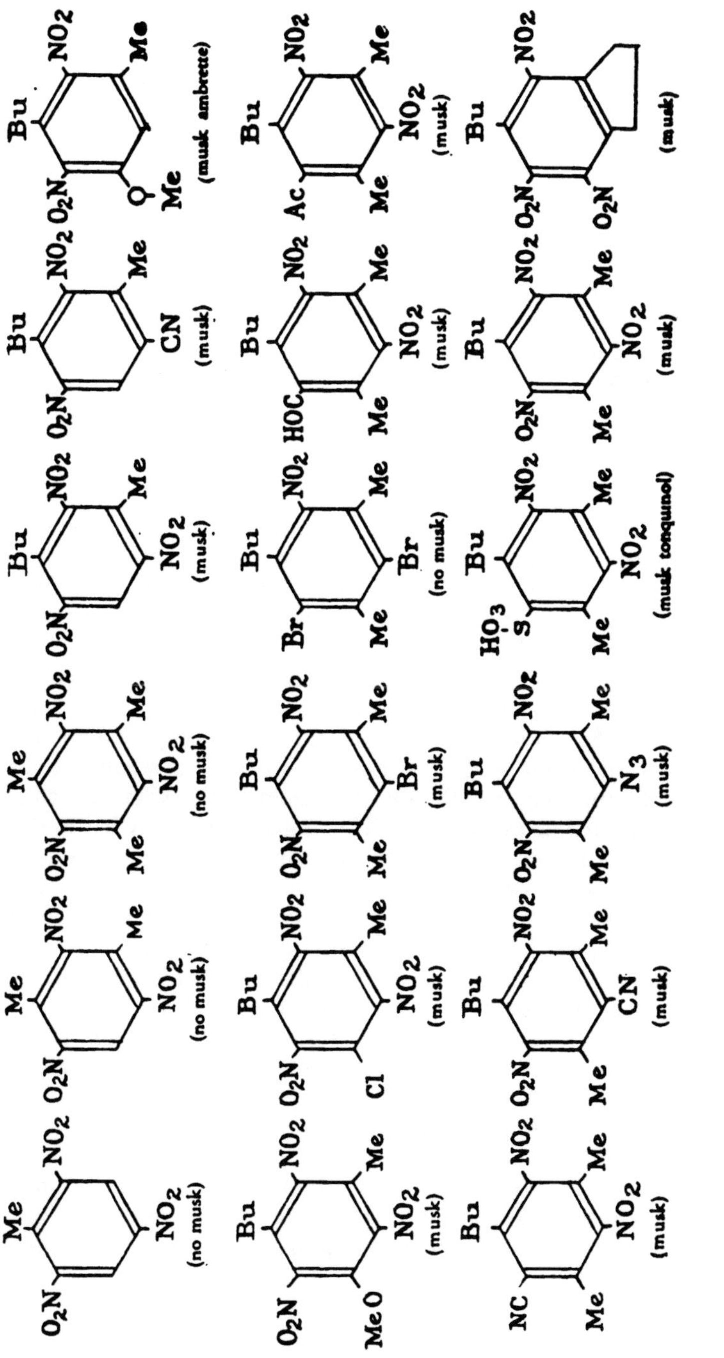

Position 3 = methyl group (or an equivalent hydrocarbon radical).

Position 4 = NO_2; or, when NO_2 at 6, H, halogen, CN or N_3.

Position 5 = H, halogen, Me or OMe.

Position 6 = NO_2; or, when NO_2 at 4, COH, $COCH_3$, CN or SO_3H.

Trinitro (1,3,5)-toluene, -*m*-xylene or -mesitylene, have no true musk odor.

Substitution of one of the essential nitro groups by any of the following, destroys the musk odor: —NH_2, —COOEt, —COPh, —CH : NOH or —CH : $CHNO_2$, or if there are halogens at both 2 and 4.

Benzothiazoles

In the course of some of our own studies in the benzothiazole and benzoselenazole groups, undertaken for the discovery of new synthetic drugs, we have encountered a striking instance of persistence of odor type as an attribute of a particular structural configuration.[16]

When 2-phenyl-benzothiazole was discovered by Hofmann,[17] 46 years ago, he was immediately struck by the remarkable fact that, although it was an unoxidized sulfur compound and of fairly high molecular weight, it possessed a very agreeable tea rose or geranium odor, which was more noticeable when the compound was heated or in solution. This unusual property led to the name "Rosenkoerper" for the product, by which it is still known.

Our experiments have disclosed the fact that this odor is in some way associated with the particular molecular structure of the compound, since the 2-phenyl group may be replaced by the furyl or thienyl groups, or the sulfur itself by selenium, without change of odor type. A methyl group at 4′ on the 2-phenyl weakens the odor, and nitro or amino groups in any part of the molecule, so far as our experiments have gone, eliminate it completely. When the 2-phenyl carries the characteristic groups of vanillin or of piperonal, the odor inclines toward that of these aldehydes.

On the other hand, benzothiazole itself has an unpleasant quinoline-like smell, and *p*-phenylene-*bis*-benzothiazole, probably because of its high molecular weight and low vapor pressure, is odorless.

16 Bogert and Stull, *Am. Perf. Ess. Oil Review*, **20**, 453 (1925).
17 Hofmann, *Ber.*, **13**, 8, 1223 (1880).

FIG. 9.

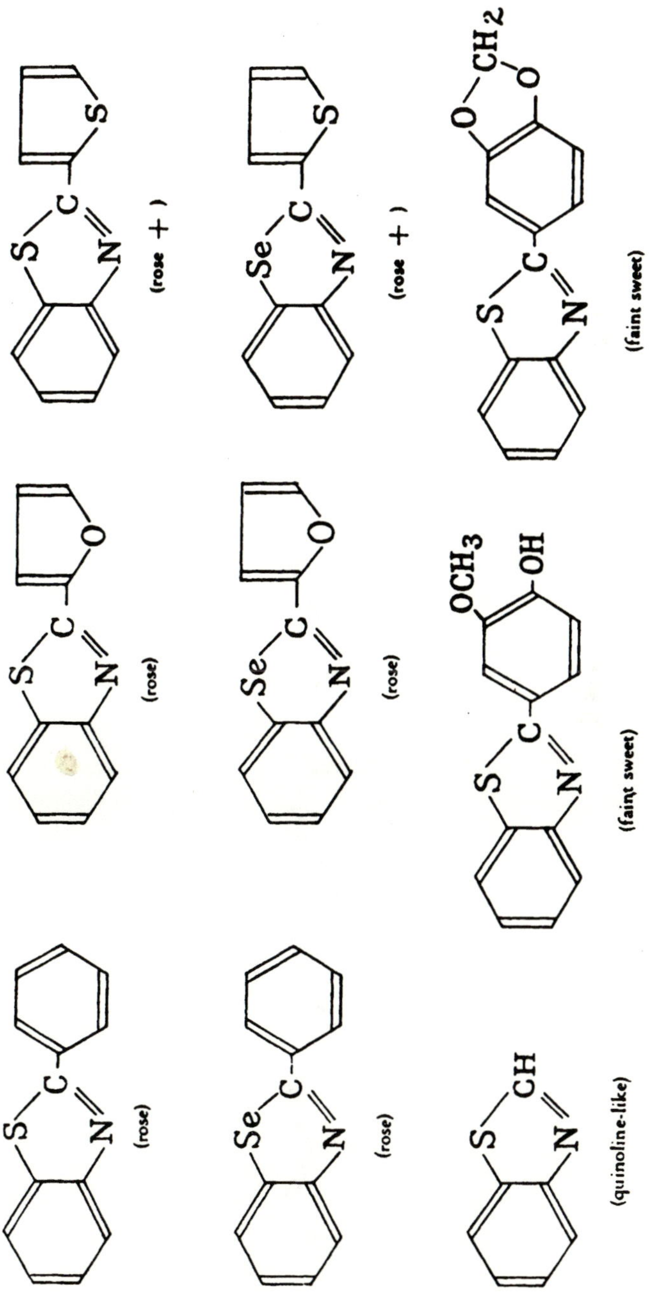

S
C
N
(rose)
Se
C
N
(rose)
S
CH
N
(quinoline-like)
S
C
N
O
(rose)
Se
C
N
O
(rose)
S
C
N
OCH3
OH
(faint sweet)
S
C
N
S
(rose +)
Se
C
N
S
(rose +)
S
C
N
O
CH2
O
(faint sweet)

FIG. 10.

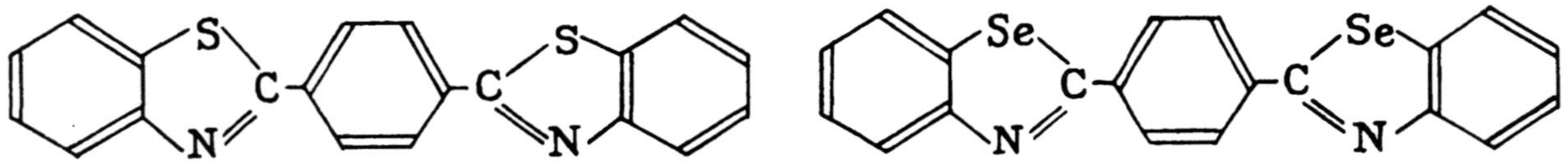

(odorless) (odorless)

In 2-phenyl-benzothiazole, Me group at 4′ weakens odor.
Nitro and amino benzothiazoles and benzoselenazoles are odorless, so far as kno wn compounds are concerned.

CHEMICAL REACTIVITY

BY

JAMES F. NORRIS
PROFESSOR OF ORGANIC CHEMISTRY
MASSACHUSETTS INSTITUTE OF TECHNOLOGY

Lecture delivered at Columbia University in the special course on Contemporary Developments in Chemistry given in the Summer Session of 1926 on the occasion of the opening of the Chandler Chemical Laboratories

COLUMBIA UNIVERSITY PRESS
NEW YORK
1927

Printed from type. Published March, 1927

Printed in the United States of America

CHEMICAL REACTIVITY

By James F. Norris

This lecture offers an excellent opportunity to sum up the work accomplished during the last few years at the Massachusetts Institute of Technology in an endeavor to get a more definite and, if possible, a quantitative idea of chemical reactivity. At the outset it is well to make clear what is meant by the term as applied in organic chemistry.

All chemists in this field recognize the fact that the members of a group of compounds containing the same type of bonding between the atoms, differ markedly in the ease with which the several members react with some fixed substance. We know, for example, that benzyl chloride is more reactive than ethyl chloride; the replacement of a phenyl by a methyl radical reduces the liability of the chlorine atom. The marked difference in the behavior of ethyl chloride and that of acetyl chloride we attribute to the replacement of two hydrogen atoms in the former by an oxygen atom. The difference is so great that we put the two compounds in distinct classes—alkyl halides and acyl halides—and rarely consider them as analogous compounds. We shall see, later, that this differentiation is an entirely arbitrary one; because the reactivity of the chlorine atom in a compound of the alkyl type can be made so great by proper substitution, that it exceeds the reactivity of the chlorine atom in certain acyl chlorides.

The fundamental property in all the cases cited is the lability of the carbon-chlorine bond; and this can be varied within the widest limits in compounds of the same type by effecting changes within the molecule.

The ease with which any compound enters into reaction with other substances is a fact of the greatest significance in the consideration of what we call chemical properties. All transformations between organic compounds are based upon the ability of the atoms in the reacting substances to enter into new combinations. Whether or not a transformation takes place in a reasonable time and the desired synthesis is effected, is determined by the reactivities of the atoms involved.

The energy change that takes place in the transformation is, of course, of great importance but it must be remembered that the free energy change in a reaction is a measure of the so-called tendency of the reaction to take place; it does not tell us whether or not the reaction will proceed under the conditions of the experiment. Much light will be thrown on organic chemistry when we are able to study reactions from the standpoint of thermodynamics, but the results will not be so immediately applicable as those to be obtained from the closer examination of the factors influencing chemical reactivity.

The Quantitative Measurement of Chemical Reactivity

About ten years ago we undertook to develop a method of studying chemical reactivity from a quantitative point of view. Any property can be appreciated definitely and used scientifically only when it can be expressed as a number which can be compared with another number that is the measure of the same property of another substance. Physical properties can be readily measured, and, as a result, we have a wealth of data of the greatest value.

The quantitative measure of chemical properties is more difficult, but when such measurements have been made the results have proved to be of the greatest importance in the development of chemistry as a science. The modern conception of the changes involved in inorganic chemical reactions are based on the quantitative study of the energy changes in these reactions. The measurements of the heats of formation of compounds and of the free energy change in reactions evaluated in thermochemical or electrochemical units have given chemistry a high place among the sciences. The facts summed up in Faraday's law and in the electromotive series of the elements made possible the broadest generalizations in regard to the chemical behavior of the elements and their compounds.

Organic chemistry lacks such broad generalizations. The radicals in these compounds correspond to the elements in inorganic chemistry. Can these radicals be studied in some way to give results of a general nature that are as applicable in this field in the interpretation of reactions, as is the electromotive series of the elements in inorganic chemistry?

As has been indicated, the reactivities of the several atoms and groups in an organic compound are the factors that determine the use of the compound when any chemical transformations are involved. No absolute measure of reactivity can be made, but under well defined and fixed conditions it can be expressed as the rate at which a definite transformation takes place. For example, we can determine the rate at which benzoyl chloride reacts with ethyl alcohol; and we can determine the rates at which a number of substitution-products of the chloride react under the same conditions. The several results obtained will show clearly to what extent the replacement of a hydrogen atom in the benzene ring by a substituent affects the lability of the chlorine atom under the conditions used. We have learned a fact of importance when we discover, for example, that the replacement of the meta hydrogen in benzoyl chloride by a nitro group so affects the bonding between the carbon and chlorine atoms that the halogen reacts seventeen times as fast in one case as in the other—the substitution brings about increased lability. Such quantitative facts must be considered and will prove of the greatest significance when we attempt to study and interpret the forces and stresses between the atoms in molecules.

The speed with which any chemical transformation takes place between two molecules is profoundly affected by the conditions under which the reaction takes place. In order to draw trustworthy conclusions as to how changes in structure in the members of a series of analogous compounds affect the lability of a definite atom present in all—in order that the results obtained may be comparable—it is necessary to know what factors outside the molecules studied affect the rate of the transformation.

Evidently attention must be paid to the temperature, the concentration, the presence of catalytic agents, and the effects of the products of the reaction and the solvent if one is used. All the reactions in the same series must take place according to the same mechanism.

When a number of similar reactions are studied under identical conditions, the results can be used to determine the relative effects of changes within the molecule on the lability of a bond that is present in all the compounds studied.

The changes in lability measured in this way are great; as a consequence it is possible to compare the effect of the several properties of the substituents on the tightening or loosening of the bond studied. The differences in lability resulting from the changing of the position of a particular substituent are also great, and it is consequently possible to study the effect of a substituent through a lengthening chain of carbon atoms.

Experimental Results

The atomic bondings that are now under investigation are the hydrogen-oxygen and the carbon-oxygen bonds in alcohols and the carbon-chlorine bond in alkyl and acyl chlorides. It is proposed to extend the work to other types of compounds.

The reaction selected for the study of the carbon-chlorine bond was that between the halogen derivatives and ethyl alcohols. The typical reaction in the case of the acyl chlorides is as follows:

$$C_6H_5COCl + C_2H_5OH = C_6H_5COOC_2H_5 + HCl$$

In the determination of the velocity-constant, the amounts of hydrochloric acid formed at definite time intervals were measured. Dilute solutions of the acyl chlorides—approximately 0.1 molar—were used, in order to have the conditions for a reaction of the first order.

The progress of the reaction was measured by determining the electrical conductivity of the solution from time to time. This method has many advantages: the solution is not brought in contact with air after it has been placed in the conductivity-cell; it avoids any side-reactions, such as hydrolysis, that would be involved if portions were removed for analysis; and it makes possible the measurement of very rapid reactions that could not be studied in the usual way. The reliability of the method was thoroughly tested. The results of the study of the conductivity of hydrogen chloride in ethyl alcohol checked with those reported in the literature. The effect of the presence of non-electrolytes was measured and taken into account. The details will be discussed fully in a future publication.

The alkyl chloride which would give the most interesting results when compared with benzoyl chloride is benzyl chloride: C_6H_5COCl and $C_6H_5CH_2Cl$. The results would show the effect of the replacement of one oxygen by two hydrogen atoms on the lability of the carbon-chlorine bond. This effect, however, is so great that it is impossible to study the two compounds under the same conditions in order to get comparable results. For this reason diphenylchloromethane, $(C_6H_5)_2CHCl$, was used as the typical member of the series of alkyl chlorides. The presence of the second phenyl radical labilized the carbon-chlorine bond to such an extent that the reaction could be studied at 25°.

After a detailed study of the method as applied to the two type compounds had been completed, a number of similar substitution-products were investigated. In the case of most of the acyl chlorides the reactions were allowed to take place at 0°. The derivatives of diphenylchloromethane were measured at 25°. The temperature coefficients of the two types of reactions were measured in order to compare the results in one series with those in the other.

The first order velocity-constants of the reaction between

diphenylchloromethane and ethyl alcohol were found to be 0.00288 at 25° and 0.00011 at 0°. For benzoyl chloride the values are 0.0493 at 25°, and 0.00549 at 0°. The significance of the relationship between these numbers will be considered later.

The results are given in the following table.

TABLE I

RELATIVE VELOCITY-CONSTANTS OF DERIVATIVES OF BENZOYL CHLORIDE AND OF DIPHENYLCHLOROMETHANE

The values given are relative, the typical member in each series being assigned the value 1. The velocity-constants used in the calculations were those determined at 0° for the acyl chlorides and at 25° for the alkyl chlorides. The concentration in the two cases was approximately 0.1 molar.

Substituent	Derivatives of C_6H_5COCl	Derivatives of $(C_6H_5)_2CHCl$
No substituent	1	1
o-Chloro	3.44	0.0095
m-Chloro	3.17	0.0451
p-Chloro	1.74	0.421
o-Methyl	3.32	2.86
m-Methyl	......	2.06
p-Methyl	0.567	18.76
p, p'-Dichloro	......	0.146
p, p'-Dimethyl	......	413
p-Ethyl	......	20.9
m-Bromo	4.08	
p-Bromo	1.93	0.331
p-Phenyl	......	12.8
o-Methoxy	18.7	92.5
p-Methoxy	0.79	3200
p-Phenoxy	......	31.6
α-Naphthyl	......	7.23
m-Nitro	17.2	
p-Nitro	17.7	

Phenyl acetyl chloride, 306.0
Benzyl phenyl chloromethane, 0.00042

Conclusions from the Study of the Carbon-Chlorine Bond

A number of other derivatives are now being studied, but the results tabulated above lead to several interesting conclusions.

(1) In the two series, the effects can be traced to at least two causes—the chemical nature of the substituent and the position of the substituent in the molecule relative to the carbon-chlorine bond.

(2) In the case of the acyl chlorides, position appears to have greater effect than the nature of the substituent. For example, the methoxyl group in the ortho position increases the constant to a value 18.7 times that of benzoyl chloride; whereas in the

para position the same group leads to reduced reactivity, the constant being 0.79 that of benzoyl chloride.

The relative effects of a positive methyl group and a negative chlorine atom on the reactivity of the carbon-chlorine bond is a further illustration of the importance of position. In the ortho position the effects of the two substituents are approximately the same, the constants being 3.44 for the chloro derivative and 3.32 for the methyl derivative. In the para compounds, where the effect of position is much less, the difference in the chemical nature of the substituent becomes evident: chlorine raised the constant (1.74) and the methyl group lowers it (0.567). In the case of the alkyl chlorides the effect of position is great, but the nature of the substituent is not overcome to as great an extent as with the acyl chlorides. The replacement of hydrogen in the ortho position in diphenylchloromethane by methyl raises the constant and by chlorine lowers it.

(3) The effect of the position of the substituent in the acyl chlorides decreases in the order ortho, meta, para.

(4) If we arrange substituents in the order of their decreasing effect on the lability of the carbon-chlorine bond, this order is different if we consider in one case ortho compounds and in the other para compounds. This fact again shows the importance of position—a fact overlooked in the past in the consideration of the relative effects of groups on the chemical properties of analogous compounds.

(5) In the acyl chlorides studied so far negative groups in all positions increase lability. Positive groups tend to decrease reactivity but this tendency is overcome when the group is in the ortho position.

(6) In the series of alkyl compounds the effect of the introduction of substituents is much greater than in the case of the acyl chlorides. The replacement of a hydrogen atom in the ortho position in diphenylchloromethane by chlorine reduces the velocity constant to about 0.01 of its former value; in the case of benzoyl chloride a similar substitution multiplied the constant by only 3.4.

(7) The effects of substituents on the two classes of compounds are in general, opposite. *m*-Chlorobenzoyl chloride reacts with alcohol 3.2 times as fast as does benzoyl chloride, whereas *m*-chlorodiphenylchloromethane under similar conditions reacts 0.04 as rapidly as diphenylchloromethane. The important fact illustrated by this example has been overlooked. The activating influence of a substituent is not a property of the substituent itself. The result of any substitution on the reactivity of an atom in the compound is determined by both the nature of the

substituent and the structure of the molecule into which it is introduced. The case cited above shows that the introduction of chlorine in benzoyl chloride makes the compound more reactive, but a similar substitution in diphenylchloromethane markedly stabilizes the carbon-chlorine bond.

The opposite effect of substituents in the two classes of compounds is illustrated further by the following examples: The introduction of bromine into benzoyl chloride labilizes the carbon-chlorine bond, whereas the analogous substitution-product of diphenylchloromethane is less reactive than the latter. The bromo-derivative of benzoyl chloride is more reactive than the chloro-derivative; the relationship is reversed in the corresponding derivatives of diphenylchloromethane.

The differences are even more marked when CH_2 is placed between the phenyl radical and the carbon linked to chlorine. The constant for phenylacetylchloride, $C_6H_5CH_2COCl$, is 306 times that of benzoyl chloride, C_6H_5COCl; the constant of benzyl phenylchloromethane, $\left.\begin{matrix}C_6H_5{-}CH_2\\ C_6H_5\end{matrix}\right\rangle CHCl$ is 0.00042 that of diphenylchloromethane, $\left.\begin{matrix}C_6H_5\\ C_6H_5\end{matrix}\right\rangle CHCl$. In these cases, again the effect is much greater in the alkyl derivatives.

(8) In the case of the derivatives of diphenylchloromethane it was possible to study the effect of the introduction of the same substituent in the same relative position in the two benzene rings. The constant of the *p*-chloro-derivative is 0.421 and of the *p,p'*-dichloro-derivative 0.146. In the case of the *p*-methyl and *p,p'*-dimethyl derivatives the constants are 18.7 and 413 respectively. The results are what might have been predicted, namely, the constant of the disubstituted derivative is of the order of magnitude of the square of the constant of the mono-derivative.

The Effect of Change in Temperature on Reactivity

The experimental results show clearly that diphenylchloromethane is a much more labile molecule than benzoyl chloride. Similar changes brought about in the two compounds result in effecting more profoundly the lability of the carbon-chlorine bond in the alkyl chloride. This fact is of particular interest in the light of the effect of change in temperature on the lability of the bond. The value of the velocity-constant for the acyl chloride at 25° is 9 times as great as at 0°, whereas in the case of the alkyl chloride the same rise in temperature increases the constant 26 times.

The temperature-coefficient of the reaction between benzoyl chloride and alcohol is approximately that of many reactions: the rate increases about 2.4 times for a rise in temperature of 10°. In the case of diphenylchloromethane and alcohol the increase is approximately 3.6 times for the same interval of temperature. This relationship will be studied farther in connection with an investigation of the specific heats of the two classes of compounds.

The Influence of Negativity on Chemical Reactivity

The replacement of hydrogen by a negative atom or group leads in the case of benzoyl chloride to increased reactivity of the carbon-chlorine bond and in the case of diphenylchloromethane to decreased reactivity. This fact points to the conclusion that the reactivity of the bond does not change progressively as the negativity of the group joined to the chlorine atom is increased.

From the available facts in regard to the reactivity of halogen compounds, it appears that as the negativity of the groups joined to chlorine increases, marked alternations in reactivity occur. If reactivity is plotted against negativity it is highly probable that the curve obtained would pass through a series of maxima and minima. It will be shown later in this paper that similar alternations in the reactivity of the hydrogen of the hydroxyl group occur in alcohols when a substituent is progressively moved away from the oxygen-hydrogen bond.

The radicals play a part in organic chemistry analogous to that played by the metals in inorganic chemistry. Our present theories in regard to the nature of atoms lead to the conclusion that the elements are built up by increasing the number of positive and negative charges of electricity they contain. This process carried out progressively results in the formation of elements whose physical and chemical properties change periodically. The negative nature of the element increases and then disappears; positive properties appear next, which progressively change to negative properties. It seems highly probable that similar alternations occur among the organic radicals as the negativity of the radical is progressively increased.

The Reactivity of the Hydroxyl-Hydrogen in Alcohols

In studying the influence of structure on the reactivity of the hydroxyl-hydrogen in alcohols, the rates at which the several alcohols reacted with *p*-nitrobenzoyl chloride were determined. The results obtained are given in Table II.

TABLE II

SECOND ORDER VELOCITY-CONSTANTS FOR THE REACTIONS BETWEEN ALCOHOLS AND *p*-NITROBENZOYL CHLORIDE

Temperature, 25°. Concentration of each reactant 1 mol in 1000 grams of ether.

Normal Primary Alcohols	*k*	Secondary Alcohols	*k*
Methyl	0.184	Isopropyl	0.010
Ethyl	.085	Sec.-butyl	.0074
Propyl	.066	Pentanol-2	.0059
Butyl	.074	Pentanol-3	.0036
Amyl	.079	Hexanol-2	.0065
Hexyl	.085	Heptanol-4	.0028
Heptyl	.069		
Primary Alcohols with a Branched Chain		**Tertiary Alcohols**	
Isobutyl	0.031	Tert.-butyl	0.0027
Methyl-2-butanol-1	.036	Tert.-amyl	.0025
Methyl-3-butanol-1	.073	Methyl-3-pentanol-3	.0014
Methyl-2-pentanol-1	.034		
Methyl-3-pentanol-1	.077	**Aromatic Alcohols**	
Methyl-4-pentanol-1	.068	Benzyl	0.017
Methyl-3-hexanol-1	.074	β-Phenylethyl	.040
		γ-Phenylpropyl	.020
		α-Phenylethyl	.0005
		α-Phenylpropyl	.0005
		α-Phenylbutyl	.0005

The variations in the velocity-constants tabulated above are so great that it is possible to draw definite conclusions as to the effect of the structure of an alcohol on the reactivity of its hydroxyl-hydrogen atom. In all cases the hydrogen atom was replaced by the *p*-nitrobenzoyl group. The replacement of this atom by a different type of group would give different results, but it is probable that the relationships between the several alcohols would be of the same order of magnitude. This point will be tested later. It is of interest in this connection to note that the velocity-constant of isopropyl alcohol is 11.7 per cent that of ethyl alcohol when measured by *p*-nitrobenzoyl chloride as a second order reaction in ethereal solution. When diphenylchloromethane reacts with the two alcohols with a large excess of the alcohol acting as solvent, the velocity constant of the reaction with isopropyl alcohol is 9.4 per cent that of the constant when ethyl alcohol is used. This fact indicates that the relationships pointed out below are of general significance.

Conclusions from the Study of the Reactivity of the Hydrogen of the Hydroxyl Group in Alcohols

(1) The lengthening of the carbon chain in normal primary alcohols results in a decrease in the reactivity of the hydroxyl-hydrogen in the first three members of the series; the constant increases progressively with the next three members and then decreases again.

In the case of the secondary alcohols with the hydroxyl group in position 2 the constant decreases with the first three members of the series and then increases.

(2) The introduction of a methyl radical as a side-chain has a marked effect on the constant. This effect is greatest in position 1; it decreases as we pass to positions 2 and 3 and begins to increase in position 4. The alternations in this effect are similar to, though greater than, those brought about by lengthening the chain—the periodicity is in cycles of 3.

(3) The percentage reduction in the constant brought about by the replacement of a hydrogen atom in position 1 by a methyl radical is of the same order of magnitude whatever the length of the chain. The constant of isopropyl alcohol is 11.7 per cent of the constant of ethyl alcohol; that of secondary butyl alcohol is 11.2 per cent of the constant of normal propyl alcohol; the constant of pentanol-2 is 8 per cent that of normal butyl alcohol; and the constant of hexanol-2 is 8.2 per cent that of the normal alcohol.

(4) When a methyl radical replaces a hydrogen atom in position 2 its effect on the constant is much less. The constant of isobutyl alcohol is 46.9 per cent of that of normal propyl alcohol; the constant of inactive amyl alcohol is 48.6 per cent that of normal butyl alcohol; and the constant of methyl-2-pentanol-2 is 44.3 per cent that of normal amyl alcohol.

(5) When a methyl radical replaces a hydrogen atom in position 3 the effect almost disappears. The constant of isoamyl alcohol is 98 per cent that of normal butyl alcohol; and the constant of methyl-3-pentanol-1 is 97 per cent that of normal amyl alcohol.

(6) When a methyl radical replaces a hydrogen atom in position 4 the effect now begins to increase. The constant of methyl-4-pentanol-1 is 86 per cent that of normal amyl alcohol.

(7) The relationships between the constants of primary and secondary alcohols containing the same number of carbon atoms are as follows: The constant of the secondary alcohols containing 3, 4, 5, and 6 carbon atoms are respectively 15.8, 10.0, 7.5, and 7.6 per cent those of the corresponding primary alcohols. In

this case the effect decreases with increasing molecular weight in the case of the first three members of the series.

(8) The relationships between the constants of tertiary and primary alcohols containing four and five carbon atoms are respectively 3.7, and 3.2 per cent.

(9) The effect of the position of the hydroxyl group in a straight chain of carbon atoms is great. The constants of pentanol-1, pentanol-2, and pentanol-3 are in the ratio 1 to 0.075 to 0.045.

(10) In the case of the aromatic alcohols there is an alternation in the value of the constant as the phenyl radical is removed from the hydroxyl group. The depression of reactivity produced by the replacement of a hydrogen atom by phenyl is so great the secondary alcohols containing this group in position 1 had the smallest constants measured.

The striking relationships between the constants of alcohols of similar structure indicates clearly that the results are significant in regard to the effect of the position of a substituent on the hydrogen bond in alcohols. It is hoped to interpret later the results in terms of the electronic theory of the bonding of atoms in organic compounds. Further work must be done, however, before the explanation of the results that suggests itself can be considered well founded.

The Reactivity of the Hydroxyl Group in Alcohols

The reactivity of the hydroxyl group in a limited number of alcohols has been studied, but much work remains to be done. The reaction selected for study was that between the alcohol and hydrobromic acid. It was found that as the reaction proceeded the water formed markedly decreased the rate, but that, if a certain minimum concentration of water were present at the start of the reaction, satisfactory constants could be obtained. It was found advisable to determine the rate of the reaction when definite quantities of water and of hydrogen bromide were dissolved in an excess of the alcohol. The reaction was thus one of the first order. The fact that the solvent used in each case was different—namely, the alcohol being studied—makes it impossible to compare the results obtained unless a correction is made for the influence of the solvents on the several rates. In the figures given in the table below these corrections have been made. The way in which the correction was made and other details will be given in a paper to be published later. The results are given in the following table.

TABLE III

FIRST ORDER VELOCITY-CONSTANTS OF THE REACTIONS BETWEEN CERTAIN ALCOHOLS AND HYDROBROMIC ACID

Temperature 100°. Concentration of alcoholic solution: water, 2 molar, hydrogen bromide, 0.16 molar.

Alcohol	*k*	Alcohol	*k*
Methyl	0.000679	Isobutyl	0.000084
Ethyl	.000260	Isopropyl	.000197
Propyl	.000442	Sec.-butyl	.000384
Butyl	.000406	Pentanol-2	.000330
Amyl	.000533	Pentanol-3	.000532

The following conclusions can be drawn from the results so far obtained.

(1) Lengthening the chain in both primary and secondary alcohols results in alternations in the reactivity of the hydroxyl groups.

(2) The effect of the position of a methyl radical as a side-chain is marked. The constant of isopropyl alcohol is 76 per cent that of ethyl alcohol; secondary butyl alcohol, 65 per cent of that of propyl alcohol; and pentanol-2, 81 per cent of normal butyl alcohol. The corresponding percentages in the case of the hydrogen reactivity of these alcohols are 11.7, 11.2, and 8 per cent respectively. If the methyl radical is in position 2, the effect is reversed. The constant for the hydroxyl reactivity of isobutyl alcohol is 19 per cent of that of normal propyl alcohol. In the case of these alcohols the hydrogen reactivity of the one having the branched chain is 46.9 per cent that of the normal alcohol.

These figures indicate clearly that a methyl radical has a greater effect on the hydroxyl group when in position 2 and on the hydrogen atom when in position 1. It is a striking fact that the two positions bear the same relation to the atom or group affected. When the methyl radical is in position 1 there are three bonds between it and the hydrogen atom: and when it is in position 2 there are three bonds between it and the hydroxyl group. In general, the determinations observed in the case of the reactivity of the hydroxyl group can be traced to the number of bonds between the substituent and the hydroxyl group. The greatest effects are shown when the number is an odd one.

Additional conclusions are indicated by the results: but measurements must be made with other alcohols before they are worthy of presentation.

The relationships established between different compounds lead to definite conclusions as to the effect of structure on reactivity. In the opinion of the author the work shows that the study of the rates of reactions can be used to help in the interpretation of chemical behavior. The conditions of experimentation can be so defined and regulated that results can be obtained with different substances which can be compared with one another. Many chemists have held that the rate at which a reaction takes place is so affected by conditions, many of which are unknown, that any conclusions drawn from velocity-constants must be open to question. It is hoped that the work outlined above is evidence of the falseness of this position. It is the belief of the author that the study of rates of reactions will bring us nearer to an understanding of the bonding between atoms and of how the attractions between the atoms are modified by changes within the molecule.

The data given in this paper were determined by C. Banta, A. A. Morton, J. F. Blake, C. J. Staud, A. A. Ashdown, F. Cortese, and H. D. Hirsch.

Chemical Relationships of Sugars, Optically Active Amino Acids, Hydroxy Acids, and Halogen Acids

BY

PHOEBUS A. LEVENE

MEMBER OF THE ROCKEFELLER INSTITUTE, AND
HEAD OF ITS DEPARTMENT OF CHEMISTRY

Lecture delivered at Columbia University in the special course on Contemporary Developments in Chemistry given in the Summer Session of 1926 on the occasion of the opening of the Chandler Chemical Laboratories

COLUMBIA UNIVERSITY PRESS
NEW YORK
1927

Printed from type. Published March, 1927

Printed in the United States of America

CHEMICAL RELATIONSHIPS OF SUGARS, OPTICALLY ACTIVE AMINO ACIDS, HYDROXY ACIDS, AND HALOGEN ACIDS

By P. A. Levene

It has been known for but a little over a century that certain crystals (such as quartz) have the power to rotate polarized light. Some have the property of dextro- and others of levorotation. The direction of rotation is determined by the crystal form. When such crystals are fused or dissolved, their solutions do not possess the properties of the solid crystals. On the other hand, crystalline substances are known, such as tartaric acid, which in the crystalline state are optically inactive, but in solution rotate polarized light either to the right or to the left. There occur substances which resemble one another in elementary composition, in many physical properties, in the character of their chemically active groups, but which differ in one property, namely, in their respective powers to rotate polarized light either to the right or to the left.

Thus, dextro- and levotartaric acids possess the same elementary composition, have the same number of hydroxyl and carboxyl groups, the same melting points, etc. The same may be said about the dextro- and levorotatory lactic acids. They possess identical reactive groups and identical physical properties such as densities, boiling points, etc.

The two classes of substances, one represented by quartz and the other by tartaric acid, must owe their optical activity to different causes. In the first case, the optical activity is due entirely to the crystal structure; in the second group the optical activity is the result of the structure of the molecule.

It is fortunate that the early work on optical activity of organic compounds was conducted on tartaric rather than on lactic acid. The former and its salts particularly have the power of forming macroscopic crystals and Pasteur made the discovery that the crystal structure of the levorotatory substance was the mirror image of that of the dextrorotatory. In other words, the crystals had an asymmetric structure and hence could not be superimposed. This discovery was the foundation of the stereochemistry of our own day, for it led to the idea of molecular asymmetry and to ideas of the three-dimensional structure of the atom in general.

It was stated above that with respect to their chemical properties and to their physical constants two enantiomorphous or antipodal substances are in the main similar. However, in one respect, each form behaves individually. Their individualities are brought out by the biological test. The differences in physiological effect of each of the two forms are very striking. Thus, dextroasparagine is sweet to the taste, whereas the levorotatory form is tasteless; the same is true regarding the dextro- and levorotatory glutamic acids. The active terpenes possess a stronger odor than the inactive forms. Not less striking are the differences in pharmacological activity of the enantiomorphous forms of certain drugs; for instance, levo-nicotine and levo-adrenaline are more potent than their antipodes, etc. Further, it is a well known fact that certain microorganisms are adapted to metabolize only certain forms and not their antipodes. Such differences of behavior have been observed in substances belonging to nearly every class of organic compounds—to alcohols, glycols, acids, hydroxy acids, amino acids, sugars, etc.

Recent years have brought out the fact that not only living organisms but also enzymes exert a selective effect on optical isomers. The best known instances are those in the group of sugars. Glucosides are known to exist in two forms, the so-called α- and β-forms. Such forms may be regarded as enantiomorphous isomers with regard to carbon atom (1). In this respect two enzymes are significant. One is extracted from ordinary yeast and is called "maltase"; the other from almonds and is known as "emulsin." The first hydrolyzes all α-glucosides regardless of the configuration of the sugar but does not hydrolyze the β-glucosides; the second hydrolyzes only the β-glucosides.

Similar selective action is known with regard to lipases and more recently it became known also in the domain of proteases. The facts discovered in the selective action of proteases are perhaps the most striking. Namely, it was found that trypsin is capable of hydrolyzing only peptides which are composed of naturally occurring amino acids. If one of the amino acids is enantiomorphous to the naturally occurring form, the hydrolysis does not occur. Also, a tripeptide or a tetrapeptide composed of naturally occurring amino acids is hydrolyzed by trypsin and so are the still higher natural polypeptides. However, it is sufficient to introduce one amino acid enantiomorphous to the naturally occurring form into the chain of naturally occurring amino acids in order that it be rendered resistant to the destructive action of trypsin. It is immaterial whether the amino acid not occurring in nature is introduced at one end of the chain or at the other, or in any other place in the chain. This antagonism between trypsin and the amino acid enantiomorphous to the naturally occurring form is very remarkable.

Now, if one scrutinizes the list of naturally occurring amino acids, he will notice that some of them are dextrorotatory and others are levorotatory. Thus, for instance, natural alanine is dextrorotatory whereas leucine is levorotatory; aspartic acid is levorotatory and glutamic acid is dextrorotatory. The question naturally arises whether trypsin is endowed with so many special groups, each fitting into a group of one natural amino acid, or whether it is a composite of many enzymes. On the other hand, it is possible that all naturally occurring amino acids, regardless of the direction of their rotations, are configurationally related.

Problems of this sort, having a certain practical background, and problems arising from mere curiosities of the human mind, have stimulated chemists to search for means of correlating the configurations of substances belonging to one class with regard to one significant group, but differing with regard to the details of their structures. Thus alanine, leucine and isoleucine are all amino acids and are identical with regard to their significant groups, NH_2 and COOH.

Alanine 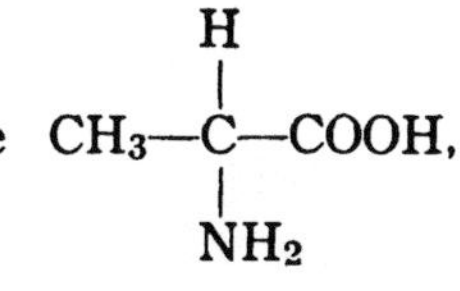 leucine 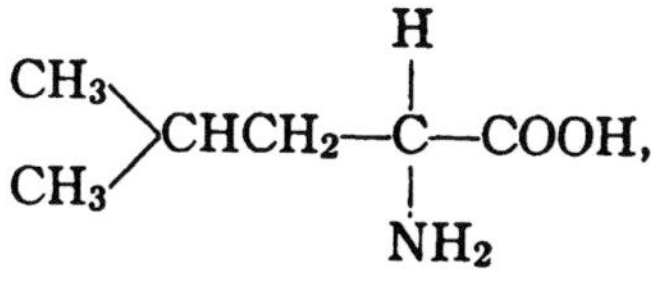

and isoleucine 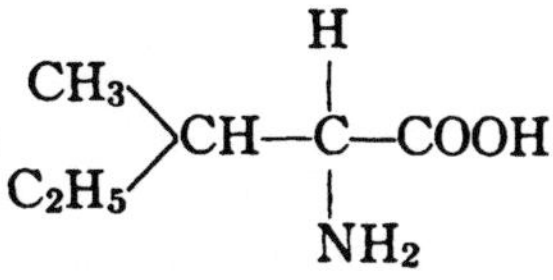 may be regarded as derivatives of aminoacetic acid. The groups entering into the aminoacetic acid have at first glance only minor differences, particularly if one compares leucine with isoleucine. In leucine the group is isobutyl, in isoleucine it is a secondary butyl, yet the natural leucine is levorotatory and the natural isoleucine is dextrorotatory. Are these substances with different rotations configurationally related or are they enantiomorphous?

The instance of amino acids is referred to first because proteins and their component amino acids in recent years have held the attention of chemists and of biochemists alike. But questions of configurational relationships arise also among individual hydrocarbons, alcohols, halides, nitro-derivatives, amines, etc. In a word, taking the simple compound

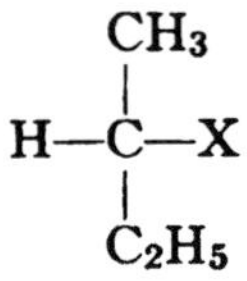

and putting in place of X any element except hydrogen or any radicle save ethyl and methyl, substances are obtained which are capable of existence in dextrorotatory and levorotatory forms and the question to be answered is "Which are configurationally related?"

When substances belong to one and the same chemical group, the solution of the problem is accessible by the methods of organic chemistry. It must be made clear that as yet there exists no way of determining the true allocation of the groups in each one of a pair of two enantiomorphous substances. That is, given two isomers of $(CH_3)(C_2H_5)CH{\cdot}X$

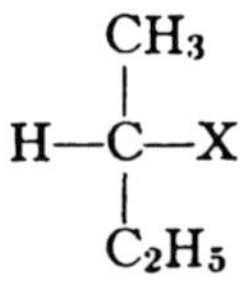

I

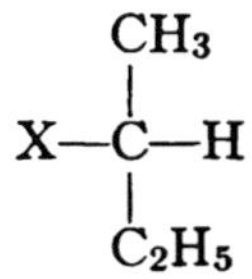

II

it is not possible to say whether the dextrorotatory form has the configuration I or II. But we may arbitrarily assign to the dextrorotatory substance the structure I and make it the substance of reference of all substances having a single asymmetric carbon atom.

The most comprehensive work on configurational relationship has been carried out in the series of hydroxy acids. The relationships are given in Chart I.

The details of the reactions leading from one substance to another cannot be presented here for lack of time. The bibliography on the subject is given at the end of the lecture. The significant fact in regard to the reactions leading to all the transformations recorded on Chart I is that they did not involve either substitution or tautomeric change on the significant asymmetric carbon atom. This condition must be observed in order to permit conclusions regarding configurational relationships.

CHART I

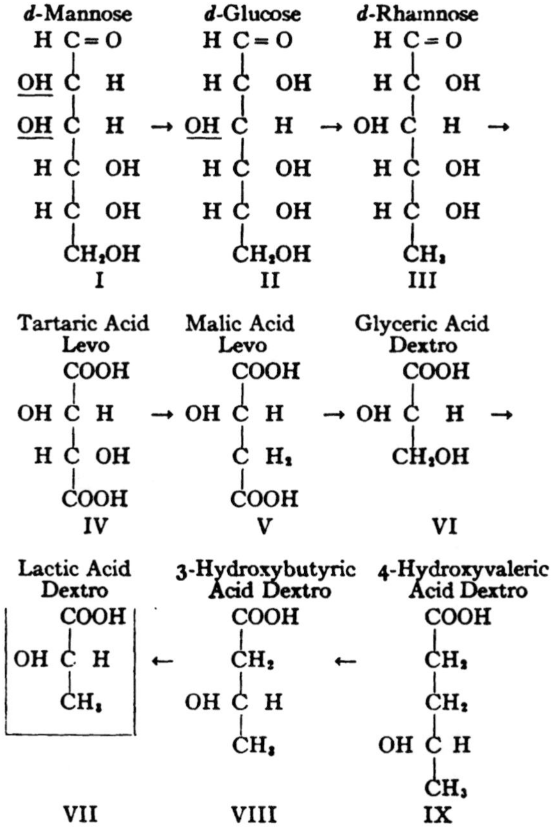

As points of reference in this series of substances may be taken, on one hand, lactic acid, as the simplest optically active hydroxy acid, on the other hand, mannose or mannonic acid, in which the hydroxyl on carbon atom (2) is in the same position as in dextro-lactic acid, namely, to the left. Because of the position of the hydroxyl to the left, it is convenient to refer to this series of monohydroxy acids as the *l*-series.

The series of amino acids is the second large group in which configurational relationships were established by methods which were of the same nature as those employed for the series of hydroxy acids. From Chart I it is seen that lactic acid may be taken as the substance of reference in the series of hydroxy acids. In a similar way aminopropionic acid (alanine) may be taken as the substance of reference in the series of a certain number of amino acids. In Chart II these relationships are given.

CHART II

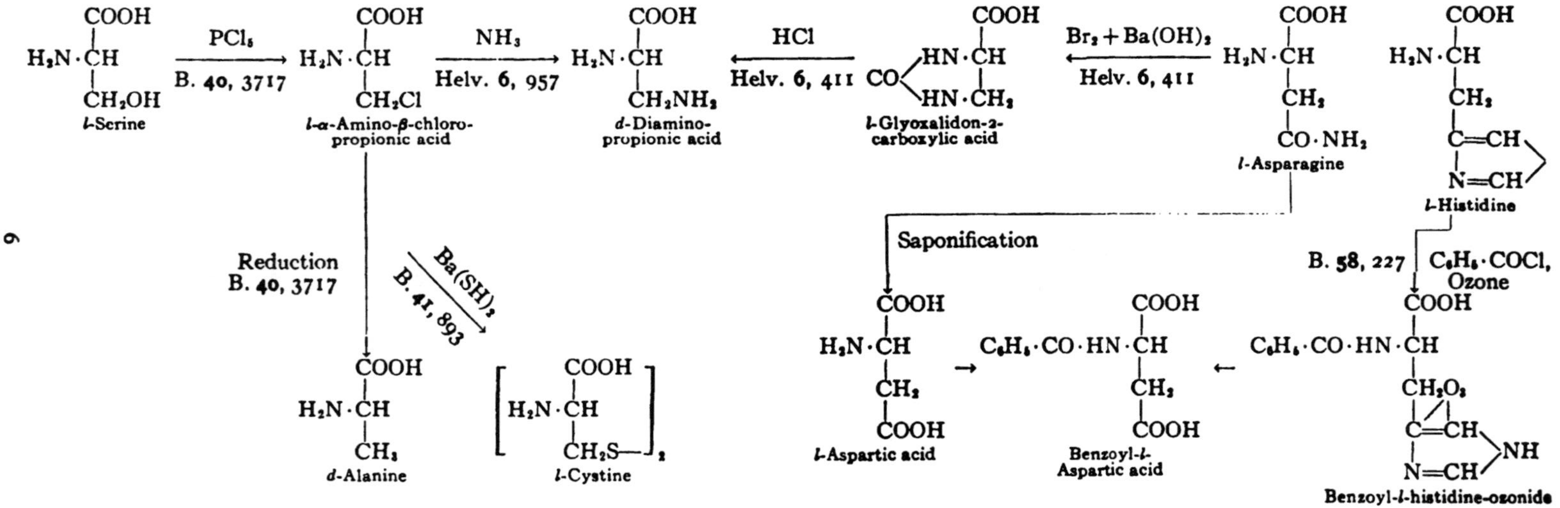

In Charts II, III and IV, and in Tables II, III, IV, V, X and XI, which are reproduced from the literature, instead of *d*- read dextro- and instead of *l*- read levo-.

Also the two naturally occurring amino acids of the aromatic series, tyrosine and phenylalanine, have been correlated, as shown in Chart III.

CHART III

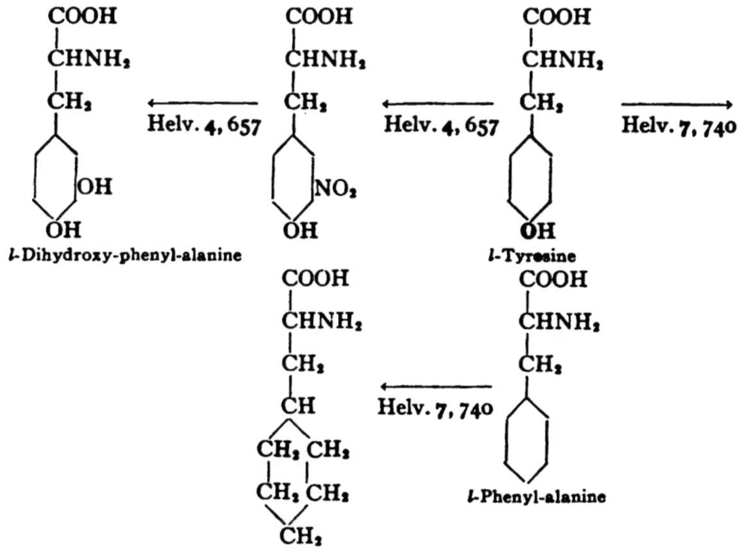

And finally, the configurational relationship was established between the naturally occurring levo-proline and the alkaloids, levo-stachydrine and levo-nicotine, as shown in Chart IV.

CHART IV

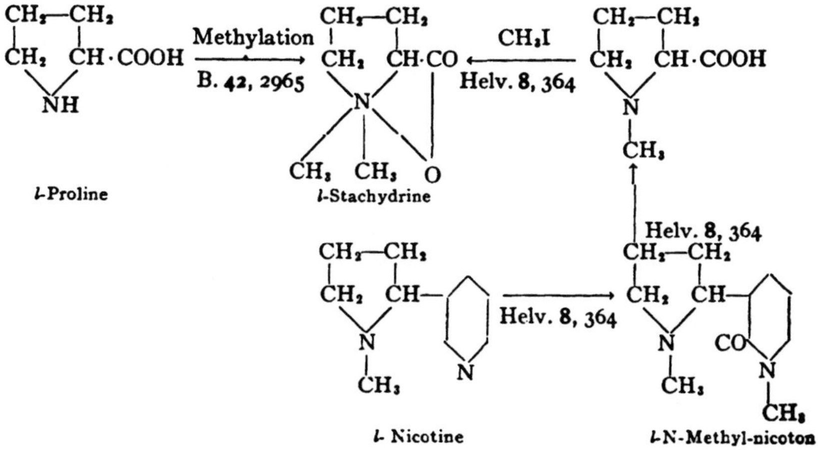

It is noteworthy also that in the series of amino acids configurationally related substances often rotate in opposite directions.

The results reviewed thus far clearly illustrate the fact that within some closely related groups of substances, configurational relationships can be established unmistakably by the direct chemical method.

The question now arises as to the possibility of correlating the configurations of amino and hydroxy acids, or of hydroxy and halogen substituted acids. *A priori* the problem seems to be readily solvable. And indeed the founders of stereochemistry did not anticipate any difficulties in this direction.

Thus, alanine is readily deaminized and converted into lactic acid and lactic acid is readily chlorinated.

$$CH_3CHNH_2COOH \xrightarrow[\text{acid}]{\text{Nitrous}} CH_3CHOHCOOH \xrightarrow{PCl_5} CH_3CHClCOOH.$$

For a time this procedure for determining configurational relationships seemed entirely dependable. Soon, however, the confidence in the method was disturbed by an unexpected result. Thus, it was found that at will either dextro- or levo-lactic acid could be obtained from alanine by the action of nitrous acid. The result depended upon the starting material. The free dextro-alanine yielded the former and its ethyl ester, the latter.

Also, two chlorides, a dextro- and a levo-, could be obtained from the same alcohol, methyl-phenyl carbinol, by acting in one case with PCl_5 and in the other with $SOCl_2$. It is evident that in the course of one of the two reactions an inversion of the original configuration took place and there exists no reliable method for discovering in which of the two reactions the inversion occurred. The inversion is called after the name of its discoverer, "Walden Inversion." This phenomenon is the cause of the difficulties in establishing the mutual configurational relationships of substances belonging to different classes of the system of organic chemistry. It necessitated the search for methods which would permit the correlation of the configurations —let us say—of an alcohol and of the corresponding halide.

There is one very important property of all optically active substances, which was made the basis of all the work on configurational relationships of substances belonging to distinct groups. This property is the variability of the numerical value and of the direction of rotation of each substance under the influence of external conditions.

The external conditions which affect optical rotation are solvent, concentration, temperature, wave length, presence in solution of other solutes, whether electrolytes or nonelectrolytes.

Much important work was done on the influence of external conditions on the optical rotation of substances, the configurational relationship of which had been established by means of the direct chemical methods. In the course of the work it was found that a given factor affected one enantiomorphous form in a certain way and the other in an opposite way. The derivatives of each form were generally affected in the same way as the parent substance. The names of Frankland, Armstrong, Patterson, Pickard and Kenyon and of Lowry are to be mentioned among the pioneers.

A very comprehensive effort to correlate the configurations of hydroxy acids among themselves by the indirect method was made by Clough. Partly from the data already in existence, and partly from those obtained by himself, this author pointed out certain significant peculiarities. Thus taking the four configurationally related hydroxy acids, dextro-lactic, dextro-glyceric, levo-malic and levo-tartaric acids, it is found that they all show a lower dextrorotation in a pure solvent (water, alcohol) than in a solution containing a metal halide; that the free acids have higher dextrorotations than the metal salts or the esters, and furthermore, that identical substituents affect the rotation of all the four acids in the same sense. Clough then postulated that all other 2-hydroxyacids reacting similarly to the four enumerated are configurationally related to them.

Table I illustrates the influence of the solvent on the optical rotations of the four acids.

TABLE I

Substance	Solvent	Concentration, Per Cent	$[\alpha]_D$
Dextro-lactic acid	Water	10	+ 2.2
	+NaBr	10	+ 7.6
Dextro-glyceric acid	Water	2.5	0.0
	+NaCl	2	+ 4.9
Levo-malic acid	Alcohol	10	− 5.9
	+NaBr	10	+21.8
Levo-tartaric acid	Alcohol	10	− 2.6
	+NaBr	10	+ 9.0
Levo-Me, lactate	Alcohol	10	− 5.4
	+NaBr	10	+ 5.8
Levo-Me, malate	Alcohol	10	− 9.4
	+NaBr	10	+ 9.1
Levo-Me, tartarate	Alcohol	10	− 2.7
	+NaBr	10	+12.6

Table II contains data on the rotations of levo-phenyl-lactic acid and its methyl ester.

TABLE II

THE INFLUENCE OF SODIUM HALIDES ON THE OPTICAL ROTATORY POWERS OF *l*-β-PHENYL-LACTIC ACID AND METHYL-*l*-β-PHENYL-LACTATE IN SOLUTION

Component	Solvent	*c.*	t°	$\alpha^{t}_{gr.}(l=2)$	$[\alpha]^{t}_{gr.}$
l-β-Phenyl lactic acid...	Water..................	2.505	20°	−1.28°	−25.5°
" " " "	Aqueous sodium chloride (4N)................	0.665	20°	−0.18°	−13.5°
" " " "	Methyl alcohol.........	10.0	20°	−4.20°	−21.0°
" " " "	Me. alcoholic sodium bromide (N)............	10.0	20°	+0.41°	+ 2.0°
Methyl *l*-Phenyl-lactate.	Methyl alcohol.........	10.0	18°	+0.90°	+ 4.5°
" " " .	Me. alcoholic sodium bromide................	10.0	18°	+ 4.46	+22.3

On the basis of these observations, Clough considers the configuration of these acids related to the four acids of Table I and classifies them all in the *l*-series.

2-*Amino acids*

As was stated above, the configurational relationships of a certain number of amino acids among themselves can be established by means of direct chemical methods. In these were observed certain regularities in the effect on optical behavior of definite external conditions.

The effect of the same conditions on other amino acids can then be observed and on the basis of these data, their configurations can be correlated with the first group of amino acids.

Some of the observations on naturally occurring amino acids are summed up in Tables III, IV and V. From these tables it is seen that all these amino acids belong to one group and furthermore comparing these data with those in Tables I and II, a great similarity in the action on the optical behavior of hydroxy and of amino acids of like external conditions is observed. On the basis of this similarity, Clough formulated the conclusion that all naturally occurring amino acids are configurationally related to dextro-lactic acid and hence belong to the *l*-series.

TABLE III

THE INFLUENCE OF INORGANIC SALTS ON THE OPTICAL ROTATORY POWER OF *d*-GLUTAMIC ACID IN AQUEOUS SOLUTION

Solvent	c	d^{25}	$\alpha^{25}_{gr.}$ ($l=4$)	$[\alpha]^{25}_{gr.}$
Water	1.50	1.003	+0.80°	+13.3°
Aqueous sodium chloride (N)	1.50	1.043	0.90	14.4
" sodium chloride (4N)	1.51	1.152	1.10	15.8
" potassium chloride (N)	1.53	1.050	0.95	14.8
" barium chloride (N)	1.53	1.092	0.96	14.4
" barium bromide (4N)	1.51	1.500	1.64	18.1
" sodium hydroxide (1 mol.)	12.25	1.075	−1.88	− 3.6
" sodium hydroxide (2 mols.)	6.55	1.050	+3.22	+11.7
" hydrogen chloride (1.5 mols.)	8.75	......	....	+37.4

TABLE IV

THE INFLUENCE OF SOLVENTS ON THE OPTICAL ROTATORY POWER OF ETHYL *l*-ASPARTATE

Solvent	c	$[\alpha]^{15}_{gr.}$ ($l=2$)	$[\alpha]^{15}_{gr.}$
None	....	−6.34° ($l=0.5$)	−11.7°
Benzene	10.92	−2.75	−12.6
Chloroform	11.02	−2.13	− 9.7
Acetone [1]	20.0	−3.50	− 8.8
Water	21.7	+1.83	+ 4.2
Water	12.25	1.03	4.2
Aqueous sodium chloride (4N)	10.81	1.44	6.7
" barium bromide (4N)	14.50	3.34	11.5
" calcium chloride (5N)	13.21	3.84	14.5
" hydrogen chloride (N)	11.37	2.73	12.0
Methyl alcohol	20.0	$[\alpha]^{20}_{gr.}$ −0.19	$[\alpha]^{20}_{gr.}$ − 0.5
Methyl alcoholic sodium bromide (N)	20.0	$[\alpha]^{20}_{gr.}$ +1.72	$[\alpha]^{20}_{gr.}$ + 4.3

[1] Ethyl *l*-aspartate reacts with acetone, the rotation changing from $\alpha_{gr} = -3.50°$ to $\alpha_{gr} - 56.6°$ ($c = 20$) in fifteen hours.

Many of the conclusions of Clough regarding the mutual relationships of the amino acids have been corroborated by Karrer and his coworkers and some of the conclusions regarding the configurational relationships of amino and hydroxy acids have been substantiated by Freudenberg and his coworkers.

The problem of the correlation of amino and of hydroxy acids was approached in our laboratory from an entirely different viewpoint. Historically it was the first attempt to correlate the

TABLE V

THE OPTICAL ROTATORY POWERS ($[\alpha]^{20}_{D.}$) OF α-AMINO-ACIDS AND OF CERTAIN DERIVATIVES FROM THEM

Amino Acid	Water	Hydro-chloric Acid	Benzoylamino Acid (in KOH, Aq.)	Aminoacyl Glycine (in Water)	Glycylamino Acid (in Water)	Hydantoin[1]
d-Alanine[2]	+ 2.7°	+14.3°	+36.5°	+50.2°	−50°	+50.6° (water)
l-Serine[3]	− 6.8	+14.4	+43.6 (in NaOH, aq., *p* nitrobenzoyl)			
l-β-Chloro-α-amino-propionic acid[4]	−15.5	+ 0.9				
d-α-Aminobutyric acid[5]	+ 8.0	+19.6	+30.7 (in NaOH, aq.)	+86.4 (K) +26.8 (A. & C.)	−20.3	
d-Valine[6]	+ 6.4	+28.7	....	+90	−19.7	−97.5 (phenylhydantoin in alcohol)
d-α-Aminohexoic acid[7]	....	+23.0	+21.9			
l-Leucine[8]	−10.3	+15.6	+6.6 (in NaOH, aq.)	+85.5	−35.0	−68.2 (in NaOH, aq.)
d-*iso*-Leucine[9]	+11.3°	+40.6°	+26.4° (in NaOH, aq.)	+33.6°	−14.7°	
l-Asparagine[10]	− 4.9	+28.5	....	....	− 6.4	
l-Aspartic acid[11]	+ 4.3	+25.7	+37.4	....	+11.1	−125 (in NaOH, aq.)
d-Glutamic acid[12]	+ 9.9	+30.8	+18.7	....	− 6.3	−79 (in NaOH, aq.)
l-Phenylalanine[13]	−35.3	− 7.1	+17.1	+54.2	+41.4	−96.4 (in 50 per cent alcohol)
l-Tyrosine[14]	....	− 8.6	+19.2	....	....	−143 (in NaOH, aq.)

1 Dakin and Dudley, Journ. Biol. Chem., 17, 29, 1913; 18, 48, 1914; T. 107, 434, 1915.
2 Fischer, Ber., 38, 2914, 1905; 39, 453, 1906; 40, 943, 1907.
3 Fischer and Jacobs, Ber., 39, 2942, 1906; Fischer and Raske, Ber., 40, 3717, 1907.
4 Fischer and Raske, loc. cit.
5 Fischer and Mouneyrat, Ber., 33, 2383, 1900; Koelker, Zeitschr. physiol Chem., 73, 312, 1911; Abderhalden and Chang, Ibid., 77, 471, 1912.
6 Fischer, Ber., 29, 2320, 1906; Fischer and Scheibler, Annalen, 363, 136, 1908.
7 Fischer and Hagenbach, Ber., 34, 3764, 1901.
8 Fischer, Ber., 33, 2370, 1900; Ber., 39, 2893, 1906; Fischer and Steingroever, Annalen, 365, 167, 1909.
9 Locquin, Bull. Soc. chim., 1, 595, 1907 [iv]; Abderhalden, Hirsch and Schuler, Ber., 42, 3394, 1909.
10 Fischer and Koenigs, Ber., 37, 4585, 1904.
11 Fischer, Ber., 32, 2451, 1899; Fischer and Fiedler, Annalen, 375, 181, 1910.
12 Fischer, Kropp and Stahlschmidt, Annalen, 365, 189, 1909.
13 Fischer and Mouneyrat, Loc. cit.; Fischer and Schoeller, Annalen, 357, l. 1907.
14 Fischer, Ber., 32, 3638, 1899.

configuration of amino and of hydroxy acids. The plan was to prepare a series of 2-aminosugar acids and to compare their properties with those of simple sugar acids. It is permissible to deal with sugar acids as if they were simple 2-hydroxy acids. This statement is best illustrated by taking a pair of epimeric acids such as gluconic and mannonic.

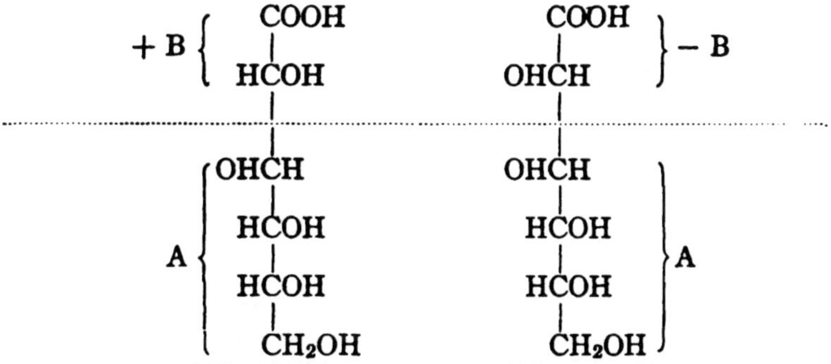

d-Gluconic Acid. $[M]_D = m$ *d*-Mannonic Acid $[M]_D = n$

It is obvious that the parts A of each molecule are identical and may be substituted by the letter R. Thus, the relationship between the two sugar acids is the same as between levo- and dextro-lactic acids, which may be represented in the following way:

Levo-lactic Acid. (*d*-series) Dextro-lactic Acid. (*l*-series)

Indeed it is possible from the numerical values of the rotations of each acid to determine for carbon atoms (2) the direction of their rotation and the numerical value of the rotation by two simple equations.

$$(+ B) + (A) = m$$

$$(- B) + (A) = n$$

Hence,

$$B = \frac{m - n}{2} \quad \text{and} \quad A = \frac{m + n}{2}.$$

In a like manner 2-aminosugar acids may be dealt with as simple 2-amino acids. The rotations of the carbon atoms (2) have actually been calculated for the entire series of 2-aminohexonic acids and the results are given in the following table (Table VI).

TABLE VI

Acid	$[\alpha]_D$ of Carbon atom (2)	$[M]^{20}_D$.	Phenyl-hydrazide	$[\alpha]_D$ of Carbon atom (2)	$[M]^{20}_D$.
Epichitosaminic	+12.5	+24.37	Gluconic	+14.25	+42.18
Chitosaminic	−12.5	−24.37	Mannonic	−14.25	−42.18
Dextro-xylohexosaminic	+12.5	+24.37	Gulonic	+14.25	+42.18
Levo-xylohexosaminic	−12.5	−24.37	Idonic	−14.25	−42.18
Epichondrosaminic	+12.5	+24.37	Galactonic	+ 8.25	+24.42
Chondrosaminic	−12.5	−24.37	Talonic	− 8.25	−24.42
Dextro-ribohexosaminic	+19.12	+37.28	Allonic	+20.8	+61.56
Levo-ribohexosaminic	−19.12	−37.28	Altronic	−20.8	−61.56

Thus, 2-aminohexonic acid and the corresponding hexonic acid could be used for the study of correlations between 2-amino and 2-hydroxy acids. The sugar acids, however, offer a great advantage for the following reasons:

Two enantiomorphous monohydroxy acids differ one from the other only in one physical property, in their rotations. Their melting points, boiling points, densities, capacities to form salts, etc., are identical.

Two epimeric sugar acids, on the other hand, may have distinctive individual properties in addition to the differences in rotatory powers. Besides, each sugar acid may be reduced to the sugar, and again two epimeric sugars may differ from one another in solubilities, melting points, rates of oxidation, etc. Thus, it is often possible to differentiate one sugar acid from the other without resorting to the differences in their optical properties. All these advantages belong also to 2-aminosugar acids.

By way of illustration, the pair of gluconic and mannonic acids may be compared with the pair of chitosaminic and epichitosaminic acids.

A striking difference between gluconic and mannonic acids is the readiness with which the latter forms a crystalline lactone. Also it is noted that the salts of gluconic acid are more insoluble than the corresponding salts of mannonic acid. Still greater are the differences in the physical properties of the corresponding sugars. Glucose is more insoluble in water and in alcohol than mannose. The melting point of glucose is 146° C., that of mannose is 132° C. Finally, the difference of the numerical values of the rotations of the α- and β-forms of glucose is normal, that of mannose is abnormal.

If the properties of chitosaminic and epichitosaminic acids are compared, one finds that chitosaminic acid is the more insoluble. Of the two sugars, chitose and epichitose, the latter has the lower melting point. Furthermore, the numerical value of the difference in the rotations of the α- and β-forms of chitosamine is normal, and that of epichitosamine is abnormal. Thus, of the four substances, gluconic and chitosaminic acids seem to belong to one class, and mannonic and epichitosaminic acids to another class.

Now, one may turn to a comparison of the influence of external conditions on the rotatory powers of the 2-aminohexonic and of the simple hexonic acids.

In the following tables (VII and VIII) are given the rotations of the free sugar acids and of their metal salts.

TABLE VII

SHOWING THE SPECIFIC ROTATIONS OF THE FREE ACIDS AND OF THEIR DERIVATIVES (ALL OF THE *d*-SERIES)

	Free acids[1] $[\alpha]^{20}_{D.}$	Na Salts[2] $[\alpha]^{20}_{D.}$	Phenyl-hydrazides[3] $[\alpha]^{20}_{D.}$	Amides[4] $[\alpha]^{20}_{D.}$
Gluconic	0.0	+11.78	+18.0	+31.2
Mannonic	+15.6	− 8.82	−10.5	−17.3
Idonic	+	− 2.52	−15.1	
Gulonic	− 1.6	+12.68	+13.45	+15.2
Galactonic	− 8.0	+ 0.40	+12.2	+30.0
Talonic	?	?	+ 4.35	
Allonic	−10.0	+ 4.30	+25.88	
Altronic	+ 8.0	− 4.05	−15.8	

[1] Observations by Levene (Levene, P. A., Jour. Biol. Chem., 59, 123, 1924).

[2] Observations by Levene and Meyer (Levene, P. A. and G. M. Meyer, Journ. Biol. Chem., 26, 365, 1916).

[3] Observations by Levene and Meyer (Levene, P. A. and G. M. Meyer, Journ. Biol. Chem., 31, 625, 1917).

[4] Observations by Hudson and Komatsu (Hudson, C. S. and S. Komatsu, Journ. Am. Chem. Soc., 16, 1142, 1919).

TABLE VIII

Acids	In 5 per cent NaOH $[\alpha]_D^0$		In 2.5 per cent HCl $[\alpha]_D^0$
Chitosaminic	+ 1.3°	c=5.0	−15°
Epichitosaminic	− 5.0°	c=5.0	+10.0°
Dextro-*d*-xylo-2-aminohexonic	−16.0°	c=2.5	+14.0°
Levo-*d*-xylo-2-aminohexonic	+ 2.0°	c=2.5	−11.0°
Chondrosaminic	−15.0°	c=2.5	−17.0°
Epichondrosaminic	+ 1.8°	c=2.5	+ 8.0°
Dextro-*d*-ribo-2-aminohexonic	+ 2.0°	c=2.5	+12.5°
Levo-*d*-ribo-2-aminohexonic	−15.0°	c=2.5	−26.0°

From these tables it is seen that each class of substances falls into two groups.

TABLE IX

First Group

Salt (ion) − free acid = + sign

Chitosaminic acid	Gluconic acid
Chondrosaminic acid	Galactonic acid
d-Levo-xylohexosaminic acid	Gulonic acid
d-Levo-ribohexosaminic acid	Allonic acid

Second Group

Salt (ion) − free acid = − sign

Epichitosaminic acid	Mannonic acid
Epichondrosaminic acid	Talonic acid
d-Dextro-xylohexosaminic acid	Idonic acid
d-Dextro-ribohexosaminic acid	Altronic acid

Of the first group, gluconic and chitosaminic acids are the representative members, and mannonic and epichitosaminic acids are the representative members of the second. The former behave as 2-hydroxy acids having the hydroxyl to the right; the latter, as those having the hydroxyl to the left. Thus, in the sugar series, the deductions made on the basis of the differences in rotation between salt and free acid are supported by the evidence found in the other physical properties of the sugar acids and of the corresponding sugars. Thus, the observations on the sugar acids furnish very significant support for the rule postulating that the configuration of 2-hydroxy and 2-amino acids may be recognized by the differences in the rotations of the metal salt and of the free acid. When the difference (Salt — FA) has a + sign the position of the hydroxyl or of the amino group is to the right, (*d*-series); when it has a — sign, the position is to the left (*l*-series).

Halogen Acids.—One encounters still greater difficulties in the attempt to correlate the configurations of this group of acids with those of the hydroxy and amino acids for the reason that there are fewer data on the influence of the external conditions on the optical behavior of the halogen acids. No attempt at all has been made to correlate the individual members of this group among themselves by the purely chemical method. The first attempt to correlate the configurations of some halogen acids by the indirect method was made by Clough. Table X, page 18 records the data on which his conclusions were based.

Table X is an illustration of the method which had been used by many authors for the purpose of correlating configurations of substances differing from one another by the character of one group or one element only. The method is based on the comparison of the influence of identical substituents on the rotations of the related substances. On the basis of the data of the above table, all the dextro-2-halogen acids were classified in one group. It remained to correlate them with the configurations of the 2-amino and 2-hydroxy acids. For this purpose also the data were scanty.

Table XI, page 19, contains the data on which the configurations of the halogen acids were correlated with those of the amino acids.

Here again the method of comparison of the influence of substituents on rotation was employed. On the basis of the data recorded in this table, Clough correlated the dextro-halogen acids with the group of amino acids previously classified as the *l*-acids. It was, however, realized by Clough himself and by others that in order to establish the relationship firmly, further data were required.

We have approached the problem in a different manner. Our argument was the following: When a hydroxyl on an asymmetric carbon is replaced by a halogen, then the electric fields around that atom are altered when the substances are brought into solution. We therefore thought it desirable to alter the polarity of a radicle attached to the asymmetric carbon atom without resorting to the reaction of substitution and then to follow the influence of that change on the optical behavior of the substance. For experimental reasons it is most convenient to bring about a change of polarity by converting $—CH_2OH$ into —COOH and —SH into $—SO_2OH$. Both methods are

TABLE X

THE SPECIFIC ROTATIONS ($[\alpha]^{20}_{D}$) OF CERTAIN *d*-α-HALOGEN ACIDS AND THEIR DERIVATIVES

	Acid	Methyl Ester	Ethyl Ester	Propyl Ester	Bromoacyl-glycine	References
d-α-Chloropropionic acid	+14°*	+26.8° (at 5°)	+19.9 (at 5°)	+11.0° (at 6°)	....	Frankland and Garner, T., 1914, **105**, 1101; J. W. Walker, T., 1895, **67**, 918.
d-α-Bromopropionic acid	+29.0	+42.6	+35.5	+22	+35.3 (water)	J. W. Walker, *loc. cit.*, Ramberg, *Annalen*, 1909, **370**, 234; Fischer, *Ber.*, 1907, **40**, 489.
d-α-Bromo*iso*valeric acid	+9.0 (water); 22.8 (benzene)	......	....	......	+47.5 (alcohol)	Fischer and Scheibler, *Ber.*, 1908, **41**, 889, 2891.
d-α-Bromo*iso*hexoic acid	+49.4	......	+50	......	+62 (alcohol)	Fischer and Carl, *Ber.*, 1906, **39**, 3998.
d-α-Bromo-β-methylvaleric acid	+26.5 (benzene)	......	....	......	+64.4 (alcohol)	Abderhalden, Hirsch, and Schuler, *Ber.*, 1909, **42**, 3394.
d-α-Bromo-β-phenylpropionic acid	+10.4	......	+9	......	+14.6 (alcohol)	Senter, Drew, and Martin, T., 1918, **113**, 158; Fischer and Carl. *loc. cit.*; Fischer and Schoeller, *Annalen*, 1907, **357**, 1.
d-Chlorosuccinic acid	+21.7 (water); 55.7 (ethyl acetate)	+42.3	+32.7	+25	....	Walden, *Ber.*, 1896, **29**, 1689; McKenzie and Barrow, T., 1911, **99**, 1919.
d-Bromosuccinic acid	+41.8 (water); 76.5 (ethyl acetate)	+51.2	+41.0	+38.0	....	Walden, *Ber.*, 1895, **28**, 1290, *Zeitsch. physikal. chem.*, 1895, **17**, 264.
d-Dichlorosuccinic acid	+80.4 (ethyl acetate)	......	+66 (chloroform)	......	....	Holmberg, *Svensk. Kem. Tid.*, 1912; Darzens and Séjourné, *Compt. rend.*, 1912, **154**, 1615.
d-Dibromosuccinic acid	+70 (water)*; 148 (ethyl acetate)	......	....	......	....	Holmberg, *Svensk. Kem. Tid.*, 1911; McKenzie, T., 1912, **101**, 1196.

*Approximate values by calculation.

TABLE XI

THE OPTICAL ROTATORY POWERS ($[\alpha]_{D.}^{20}$) OF CERTAIN α-BROMOACYL AND α-AMINOACYL DERIVATIVES OF THE α-AMINO ACIDS

Amino Acid	Amino Acid	Chloro-acetyl-amino Acid	Glycyl-amino Acid	*d*-α-Bromo propionyl-amino Acid	*d*-Alanyl-amino Acid	*d*-α-Bromo-*iso*hexoylamino Acid	*l*-Leucylamino Acid	References
Glycine......	....	....	....	+38.3°	+50.2°	+62.0° (in alcohol)	+86.0°	Fischer, *Ber.*, 1908, **41**, 850; 1906, **39**, 2911.
d-Alanine......	+2.7°	−45°	−50°	−16.5	−21.6	+23.3 (in alcohol)	+23.5 (in methyl alcohol)	Fischer, *Ber.*, 1906, **39**, 453; Fischer and Schulze, *Ber.*, 1907, **40**, 943.
l-Alanine......	−2.7	+45	+50°	+68.2	+68.9	....	....	Fischer and Raske, *Ber.*, 1906, **39**, 3989.
l-Leucine......	−10.3	−13.8	−35.2	+ 2.0 (in alcohol)	−17.0	+16.2 (in ethyl acetate)	−13.4 (in *N*-NaOH)	Fischer, *Ber.*, 1907, **40**, 1754.
d-Leucine.....	+10.3	+13.8	+35.2	....	....	+35 (in ethyl acetate)	+68.9 (in *N*-HCl)	Fischer and Koelker, *Annalen*, 1907, **354**, 39.
d-*iso*Leucine...	+10.5	+25.0 (in alcohol)	−14.7	+24.5 (in alcohol)	+ 6.1 (in HCl)	+49 (in ethyl acetate)	+18.1	Abderhalden, Hirsch, and Schuler, *Ber.*, 1909, **42**, 3394.
l-*iso*Leucine....	−10.5	−25.0 (in alcohol)	+14.7	....	....	+19 (in alcohol)	+53.1	Abderhalden and Schuler, *Ber.*, 1910, **43**, 907.

The values in this table are for specific rotations in aqueous solution unless stated otherwise.

being made use of in our laboratory but more progress has been made by the second method and the results will now be presented.

The following thio-derivatives have been studied: secondary mercaptans, 2-thio-monocarboxylic acids, thio-dicarbocyclic acids, and 3-thio-monocarboxylic acids.

The results are as follows:

In the series of secondary mercaptans, all showed a reversion of their optical rotation on oxidation to the corresponding sulfonic acids. The mercaptans thus far analyzed in this respect are, 2-mercaptobutane, 2-mercaptoisohexane, 2-mercaptooctane and benzy-phenyl-mercaptomethane. On oxidation to the corresponding sulfonic acids, the rotations of all the mercaptans were reversed in their directions. From these observations it is evident that derivatives of secondary alkyls analogous in configurations but differing markedly in the polarity of the significant groups rotate polarized light in opposite directions. On the basis of these considerations the conclusion was drawn that secondary alcohols and secondary halides rotating in opposite directions are configurationally related.

Monocarboxylic acids substituted in position (2). Of this series only one acid has been observed thus far, namely, thiolactic acid. The sulfoacid prepared from it rotated in the same direction as the parent substance. From this observation one could conclude that dextro-chloropropionic and dextro-lactic acids are configurationally related. Further analysis brought additional evidence in favor of this assumption.

It is characteristic for dextro-lactic acid, for dextro-alanine and for other 2-hydroxy and 2-amino acids of the *l*-series, that the differences in the rotations of ion-free acids has a — sign, as seen from the following table (Table XII).

TABLE XII

	C	NaOH$[M]_D$ 2 Eq.	NaOH$[M]_D$ 1 Eq.	Free Acid$[M]_D$	HCl$[M]_D$
1. Dextro-lactic acid[1]	5%	....	−11.88	+ 1.80	
2. Dextro-alanine[2]	....	....	....	+ 2.40*	+13.60
3. Dextro-2-bromopropionic acid[3]	....	....	− 9.34	+37.96	
4. Dextro-2-thiopropionic acid[4]	....	+ 7.32	− 5.58	+58.98	
5. Dextro-2-sulfopropionic acid†	....	− 3.28	+12.84	+13.76	

1 Purdie and Walker, Trans. Chem. Soc., 67, 630 (1895).
2 Clough, G. M., Trans. Chem. Soc., 113, 540 (1918) and observations in this laboratory.
3 Levene, P. A., and Mikeska, L. A., Unpublished.
4 Levene, P. A., and Mikeska, L. A., J. Biol. Chem., 53, 86 (1925).
* Ionized condition.
† Partially racemized, unracemized $[M]_D$ = + 49.50.

Thus the presence of sodium hydroxide in the solutions of these acids affects their rotations in the same sense. Of the five substances two (4 and 5) are definitely known to be configurationally related on the grounds of method of preparation, and two (1 and 2) are regarded configurationally related on the basis of indirect evidence. It now seems justifiable to classify the dextro-bromopropionic acid in the same series with dextro-lactic acid, namely, in the *l*-series.

Substituted dicarboxylic acids.—Of this class only the derivatives of succinic acid were analyzed. In the following table (XIII) the rotations of the free acids and of the salts of the various derivatives are compared.

TABLE XIII

	C	HCl Solution $[M]_D$	Free Acid $[M]_D$	1 Eq. NaOH$[M]_D$	2 Eq. NaOH$[M]_D$	3 Eq. NaOH$[M]_D$	Amide $[M]_D$
1. Levo-malic acid[1]	10%	......	− 3.59	− 7.60	− 9.22		
2. Dextro-bromo-succinic acid[2]	....	......	+142.00 (in eth. acet.)	....	....	....	+133.10 (in eth. acet.)
3. Dextro-amino-succinic acid[3] (aspartic) ac. 2..........	....	+42.60	+ 7.80	− 6.65	+14.63		
4. Dextro-thiosuccinic acid[4]...	....	......	+73.06	+38.04	+48.57	+41.09	
5. Dextro-sulfo-succinic acid.	....	......	+51.53	+49.27	+37.89	+25.67	
6. Dextro-thiosuccinamide.....	....	......	+89.40	+36.25	+28.35		
7. Dextro-sulfo-succinamide..	....	......	+49.30	+46.56	+ 2.26		

1 Schneider, G., Ann. Chem., 207, 266 (1881).
2 Walden, P., Z. physiol. Chem., 17, 249 (1895).
3 Wood, J. V., Trans. Chem. Soc., 105, 1992–93 (1914).
4 Levene, P. A., and Mikeska, L. A., J. Biol. Chem., 63, 86 (1925).

In order to interpret the above data it is necessary to bear in mind the progress of ionization of the individual groups in the substituted succinic acids with the decrease in hydrion concentration (increase in *p*H values). It is as follows:

CHART V

MALIC ACID

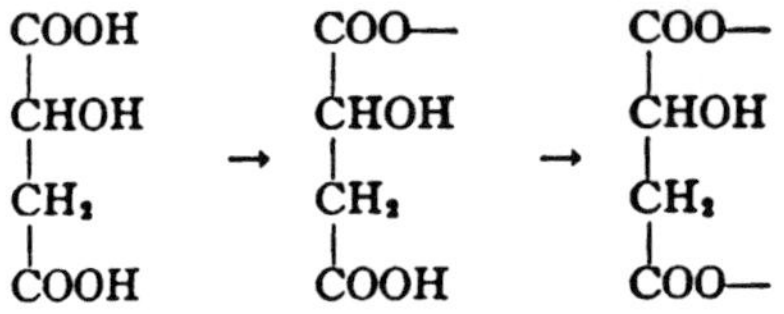

ASPARTIC ACID

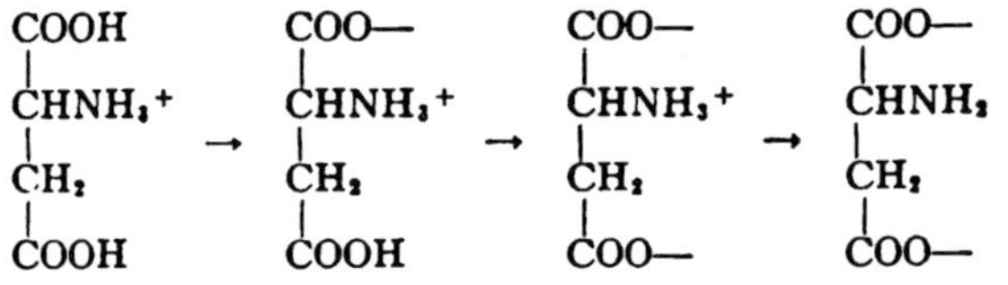

THIOSUCCINIC ACID

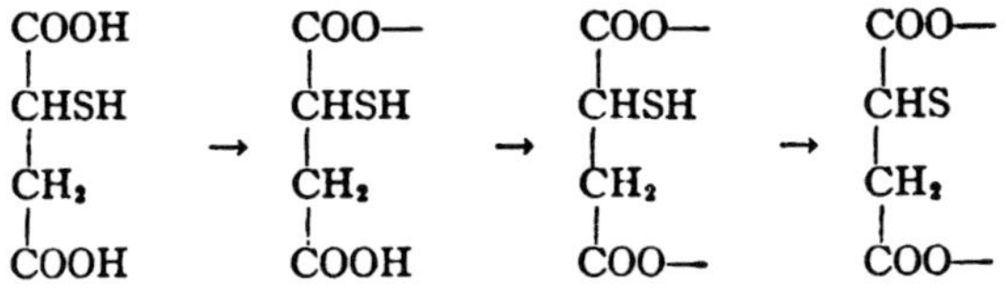

SULFOSUCCINIC ACID

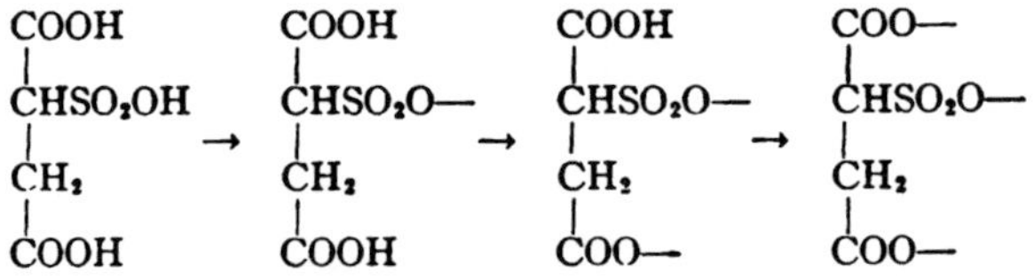

Comparing Chart V with Table XIII, one will notice that in malic and aspartic acids the ionization of the carboxyls in positions 2 and 3 leads to a change in rotation towards the left. In the thioacid, a peculiarity is noted, namely, the ionization of the carboxyl in position 2 leads to a change of rotation towards the left as in the other two instances. The ionization of the carboxyl in position 3, on the other hand, leads to a change of rotation to the right from that of the rotation of the monion.

In the sulfoacid the optical change with the ionization of individual groups is the same as in malic acid. That the change in rotation with the ionization of the carboxyl in position (2) proceeds as in the monocarboxylic acids is seen from the behavior of thio and sulfo-succinamide (NH_2 group on carboxyl in position) (3). Thus, on the basis of the above considerations, the conclusion is warranted that all the above-mentioned substituted succinic acids belong to the *l*-series.

3-Halogen acids.—There are fewer data on the basis of which to correlate the configuration of acids substituted in position (3). Dextro-3-hydroxybutyric acid was found configurationally related to dextro-lactic acid, and therefore was classified as an *l* acid. The rotation of this acid with the progress of ionization showed a shift to the left. Whether the same rule holds for acids substituted in position (3) by other groups can not as yet be stated definitely. The existing data, however, seem to indicate that in this group of acids, the direction of the change in rotation with the progress of ionization seems to depend on the polarity of the substituting group. Thus it was shown above that in the substituted succinic acids the change of rotation with the progress of ionization of the carboxyl in position (3) was to the left for dextro-malic, dextro-aspartic and dextro-sulfosuccinic acids and to the right for dextro-thio-succinic acid. For the group of 3-hydroxy substituted acids, the following data are available (Table XIV).

TABLE XIV

	C	HCl[M]D	2 Eq. NaOH[M]D	1 Eq. NaOH[M]D	Free Acid [M]D
Dextro-3-hydroxybutyric acid[1]	....	....	....	+14.80	+24.60
Levo-3-aminobutyric acid[2].....	10%	+29.7	....	+14.70	−35.00
Levo-3-chlorobutyric acid[2].....	10%	....	....	−50.59	−61.00
Dextro-3-thiobutyric acid[3].....	....	....	+23.05	+39.33	+49.26
Dextro-3-sulfobutyric acid.....	....	....	+19.59	+ 7.04	+ 6.90

1 McKenzie, Trans. Chem. Soc., **81**, 1402 (1902) and Levene, P. A. and Haller, H. L., J. Biol. Chem., **65**, 49 (1925).
2 Fischer, E., and Scheibler, H., Ann. Chem., **383**, 349 (1911).
3 Levene, P. A., and Mikeska, L. A., Unpublished.

Taking it for granted that the oxidation of the 3-thioacid to the sulfonic acid proceeds without change of configuration, one may assume that in 3-hydroxy and in 3-chlorobutyric acids the changes in rotation with progress of ionization will proceed in different directions as is the case in the 3-thio and 3-sulfobutyric acids. It would then follow that levo-3-chlorobutyric acid is configurationally related to dextro-3-hydroxybutyric acid. From the behavior of malic and aspartic acids, on ionization of the carboxyl in position (3) it may be concluded that configurationally related 3-hydroxybutryric acid and 3-aminobutyric acid should show similar optical change with the progress of ionization of the carboxyl and hence, dextro-3-hydroxybutyric and levo-3-aminobutyric acids should be configurationally related. Hence the first three substances given in Table XIV should all be configurationally related and should all belong to the *l*-series.

SUMMARY

Thus, the two different methods, the older one based on the comparison of the influence of external conditions and the one used by us, consisting of changing the polarity of a group attached to the asymmetric carbon atom, lead to the same conclusions, namely:

1. All naturally occurring amino acids have the same configuration as *dextro-lactic acid.*
2. All hydroxy acids, in which the progress of ionization leads to a change of rotation to the left are configurationally related to *dextro-lactic acid.*
3. All 2- and 3-substituted amino acids which behave as (2) are configurationally related to *dextro-lactic acid.*
4. Dextro-chloropropionic and levo-3-chlorobutyric acids are configurationally related to *dextro-lactic acid.*
5. In dextro-lactic acid, the asymmetric carbon atom has the configuration of carbon atom (2) of *d*-mannonic acid.
6. All substances having this configuration are classified in the *l*-series.

In conclusion, I wish to add that the intimate knowledge of the correlation of chemical structure and optical behavior may have, in addition to the significance outlined above, still another more general and more comprehensive interest. Optical activity in organic compounds is most generally brought about by the distortion of the tetrahedron of the carbon atom. Also reactivity of a molecule is undoubtedly caused by the same factor. Stability of any structure and naturally also of molecular structure depends on symmetry; instability is synonymous with reactivity. More intimate knowledge of phenomena associated with optical activity, such as Walden Inversion, racemization, etc., will contribute much to our knowledge of the inner mechanism of chemical reactions.

BIBLIOGRAPHY

On Configurational Relationships of Hydroxy Acids.

Abderhalden, E., and Eichwald, E., *Ber. d. deut. chem. Ges.*, 51, 1312 (1918).
Bremer, G. J., *Ber. d. deut. chem. Ges.*, 8, 861 (1875).
Fischer, E., *Ber. d. deut. chem. Ges.*, 29, 1377 (1896). (Configuration of tartaric acid.)
Freudenberg, K., *Ber. d. deut. chem. Ges.*, 47, 2027 (1914).
Freudenberg, K., and Braun, Fritz, *Ber. d. deut. chem. Ges.*, 55, 1339 (1922).
Karrer, P., and Klarer, W., *Helv. Chim. Acta*, 8, 393, (1925).
Levene, P. A., and Haller, H. L., *Jour. Biol. Chem.*, 65, 49, (1925); *Jour. Biol. Chem.*, 67, 329, (1926).
Levene, P. A., and Walti, A., *Jour. Biol. Chem.*, 68, 415, (1926).
Walden, P., Optische Umkehrerscheinungen, Fried. Vieweg und Sohn Braunschweig, 1919.
Wohl, A., and Momber, F., *Ber. d. deut. chem. Ges.*, 50, 455 (1917).
Wohl, A., and Schellenberg, *Ber. d. deut. chem. Ges.*, 55, 1404 (1922).

On Configurational Relationships of Amino Acids.

Fischer, E., and Raske, K., *Ber. d. deut. chem. Ges.*, 40, 3717 (1907); *Ber. d. deut. chem. Ges.*, 41, 893 (1908).
Freudenberg, K., and Noe, A., *Ber. d. deut. chem. Ges.*, 58, 2399 (1925).
Karrer, P., and Ehrensheim, *Helv. Chim. Acta*, 8, 323 (1926).
Karrer, P., and Kaase, W., *Helv. Chim. Acta*, 6, 411, (1923); 6, 957, (1923).
Karrer, P., Jaggi, W., and Takahoshi, T., *Helv. Chim. Acta*, 8, 360 (1925).
Karrer, P., and Schlosser, *Helv. Chim. Acta*, 6, 411 (1923).
Langenbeck, W., *Ber. d. deut. chem. Ges.*, 58, 227 (1925).
Waser, E., and Brauchli, E., *Helv. Chim. Acta*, 7, 740 (1924).

The Influence of External Conditions on Configurationally Related Substances.

Armstrong, H. E., and Walker, E. E., *Proc. R. Soc. A.*, 88, 388 (1913).
Clough, G. W., *Jour. Chem. Soc.*, 113, 526 (1918).
Frankland, P., and Wharton, F. M., *Jour. Chem. Soc.*, 69, 1587 (1896).
Freudenberg, K., and Rhino, F., *Ber. d. deut. chem. Ges.*, 57, 1547 (1924).
Karrer, P., and Kaase, W., *Helv. Chim. Acta*, 2, 436 (1919); 3, 248 (1920).
Lowry, T. M., *Jour. Chem. Soc.*, 103, 1067 (1913).
Patterson, T. S., *Jour. Chem. Soc.*, 77, 1106 (1900); 85, 1142 (1904); 93, 1843 (1908); 95, 323 (1909); 109, 1145 (1916); etc.
Pickard, R. H., and Kenyon, J., *Jour. Chem. Soc.*, 99, 45 (1911); 101, 620, 1427 (1912); 103, 1923 (1913); 105, 830 (1914); 105, 1115, 2226 (1914); etc.
Rupe, H., and Ackermann, *Ann. d. Chem.*, 420, 1 (1919).

Influence on the Optical Rotation of the Change of Polarity of One of the Groups Attached to the Asymmetric Carbon Atom.

Levene, P. A., and Mikeska, L. A., *J. Biol. Chem.*, 59, 473 (1924); 60, 1, 685 (1924); 63, 85 (1924).

Configurational Relationships in 2-Aminohexoses and 2-Aminohexonic Acids.

Levene, P. A., and co-workers; Monographs of The Rockefeller Institute for Medical Research, No. 18 (1922).
Levene, P. A., *J. Biol. Chem.*, 59, 123 (1924); 63, 95 (1924).

CONFIGURATIONAL RELATIONSHIPS OF SIMPLE SUGARS AND SUGAR ACIDS ESTABLISHED ON THE BASIS OF VAN'T HOFF'S SUPERPOSITION THEORY.

Freudenberg, K., Brauns, F., and Siegel, H., *Ber. d. deut. chem. Ges.*, 56, 193, (1923).

Hudson, C. S., *J. Amer. Chem. Soc.*, 31, 66 (1909); 38, 1566 (1916); 39, 462 (1917); 40, 813 (1918).

Hudson, C. S., and Yanowski, E., *J. Amer. Chem. Soc.*, 39, 1013 (1917).

Hudson, C. S., and Komatsu, S., *J. Amer. Chem. Soc.*, 41, 1141 (1919).

Levene, P. A., *J. Biol. Chem.*, 23, 145 (1915).

Levene, P. A., and Meyer, G. M., *J. Biol. Chem.*, 26, 355 (1916); 31, 623, (1917).

Weerman, R. A., Dissertation published by A. Kruyt, Amsterdam, (1916).

Reversible Oxidation-Reduction Reactions in Organic Systems

BY

W. MANSFIELD CLARK
PROFESSOR OF CHEMISTRY, HYGIENIC LABORATORY
U. S. PUBLIC HEALTH SERVICE

Lecture delivered at Columbia University in the special course on Contemporary Developments in Chemistry given in the Summer Session of 1926 on the occasion of the opening of the Chandler Chemical Laboratories

COLUMBIA UNIVERSITY PRESS
NEW YORK
1927

Printed from type. Published March, 1927

Printed in the United States of America

REVERSIBLE OXIDATION-REDUCTION REACTIONS IN ORGANIC SYSTEMS

W. MANSFIELD CLARK

The term "oxidation" has an obvious meaning when applied to the burning of carbon. Since CO_2 is a "higher" oxid than CO, the reversal of the process of oxidation, namely reduction in degree of oxidation, is rather naturally called *reduction.*

To appreciate why the terms "oxidation" and "reduction" have been extended to include processes such as:

$$Fe^{++} \rightleftharpoons Fe^{+++}$$

or

$$\text{Indigo white} \rightleftharpoons \text{Indigo blue} + H_2$$

requires historical perspective. It is trivial to cite as origins the ingenious devices invented to incriminate oxygen in all such transformations. The incentive to these inventions, the real origin of the classification and of its terms, we shall appreciate only as we recapture the fresh wonderment of the early chemists who revealed the engines of industry and the pulse of life to be dependent literally from minute to minute upon one element of our atmosphere. These were the large phenomena that established the category. Into it fell quite naturally the rusting of iron and the bluing of the cloth as it came from the indigo vat.

It was a long process of dissection at the hands of chemists that isolated particular aspects of specific oxidation-reduction reactions. As a consequence of emphasis upon one or another of these isolated aspects we have in modern thought three leading tendencies—that of the physical chemist who describes in terms of electron-transfer, that of the organic chemist who describes in terms of hydrogen- or oxygen-transfer and that of the biochemist who is conscious that a century and a half of research, including the rise and development of physical chemistry and of organic chemistry, has left us very hazy in our concepts regarding the more intimate relation of oxygen itself to the chemistry of our tissues. His natural tendency is to emphasize the importance of the element which determines consciousness.

Now as affairs scientific go each viewpoint has its too ardent advocates, but again as affairs scientific go the problem is being recast into new forms in which mechanism plays a minor part. Under the safe guidance of thermodynamics there are being formulated and experimentally evaluated the energy relations of different systems. The significance of this will be revealed as the story unfolds.

It is well known that when the free energy of a chemical process can be made to flow in a purely electrical channel its measurement can be made elegantly exact. Consider for instance the classic case of the reduction of ferric ions to ferrous ions at a definite acidity by hydrogen. The device used in the study of this case is the following:

A hydrochloric acid solution is divided (see Fig. 1) and its

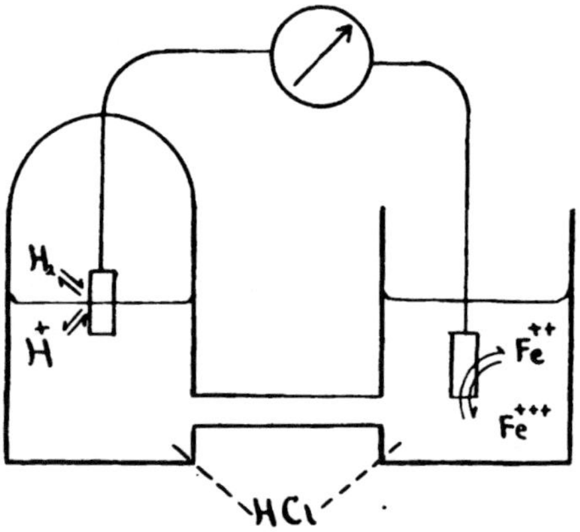

Fig. 1.

two containers are connected by a tube made comparatively narrow in order that two processes may be approximately isolated. To one vessel is added a definite mixture of ferrous and ferric chlorides. In the other is placed hydrogen gas at a definite pressure. In each is immersed a platinum electrode. This arrangement constitutes an electric cell. Accompanying the withdrawal of current from this cell there occurs in one of the half-cells the oxidation of hydrogen to hydrions and in the other half-cell the reduction of ferric to ferrous ions. It is an example of what Ostwald described by the spectacular phrase "chemical action at a distance." The reaction may be reversed by driving current from an external source against the electromotive force of the cell. In short, the cell is reversible. If now at constant temperature the external electromotive force is nicely balanced against that of the cell there is experimentally attained one of the nearest approaches to the ideal condition for maximum work. Thermodynamics then furnishes the equation relating the free energy of the reaction to its equilibrium constant.

The free energy in this instance is measured in electrical units and since we know that one faraday of quantity is associated with the transformation of one gram mol of ferrous to ferric ions, it is useful to center attention upon the intensity factor of the work term, namely, the potential difference or the electromotive force of the cell. It describes the driving force with which a definite hydrogen pressure, restrained by a definite hydrion concentration, tends to transform an equi-molecular mixture of ferrous and ferric ions toward the completely ferrous state.

For many years this method of study has been applied to inorganic systems and successfully applied when the systems studied have been well chosen and adequately formulated. These conditions for success are important, and are no small items of the undertaking; for it is to be emphasized that the thermodynamic equation merely formulates the relation of energy change to *a* process in general terms and leaves it to the user first to determine whether the energy change in question is susceptible to measurement by a particular device and, if it is, to discover in the second place what relations among the components of the system will furnish a successful solution.

As a guide in the latter task, it is more or less immaterial what scheme of mechanism is postulated but consistency is essential. Elsewhere I have described in more detail the following scheme.

The difference of potential, E, between an electrode and a solution is assumed to originate in the difference in the escaping tendencies of the electrons in the two phases. This leads to the "fundamental" equation (1).

$$E = C - \frac{RT}{F} ln\, e, \quad . \quad . \quad . \quad . \quad . \quad . \quad . \quad (1)$$

where *e* represents the escaping tendency of the electrons in the solution system described by

$$Ox + ne \rightleftharpoons Red^{n-}$$

$$Ox^{n+} + ne = Red,$$

or by intermediate cases.

For purposes of simplicity, it will be assumed that it is legitimate to proceed with the argument by employing concentrations in place of "activities" and that the equilibrium state for one of the above reactions can be written as in (2):

$$e = \sqrt[n]{K\frac{[\text{Red}^{n-}]}{[\text{Ox}]}}. \quad \ldots \ldots \quad (2)$$

Substitution of (2) in (1) yields (3)

$$E_h = C' - \frac{RT}{nF} ln \frac{[\text{Red}^{n-}]}{[\text{Ox}]}, \quad \ldots \ldots \quad (3)$$

by a collection of the constants in the term C′ and the reference of E to the hydrogen standard as designated by E_h.

Now the important consequence of this formulation is that the arbitrarily assumed difference between oxidant and reductant is one which makes the reductant, in the case written into (3), an anion. Obviously, the oxidant could be a cation and the reductant neutral or the charges could be otherwise distributed so long as the difference is equivalent to a gain of electrons by reductant. In any case the reductant is less basic or more acidic than the oxidant and thus it is evident that the hydrion concentration of the solution is a fundamentally important factor.

This becomes more definitely evident when the equation is developed.

Consider the simple case

$$\text{Ox} + e \rightleftharpoons \text{Red}^-$$

for which the electrode potential equation is

$$E_h = C' - \frac{RT}{F} ln \frac{[\text{Red}^-]}{[\text{Ox}]}. \quad \ldots \ldots \quad (4)$$

Since Ox is assumed not to dissociate, [Ox] is the total oxidant $[S_o]$.

To include the total reductant $[S_r]$ write

$$[S_r] = [\text{Red}^-] + [\text{H Red}]$$

and

$$\frac{[\text{Red}^-][\text{H}^+]}{[\text{H Red}]} = K_a.$$

Hence

$$[\text{Red}^-] = \frac{K_a[S_r]}{K_a + [\text{H}^+]}. \quad \ldots \ldots \quad (5)$$

The substitution of (5) in (4) and the collection of constants which can be collected gives:

$$E_h = E_0 - \frac{RT}{nF} ln\frac{[S_r]}{[S_o]} + \frac{RT}{nF} ln[K_a + [H^+]]. \quad . \quad . \quad (6)$$

It is only a slight elaboration of this process which leads to the formulation of the case of a two-electron transfer described by:

$$Ox + 2e \rightleftharpoons Red^{--}$$

$$E_h = E_0 - \frac{RT}{nF} ln\frac{[S_r]}{[S_o]} + \frac{RT}{nF} ln[K_1K_2 + K_1[H^+] + [H^+]^2]. \quad (7)$$

Furthermore it is merely elaboration of the simple principle which leads to rather formable equations in special cases.

I shall try not to introduce many more of these tiresome equations but I do believe one example is worth study since its analysis in a specific case will clearly indicate experimental procedure.

In the experimental study of a specific case to which equation (7) applies, the procedure will be in two distinct steps. In the first the system is buffered so that $[H^+]$ remains at a particular value. The constant value of the last term of (7) when combined with E_o, reduces the equation to the simple form (8)

$$E_h = E'_0 - \frac{RT}{nF} ln\frac{[S_r]}{[S_o]}. \quad . \quad . \quad . \quad . \quad . \quad (8)$$

When $\frac{[S_r]}{[S_o]} = 1.$ $E_h = E'_0.$

Since the value of R, the gas constant, of T, the absolute temperature, and of F, the faraday, are known, it now remains to determine n, if it is not already known from chemical evidence. This is done by varying the ratio $\frac{[S_r]}{[S_o]}$. In the case at hand $n = 2$.

The second process is to keep the ratio $\frac{[S_r]}{[S_o]} = 1$ when the term in which it appears becomes zero and the equation reduces to (9).

$$E_h = E_0 + \frac{RT}{2F} ln[K_1K_2 + K_1[H^+] + [H^+]^2]. \quad . \quad . \quad (9)$$

If we are fortunate enough to deal with a case where K_1 and

K_2 are small $[H^+]$ can be made large in relation to K_1 and K_2 and (9) is practically

$$E_h = E_0 + \frac{RT}{F} ln[H^+] \quad . \quad . \quad . \quad . \quad . \quad . \quad (10)$$

When pH = 0. $E_h = E_0$.

If $[H^+]$ can next be made small in relation to K_1K_2

$$E_h = E_0 + \frac{RT}{nF} ln[K_1K_2].$$

Hence K_1K_2 is determined.

Intermediate values of $[H^+]$ reveal the values of K_1 and K_2 separately as will be shown graphically.

Returning now to the first step we may illustrate it by means of Figs. 2 and 3. In each instance *n*, the number of equivalents involved in the transformation of oxidant to reductant, is 2. Consequently the same type-curve applies to all these cases, and interest centers upon the relative position of each specific curve along the E_h axis.

By the equation or from the fact that the curve is asymptotic to the potential axis in its upper and lower reaches, it is evident that the potential would be infinitely positive were the pure oxidant alone present or infinitely negative were the pure reductant alone present. However, to establish a pure oxidant in the conditions specified, it would be necessary not only to make use of a reagent of infinite positive potential but also to guard the material against interaction with the solvent. We therefore have no interest in "perfectly pure" oxidants or reductants but only in those having impurity below the limits of detection. Then, since a mere trace of an extraneous reagent would shift the system definitely into the more flattened portion of the curve, we choose for characterization of the system the most stable condition, namely, that of the mid-point of the curve.

If at a definite pH value the curve of one system is positive to that of a second, the first system is oxidizing with respect to the second and the second reducing with respect to the first. This must be qualified by noting that when two systems are close to one another a high percentage oxidation of the lower could be in the position of an oxidant with respect to a low percentage oxidation of the second.

In any case the formulation of simultaneous equilibria involves the total quantities of the materials interacting.

When pH is constant the curves relating percentage reduction to potential are so monotonous in form that they are easily remembered, and we may center attention on the diversified pictures of the second aspect, the relation of the mid-point of reduction to variation in pH.

When the percentage oxidation is kept at 50 per cent the term $ln\frac{[S_r]}{[S_o]}$ becomes zero. Then variation of $[H^+]$ by means of definite

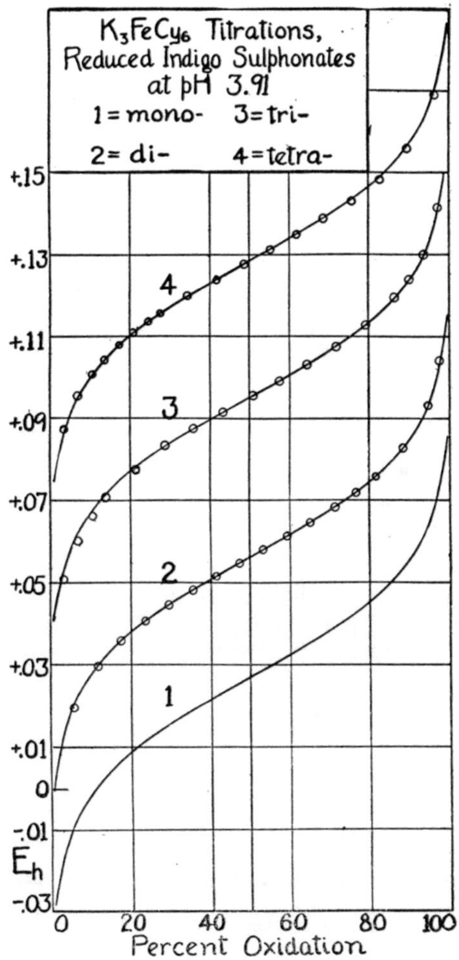

FIG. 2.

buffers leads to pictures such as Fig. 4. It is hardly necessary to examine in detail the correspondences between the algebraic and the graphical representations. It will suffice to state the following consequences.

Where the projection of the upper section of the curve intersects with the next section the value of pH is the value of log $\frac{1}{K_1}$ expressed as pK_1. The next intersection determines pK_2.

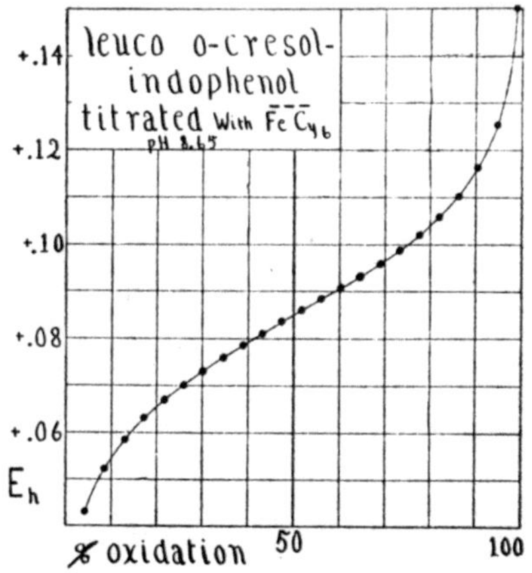

FIG. 3.

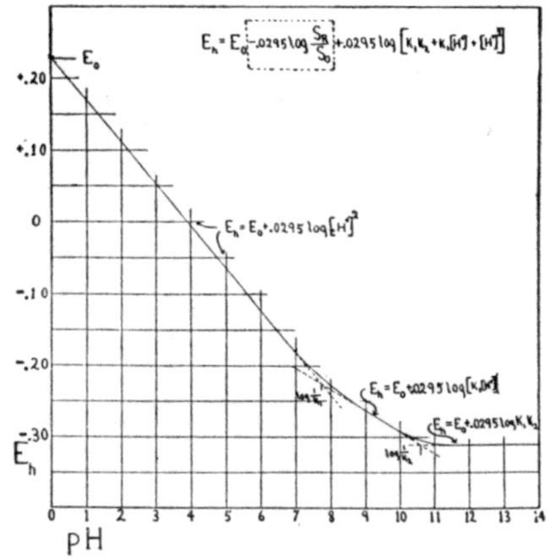

FIG. 4.—Variation of potential with pH when total oxidant and total reductant are kept constant at $\frac{S_r}{S_o} = 1$.

Anthraquinone, 2,7-disulphonic acid and its reductant at 25°.

(From Chem. Rev., Vol. II.)

At low values of pH the potential varies linearly with pH, and when pH = 0, $E_h = E_0$. E_0 is the so-called "normal potential."

The value of this normal potential is a convenient datum with which to fix the general position of the system.

Thus by **experimental control of conditions** the several items of the equation are evaluated step by step.

Somewhat more complicated than the case illustrated in figure 4 is the case of an indophenol system such as

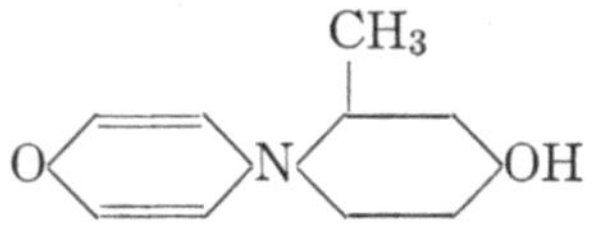
oxidant

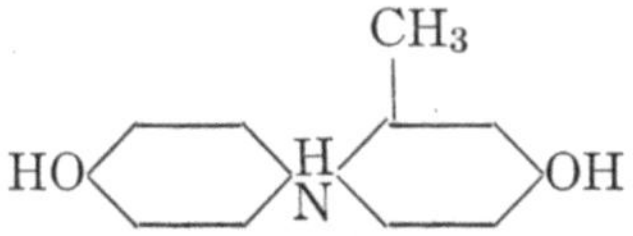
reductant

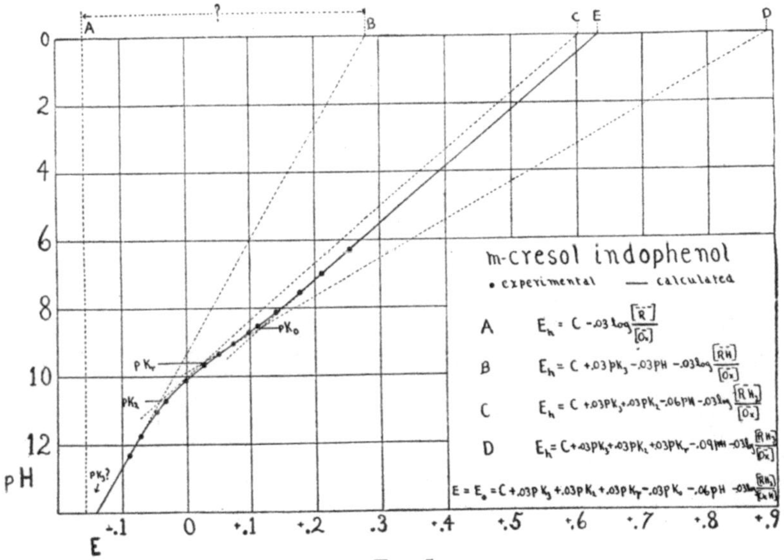

FIG. 5.

In addition to the two acidic groups created by reduction there is a phenolic group common to oxidant and reductant. Because the dissociation constant of this group changes when the molecule is oxidized or reduced the value of each constant can be determined. The change makes itself evident in a displacement of the trend of the curve in Fig. 5.

Figure 6 shows similar E'_0: pH curves for three other indophenols.

If now the results of the first type of experiment be synthesized with those of the second there results a surface such as the one representing the system of 2,6-dibromo indophenol shown in Fig. 7.

It will be remembered that in the first experiment the level of pH was maintained constant while the percentage oxidation was varied. The potential then follows one of the S-curves. In the second type of experiment the percentage oxi-

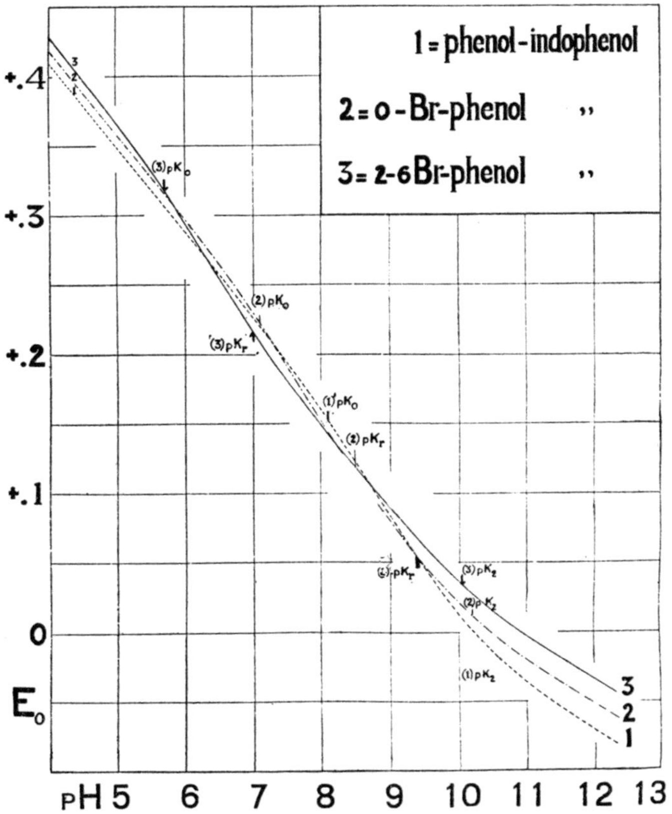

FIG. 6.

dation was kept constant, preferably at 50, and pH was varied, giving one of the curves running from the upper right-hand corner to the lower left-hand corner of the isometric drawing.

The S-curve at the upper right-hand corner is a curve of constant potential. It shows how the system would behave were it introduced as an indicator in very small relative concentration and were acted upon by another system present in large amount so as to hold the potential constant while pH is changed. Stabilizers of oxidation-reduction intensity are comparable with acid-base buffers. I call them poising systems.

A still more complicated case can be briefly reviewed by means of the formulating equations. I refer to a system of which benzidine, for example, is the reductant. In such cases the reductant has two dissociation constants Kr_1 and Kr_2 and the oxidant two dissociation constants Ko_1 and Ko_2 as defined in the original paper on this subject to which reference must be

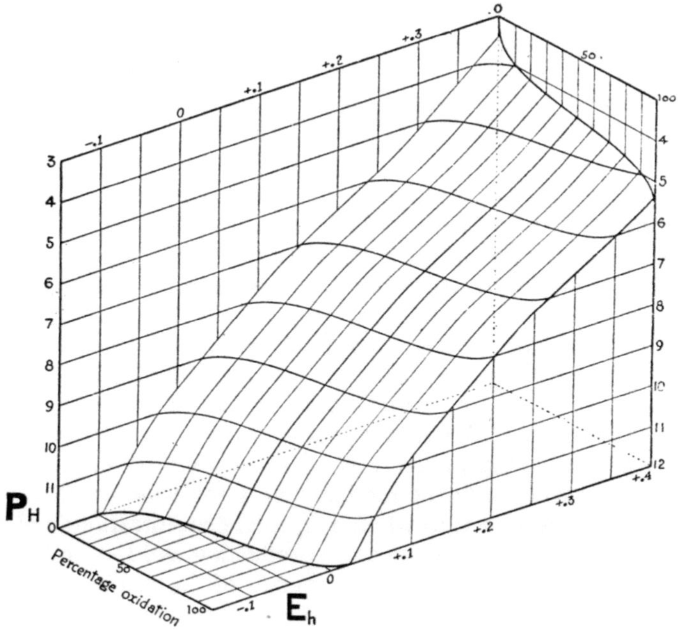

FIG. 7.

made for details. The disconcerting aspect is that certain species of these ions or undissociated residues combine to form the colored substances called meriquinones. If this meriquinone is designated by M the equation is

$$E_h = \underset{\text{I}}{E_0} - \frac{RT}{2F} ln \underset{\text{II}}{\frac{[S_r] - [M]}{[S_0] - [M]}} - \underset{\text{III}}{\frac{RT}{F} ln pH} + \frac{RT}{2F} ln \underset{\text{IV}}{\frac{Kr_1Kr_2 + Kr_2[H^+] + [H^+]^2}{Ko_1Ko_2 + Ko_2[H^+] + [H^+]^2}}.$$

If M is formed from equimolecular parts of oxidant and reductant term II vanishes when $[S_r] = [S_o]$. When also $[H^+]$ is large in relation to the constants in term IV, this term becomes zero. When pH = 0, term III vanishes. E_0 is thus found.

Maintaining $[S_r] = [S_o]$, $[H^+]$ is next varied whence by methods previously described the constants of term IV are determined.

Next [M] can be determined in a specific case at constant pH by comparing the potential at 50 per cent oxidation when

$$E_h = E'_0,$$

and at small percentage oxidation when

$$E_h = E'_0 - \frac{RT}{2F} ln \frac{[S_r] - [M]}{[S_0] - [M]}.$$

Also when [M] is determined the association constant of meriquinone formation is determined from a rather complex set of equations developed in the original paper on the subject.

Now while it is true that the benzidine, *p*-amino dimethyl aniline and similar systems are very difficult to study by reason of the rapidity of the decompositions, we believe that their main characteristics have been outlined, formulated and evaluated with sufficient accuracy to justify some important conclusions. So far as the organic chemistry of these systems is concerned much remains to be done in the way of more accurate evaluation of constants, particularly the meriquinone association constants, the detailing of the extent to which pH, total concentrations and percentage oxidation determine color production and the explanation of peculiarities in the absorption spectra of the meriquinones and haloquinones. But so far as biochemical applications of benzidine and similar reagents are concerned our conclusion is definite. These systems are too complex for practical use. We could now design, with rational instead of empirical methods, the construction and use of a benzidine reagent, but it seems not to be worth while to do so for we have traced too many pitfalls into which the user may fall.

Since the components of no two systems have the same dissociation constants, it is evident that there is no common E'_0: pH curve. For instance, the curves of Fig. 8 all differ. The ferro-ferricyanide system is oxidizing with respect to the indophenol system in alkaline, neutral and *slightly* acid solution and reducing in *distinctly* acid solution. The methylene blue curve has a remarkable slope quite out of harmony with the more usual. In short, the description of each system must be hammered out on the anvil of mathematical formulation with data hot from the fire of experiment. By this sort of hand labor, and this sort

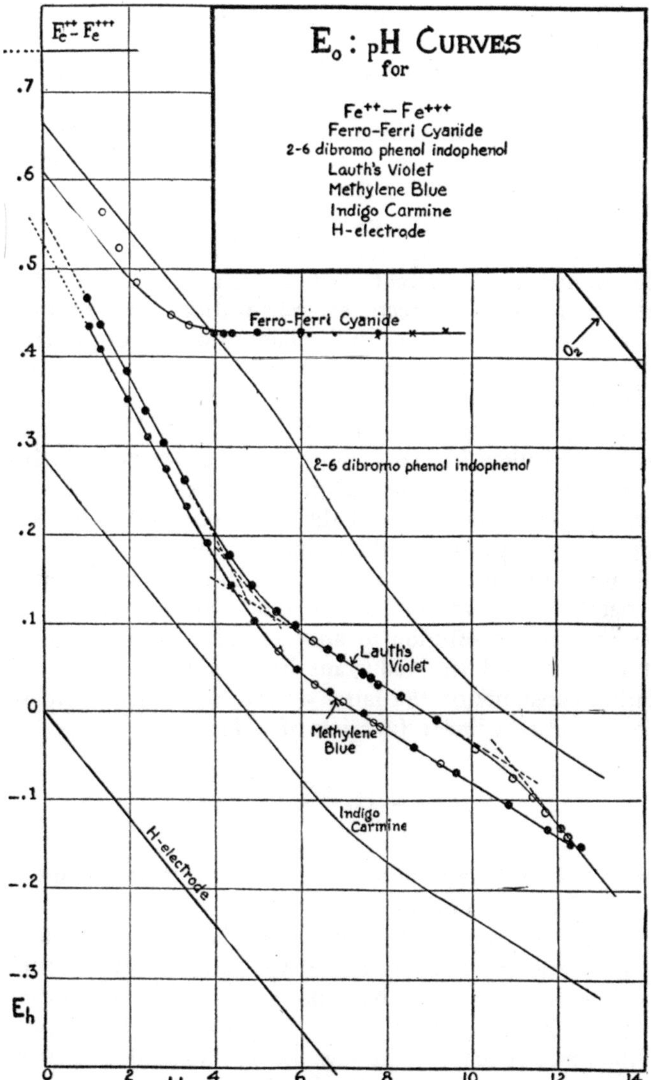

FIG. 8.

only, can there be provided stencils, as it were, with which to compare one system with another.

Now even a casual survey will reveal the need of such descriptions for purposes ranging from the more general correlations of the physical chemist to those of the cytologist in his dealing with the conduct of a specific reagent in the oxidation-reduction metabolism of the living cell. Therefore, the need of laborers is great.

We, of the Hygienic Laboratory, have been particularly interested in developing a series of indicators useful in detecting intensities of reduction in a manner comparable with the use of acid-base indicators in detecting intensities of acidity (pH value). Fig. 9 illustrates the E'_0 : pH curves of some of the more useful of such systems.

Inasmuch as so many characteristic constants of a system can be determined by the potentiometric method, data of considerable value regarding the effects of substitutions are accumulating. The work of Biilmann and his collaborators in Copenhagen, the work done by LaMer and Baker in Columbia University, that of Conant and his students at Harvard and of my associates, Dr. Cohen, Dr. Gibbs and Dr. Sullivan, at the Hygienic Laboratory, have resulted in many interesting conclusions which I shall not have time to describe. Figure 10, taken from LaMer and Baker's paper, shows some of the shifts observed when different groups are substituted in the quinone molecule.

Another interesting development is this. Quinhydrone, which is a peculiar combination of quinone and hydroquinone, dissociates in aqueous solution to an equimolecular ratio of the oxidant quinone and the reductant hydroquinone. The acid dissociation constants of the latter are so low that in acid solution the potential is a linear function of pH:

$$E_h = E_0 - 2.3 \frac{RT}{F} pH.$$

Since E_0 has been determined, a measurement of E_h and of temperature T determines pH.

The quinhydrone system had been given a preliminary study by Haber and Russ in 1904 and was more carefully studied at Columbia University by Nelson and Granger, as described in Granger's dissertation of 1920. However, it is to the credit of Biilmann (1920) that he saw the application of what he calls the quinhydrone electrode to the determination of pH. It is now a serious rival of the hydrogen electrode and is being applied extensively.

Concurrently with the extension of the quinhydrone electrode to uses of more immediate practical importance, has run the careful study of the effect upon its potential of varying concentra-

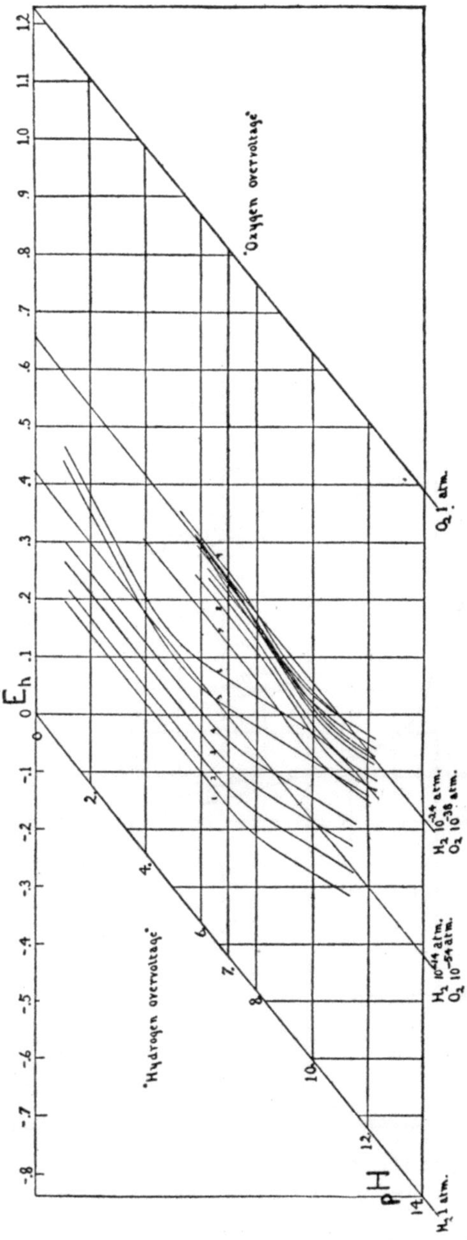

FIG. 9.—E'_0 : pH curves of a few oxidation-reduction indicators.

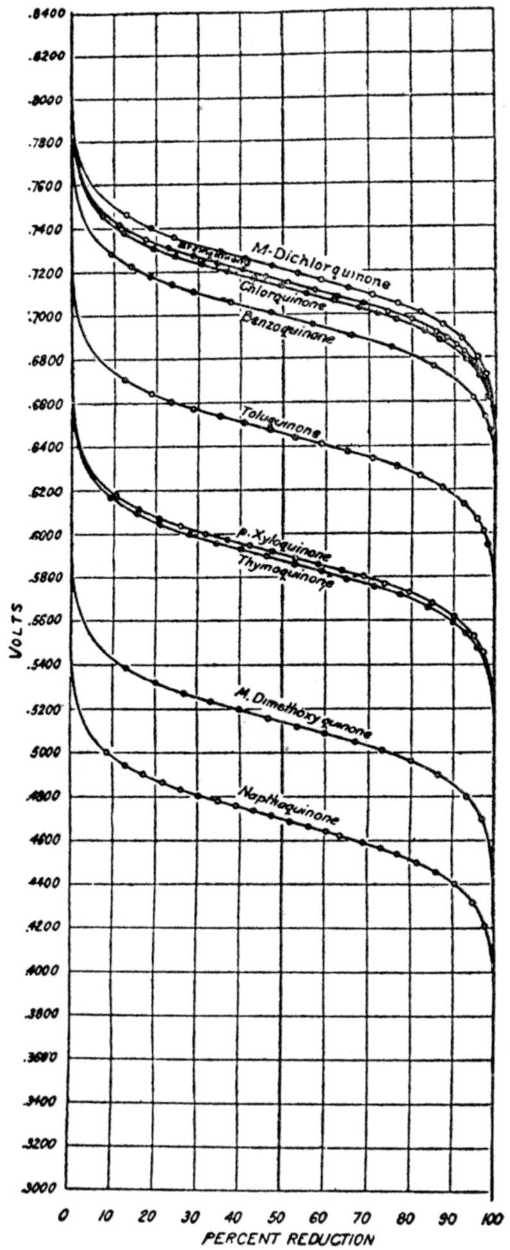

FIG. 10.—Relation of electrode potential to percentage reduction showing effects of substitution.

After Baker.

tions of different neutral salts. The formulation of the results by Sørensen and Linderstrøm-Lang has, of course, been in terms of "activities." This concept I have not used since it does not alter the general form of the equations and introduces activity coefficients which sometime complicate the equations to an extent unnecessary for purposes of elementary exposition. Without it salt-effects are described as "salt-errors."

Since organic compounds run the gamut of variation in number, strength and general character of their ionizations, it can be predicted that they will furnish excellent material for testing some of the newer theories of solution. They also provide excellent material for the comparison of aqueous with non-aqueous solutions, as Conant has found.

Let me now dwell briefly upon another fascinating aspect.

We formulated our equations with the guidance of the postulate of electron-transfer. That the resulting equations fit the experimental facts is no argument whatever that the postulate represents actuality. It can easily be shown that the same working equations can be derived with the guidance of other postulated mechanisms or derived entirely without the guidance of any mechanistic postulate whatever.

But once we have the experimental data and realize their independence of mechanistic postulate we are curious to see if they are suggestively in favor of any specific mechanism of oxidation-reduction.

In the first place it has been definitely shown that the ionization constants appertaining to groups in the reductant created by the process of reduction have distinctly different values. Now while an ionization constant is in a sense representative of a statistical affair, it may certainly be inferred that a single diacidic anion can and probably does acquire hydrions stepwise. At constant hydrion concentration where the anions would *tend* to take one hydrion and one only, the oxidation-reduction process in the cases mentioned involves two equivalents and these equivalents are paired, in the sense that the most accurate measurements made have revealed no trace of a stepwise reduction.

If the process of intuitive inference is to be permitted in science it seems to me that it is legitimate to conclude that in the class of cases referred to, the process of reduction is essentially the acquirement of an electron pair followed by or not followed by an acquirement of hydrions according to the relation of their concentration to the several dissociation constants.

After all it is an electron exchange somewhere in the experimental set-up that is measured.

But this sort of measurement is not generally applicable in

organic chemistry any more than it is in inorganic chemistry. Conant has furnished good evidence that certain ethylene linkages are hydrogenated not by any process that is susceptible to electrochemical measurement and formulation, but by the direct addition of hydrogen with the aid of catalysts.

Or consider the classic case of direct oxygen addition, namely, the oxidation of hemoglobin to oxyhemoglobin. Conant could find no evidence that this system can be measured or formulated by the methods we have been considering. On the other hand he found that the hemoglobin-methemoglobin system can be so measured and formulated.

Thus the electrochemical method by both its positive and its negative revelations is beginning to disperse the protagonists of too narrow general theories. Electron-transfer, hydrogen-transfer, oxygen-transfer are each to receive due consideration.

What then of biochemistry? Few reversible, electromotively active organic systems are found among the compounds which the biochemist has isolated from living cells and from among the products of their metabolism. Yet the favorite tool of the biochemist in his study of biochemical oxidation-reduction has been the reversible and electromotively active dye system, notably that of methylene blue.

This is a truly remarkable situation.

Now Biilmann has shown by the electrochemical method that in the reduction of certain of the azo dyes there is a reversible stage followed by an irreversible rearrangement. We have shown that in the oxidation of certain diamines, such as benzidine, there is a reversible stage followed by an irreversible process which we infer to be a certain type of autoxidation. Conant, who has recently reviewed the subject of irreversible oxidation-reduction, presents the view that there can be in many instances a reversible stage followed by the irreversible step leading to the products ordinarily dealt with, and by means of certain postulates he presents a very good rational aspect to the otherwise still useful empirical potentials defining what reagents will produce the reduction and what reagents will not. Thus in the study of irreversible reactions there is being stimulated the search for that type of intermediate, reversible change which is amenable to formulation and to definite measurement by the methods here described.

I mention this in connection with the biochemical problem because in our own work, dating from 1919 when Gillespie opened the subject (Gillespie 1920), we have found that the effects of cell suspensions upon electrodes were as if some system or systems in the cell had a definite electromotive activity. Recently Mr. Cannan of University College, London, while working with Dr. Cohen and me at the Hygienic Laboratory, obtained results which can best be explained at present on the hypothesis that the living cell activates some of its metabolites in such a way that there is produced an extremely small quantity of electromotively active material. This the cell continues to supply from the comparatively large reserve. The smallness of the quantity present at any moment accounts for the precarious nature of the potentials, e.g., the ease of polarization of electrodes. The definiteness of some of this material would account for the definite trend of potentials and the interaction between cell suspension and reversible oxidation-reduction indicators. The results entirely freed from postulates lead to a powerful experimental method of attack.

Now the reaction

Succinic acid + methylene blue $\rightleftharpoons$ fumaric acid + methylene white might presumably alone reach equilibrium as defined by

$$\frac{[\text{fumaric acid}]\ [\text{methylene white}]}{[\text{succinic acid}]\ [\text{methylene blue}]} = K,$$

but it would probably take a very long time . In the presence of muscle tissue the attainment of equilibrium is greatly hastened. Thunberg, observing the extent to which the methylene blue system is transformed and using our data for the methylene blue system, calculated the potential to which the fumaric $\rightleftharpoons$ succinic system is transformed by muscle tissue. His value is +.005 at pH 6.7.

Quastel in the same type of experiment with resting bacterial cells determines K and thereby Conant with the use of our potential data for methylene blue finds —0.009 at pH 7.3. As Conant remarks the agreement is "good" when it is remembered that neither temperature nor pH were the same in the two experiments. Conant adds: "The fact that two different investigations using different biochemical catalysts obtain essentially the same equilibrium constant is convincing evidence that the process is strictly reversible under these conditions . . . the discovery of any method of changing an irreversible system into a strictly reversible one is of much importance to the general theory of oxidation-reduction reactions quite apart from biochemistry."

Now the potential found for cell suspensions containing this succinic acid-fumaric acid system is distinctly negative to that which the Needhams on the one hand and Wurmser and Rapkine on the other are finding in aerobic cells by micro-injection of oxidation-reduction indicators. The region of potential they find is confirmed by other types of experiment and of argument brought forth by Cannan, Cohen and Clark and by some diffusion experiments of Mrs. Brooks. In short, very different types of experiment converge to the conclusion that the aerated cell maintains a potential distinctly oxidative to the methylene blue system and distinctly reductive to certain brominated indophenol systems. But this is enormously far distant from the potential of any system in equilibrium with the oxygen of our atmosphere. We can see no other conclusion than the following: In the participation of oxygen in the chemistry of the living cell there is, on the one hand, no *potentiometrically* measurable oxygen equilibrium and yet, on the other hand, there is some means by which the oxygen, contending against the reductive processes of the cell, maintains a more or less definite and potentiometrically measurable level of oxidation-reduction intensity.

Thus is resolved, we believe, two distinct aspects of the biological problem, the problem of the catalysis of oxygenation as a special aspect of oxidation applying to the aerobic metabolism of the cell and the problem of the catalysis leading to that type of oxidation-reduction which can be formulated by the electrochemical methods described, and studied when the cell is under anaerobic conditions.

This view should dispel certain controversies now current. It may lead to worse but these can be endured for a time since the experimental methods will lead to definite results.

For references to the literature see: Clark, *Chemical Reviews*, Vol. II, p. 127; Conant, *Chemical Reviews*, Vol. III, p. 1.

Crystal Structure in its Relation to Chemical Problems

BY

RALPH W. G. WYCKOFF

RESEARCH CHEMIST, GEOPHYSICAL LABORATORY
CARNEGIE INSTITUTION OF WASHINGTON

Lecture delivered at Columbia University in the special course on Contemporary Developments in Chemistry given in the Summer Session of 1926 on the occasion of the opening of the Chandler Chemical Laboratories

Papers from the
GEOPHYSICAL LABORATORY
Carnegie Institution of Washington
No. 624

COLUMBIA UNIVERSITY PRESS
NEW YORK
1927

Printed from type. Published March, 1927

Printed in the United States of America

CRYSTAL STRUCTURE IN ITS RELATION TO CHEMICAL PROBLEMS[1]

By

RALPH W. G. WYCKOFF

It is the purpose of this lecture to indicate some of the different ways in which X-ray diffraction analysis has supplied information of chemical interest and to suggest some of those paths along which this method of physical research may be expected to become increasingly helpful. These applications are already so numerous that it will be possible to do little more than list them and to cite as illustrative examples a very few of the numerous data that have been accumulated.

Although the first X-ray diffraction measurements upon crystals were carried out not more than fourteen years ago, crystal analysis has in the meantime built up for itself complicated and new methods of experimentation and interpretation. Its various experimental procedures are purely physical. Ways of analyzing the results of these experiments have been adopted from other branches of science and developed to meet the needs of this interpretation. These ways are in the main crystallographic and crystallography is in its essence and in its most extended development a branch of pure geometry. The substances which give rise to X-ray diffraction patterns are definite chemical compounds. In order to progress far in the interpretation of these patterns it is necessary to know something of the chemical characteristics of the compounds producing them. Conversely results of this investigation often make it possible to know or to infer something of chemical importance.

Thus though its methods are physical and though chemical information is essential to its progress, crystal analysis does not fit easily into either physics or chemistry as the boundaries of these sciences are now customarily drawn. It is perhaps worth while to try to form a picture of the relation between X-ray crystal analysis and these two fundamental sciences. In this way the present and the probable future significance of the Laue experiment will appear more clearly and sharply and the kind of knowledge that should properly be the product of the science founded on this experiment may be better appreciated.

In the past physics has been the science which defines and correlates the properties of individual and separate kinds of matter. Chemistry, on the other hand, has properly concerned itself with the phenomena which arise when different substances are brought into contact with one another. In its early stages of development the main goal of chemistry was the *description* of what takes place when various substances interact. When this descriptive phase had sufficiently elaborated itself emphasis in chemistry naturally shifted from the question of what takes place to the question of how and why these things happen. The product of this shift of emphasis is, of course, physical chemistry. Its materials have been chemical, its methods those of physics. Because the vast majority of chemical reactions take place among gases and liquids, it has become essentially a compounding of the physics and the chemistry of fluids and solutions. It is natural to expect that as knowledge increases this science of fluids will incorporate into itself more and more of the chemistry of the present time.

Physics in the last fifty years has not followed an analogous evolutionary development. The physics of the last century aimed to give a description and had ideas in explanation of the principal properties of matter in its different forms. With the birth and spread of atomic physics, however, this older science has practically ceased to exist as a living and growing entity.

We have already suggested that physical chemistry is the natural heir to those aspects of this knowledge which deal with fluids. Vital interest in solids flagged even before the appearance of atomic physics and physical chemistry. The reason for this is perhaps easily found. Physical chemistry became a rapidly growing branch of knowledge just as soon as, but not before, a sufficiently truthful picture had been found of the nature and behavior of the molecules and the atoms composing the fluids with which it deals.

X-ray diffraction measurements upon crystals are important not so much because they may be occasionally useful to mineralogy, to chemistry, to atomic physics, or biology, but because they do essentially the same thing for the science of solids that the development of the kinetic theory of gases and the theory of ionic dissociation accomplished for physical chemistry. By revealing the constituents of crystals and their arrangement with respect to one another this new method of experimentation permits the building of a new science which calculates the external properties of crystals from the properties of their ultimate units. Only preliminary and rather halting attempts have yet been made to develop such a theory. Nevertheless it is inevitable that knowledge of solids must multiply rapidly as experience is added and experiments are widened and improved.

The physicist undoubtedly would object to this division of the main body of older physics between physical chemistry and a new science of solids. He would assert that he has not relinquished his right to classical physics, that his interest is only temporarily transferred to atomic physics because it offers a more fruitful field for immediate investigation, and that as soon as the most important of these problems have been solved he will return to the material of the old physics with new information which will give this material a changed and greatly widened development. The fact, nevertheless, remains that the active trend at the present moment is towards a merging of the old physics and the old chemistry with the resulting growth of the complementary sciences of fluids and of solids.

The foregoing comparisons have been drawn between physical chemistry and the science of solids and their complementary character has been emphasized because it is believed that only from this standpoint can the results of crystal analysis be adequately judged. Both sciences are a compounding of physics and chemistry and thus the nature of the information that they yield is in many respects similar. They have, however, this striking difference. The most important property of fluids is their ability to react together. Thus the chemical aspects of the science of fluids far outweigh all others. On the other hand the general physical properties of solids are of much greater importance than their occasional chemical reactions. It is for this reason that in emphasizing the chemical aspects of the results of crystal analysis we shall necessarily neglect many of its most interesting conclusions.

The determination of the nature of the forces of chemical combination and an understanding of the manner in which they operate between different kinds of atoms are obviously of the most fundamental importance to chemical theory. Closely associated with these questions is the problem of the existence of molecular groupings of several atoms which may function as single units in fixing both the chemical and the physical properties of compounds. Though the determinations of atomic positions which are furnished by crystal structure results obviously cannot prove anything about the forces existing between atoms, this information when taken in connection with knowledge derived from other forms of experiment has thrown light upon the nature of chemical combination.

The development and proof of the gas laws showed unmistakably that in the gaseous state practically all compounds consist of molecules each composed of a few atoms. The extension of these gas laws to solutions indicated that similar molecular groups of atoms exist in the large class of solutions to which these laws satisfactorily apply. Furthermore the phenomena of organic chemistry are scarcely comprehensible unless the atoms of organic compounds are aggregated into molecules not only in the gaseous state and in solutions, but also when the compounds exist as liquids and as solids. Proceeding from these experimental results the theoretical chemistry of fifteen or more years ago generally assumed that the normal state of any compound is one in which it consists of such molecular aggregates of atoms. The earliest results of crystal analysis, together with the data subsequently collected, indicate that most of the crystals yet studied do not consist of molecules. The nature of this evidence can be simply illustrated by reference to sodium chloride. The structure of this compound will be remembered as a cubic one in which each atom of sodium is surrounded by six equidistant atoms of chlorine and each atom of chlorine by six equally near sodium atoms. Since therefore no pair of sodium and chlorine atoms are especially close to one another, molecules of NaCl would be possible only in case it were assumed that the distance between two atoms is independent of the intensity of the forces between them. This assumption is equivalent to an hypothesis of the absolute constancy of atomic radii which is not borne out by the existing X-ray data.

In a few instances, crystals appear to be made up of molecules. Hexamethylene tetramine and arsenious oxide are typical examples. Sub-molecular groupings of atoms, such as the carbonate and the sulfate groups, are likewise frequently encountered among the observed crystal structures. Caution must always be used in drawing conclusions from crystal structure data about the existence of molecular groupings. The following example will suffice to illustrate the difficulties that may arise. The close association of four CN groups about a zinc atom in $K_2Zn(CN)_4$ and of the four oxygen atoms about an atom of molybdenum in Ag_2MoO_4 are compatible with the existence of $Zn(CN)_4$ and MoO_4 ions in these crystals. If they were taken as proofs of the presence of these ions, however, it would be necessary likewise to conclude that crystal structure evidence proves that chemically unreasonable MgO_4 groups occur in the structurally isomorphous $MgAl_2O_4$.

It is important to try to relate this experimental suggestion of both molecular and non-molecular solids to present-day ideas concerning the nature of the forces of chemical combination. In recent years it has been customary to distinguish between polar and nonpolar chemical bonds. If the electron is recognized as the important agent in holding atoms together, then strongly electropositive and strongly electronegative atoms may conceivably be bound together through the acquisition by the electronegative atom of one or more electrons belonging to the neutral electropositive atom. Such a bond is polar. When chemical union is neither accompanied by, nor is the result of, such an electronic interchange, the bond is nonpolar. Experiment shows that all compounds involving bonds that are undoubtedly polar possess crystal structures which, like sodium chloride, are non-molecular. If polar compounds yield on solution ions which consist of several atoms, these complex groups can likewise be discerned in their crystal structures. Thus calcite is most simply imagined as a somewhat distorted sodium chloride grouping of calcium atoms and carbonate groups and $(NH_4)_2SnCl_6$ is a fluorite (CaF_2) structure with NH_4 groups replacing fluorine atoms and $SnCl_6$ groups in place of calcium atoms. Since in solution these compounds consist of ions bearing opposite charges, their non-molecular character in the solid state is readily understood if their crystals likewise are composed of charged atoms or atom groups which are held together mainly by the electro-static forces existing between these "ions." It has been asserted that the intensities of X-ray reflections prove the existence of charged atoms in these crystals. The calculations underlying these supposed proofs have assumed that the scattering powers of the atoms involved were proportional to their electron numbers. It has long been suspected that this assumption is only very roughly true. Recent experimental determinations[2] of the F-curves of various atoms confirm this suspicion. These demonstrations cannot therefore be convincing. In spite of this absence of direct proof there can be little doubt of the essentially ionic nature of polar bonds.

No satisfactory picture can be drawn of the mechanism of nonpolar linkages. It is perhaps simplest to imagine that they arise through electrons being held in common, or shared, by neighboring atoms. Such a sharing would be produced, for instance, if the path of an electron lay partly within the field of one atom and partly within the field of another. From such a standpoint as this, completely polar and completely nonpolar bonds are merely limiting cases in a continuous series of bonds.

It is natural to suppose that such more or less nonpolar bonds exist between the atoms in molecules and in complex ions. Nevertheless many investigators have assumed that nonpolar unions either never exist or are of comparative unimportance. From this point of view the carbon-to-oxygen bond in calcite, for instance, is not qualitatively different from the calcium-to-oxygen bond. This compound, hence, splits on solution into calcium and carbonate ions, rather than into any other imaginable pair, mainly on account of the predominantly great electrostatic attraction existing between carbon and oxygen. At the present time there appears to be no satisfactory way of selecting between these two interpretations. The phenomena of organic chemistry, however, are more easily understood in terms of nonpolar as well as polar linkages.

A most fruitful development of the science of solids would unquestionably follow an answer to this question of the exact character of chemical bonds. Efforts have already been made to calculate physical properties of crystals from their structures by employing the assumption that all their atoms are ions. Thus lattice energies, compressibilities and the like have been calculated [3] for the alkali halides, for ZnS, for CaF_2, for rutile (TiO_2) and for other simple types of structures. These lattice energies are of the greatest importance to chemical theory because they furnish one of the essential links in the chain of energy values through which the possibility of a chemical reaction and the progress of this reaction can eventually be calculated. Attempts productive of a certain success have also been made to calculate the refractive indices [4] and the crystallographic constants [5] of some crystals from a knowledge of their atomic arrangements. The assumption of simple ions underlying all these calculations is not yet well grounded and consequently they are liable to future revision. Nevertheless they are of the greatest importance as examples of one way in which crystal structure knowledge will add to an understanding of both the physics and the chemistry of solids.

Besides ionic compounds composed of strongly electro-positive and electro-negative elements on the one extreme and the distinctly molecular compounds on the other, there are innumerable compounds intermediate between them which involve the amphoteric elements lying midway across the periodic table. The character of the atomic bonds in these compounds is little understood and only a few of the simplest of their structures have yet been satisfactorily established. Such crystals as carborundum (SiC in its various forms),[6] the high temperature modifications of quartz [7] and cristobalite [8] (SiO_2) and the diamond show no evidence for the existence of molecules. In the truly ionic compounds there is no relationship between the valence of an atom and the arrangement of other atoms about it. In the silicon and the carbon compounds just mentioned, however, the number of other atoms immediately surrounding each atom is equal to its chemical valence. The experimental data are not yet numerous enough to show whether these numerical coincidences are affairs of chance or whether atoms in such amphoteric non-molecular crystals are bound by some form of electron sharing rather than by polar unions. From attempts that have been made to obtain the crystal structures of several of the less simple silicates and aluminates it is apparent that the structures of many of these amphoteric solids are of exceeding complexity.

Besides those compounds in which the atoms are held together by primary valences, innumerable chemical compounds exist which are formed through the operation of forces of secondary or residual valence. It is of interest to see what information concerning these compounds can be derived from a study of their atomic arrangements. Although only the simplest compounds of this type have yet been investigated, their structures are in complete accord with the demands of the coordination theory of Werner. Thus the NH_3 groups in the hexammoniate halides of nickel are arranged, six about each nickel atom, at corners of an octahedron. A similar distribution undoubtedly prevails for the NH_3 groups in $Ni(NO_3)_2{\cdot}6NH_3$ and for the H_2O groups in $Zn(BrO_3)_2{\cdot}6H_2O$ and in the alums. It is likewise in accord with this theory that $[Ni(NH_3)_6]$ in $NiCl_2{\cdot}6NH_3$ is spatially analogous to $[PtCl_6]$ in $(NH_4)_2PtCl_6$. Furthermore the beryllium atom [9] in $Be_4O(C_2H_3O_2)_6$ appears to exhibit its coordination number four and the $[Co(NH_3)_6]^{+++}$ complex in the cobalt amine [10] $CoI_3{\cdot}6NH_3$ is spatially like the $[Ni(NH_3)_6]^{++}$ complex in $NiCl_2{\cdot}6NH_3$. This nickel compound is an ammoniate rather than an amine. Crystallographic investigations have shown that the symmetry properties of some groups of amines change in a regular fashion with increases in the complexity of the substitutions within the first coordination sphere. As X-ray methods become able to handle more complicated compounds, their application to such crystallographically related amines will undoubtedly add much to the existing knowledge of the operation of the forces of secondary valence.

Various attempts have been made from time to time to define the factors which determine the atomic arrangement assumed by any crystal. Some of these factors are gradually being recognized but much more must be learned before it can be understood, for instance, why $NaClO_3$ is cubic while $KClO_3$ is monoclinic, or why $NaNO_3$ is rhombohedral while KNO_3 is orthorhombic in symmetry. It is not yet known with certainty whether the structural unit in crystals is always the atom or whether it may not sometimes be a molecular or sub-molecular grouping of atoms functioning together. In such crystals as calcite the observed reflections can be explained only if all the atoms, and not merely calcium and carbonate ions, have positions agreeing with the observed symmetry. Some evidence suggests that such groups as NH_4 and CN may act as single symmetry units. Until it has been ascertained if all the atoms in a crystal must occupy positions compatible with the observed symmetry, it cannot be known whether the shapes of atoms as well as their relative sizes are of importance in determining the nature of the observed atomic arrangement.

Neither the relative nor the absolute size of atoms—nor their spheres of influence—are as yet adequately known. The measurements resulting from crystal structure studies are of distances between atomic centers. If atoms were of invariable size their radii could be established from a knowledge of the atomic arrangements in crystals of one or more elements. The available data, however, show that this is not the case but that the effective size of an atom is very considerably influenced by the state of its chemical combination. Information in addition to that furnished by crystal structure results is therefore needed to define the atomic dimensions. Such estimates of size have been based upon measurements of the viscosity of gases, of the conductivity of ions, and their specific refractivities, and upon various ideas concerning the structures of atoms. They vary as widely as do their sources. Some are obviously more plausible than the others, but none is based upon thoroughly satisfactory grounds. The estimate of atomic size which perhaps most nearly approximates true values is the one developed from considerations of ionic [11] refractivities. Calculations [12] of radii have recently been carried out using these refractivities and the available data of crystal analysis together with various extrapolations from these data. The results of these calculations represent the most elaborate development yet offered of the idea of radii. Insofar as the nature of the chemical union between the atoms in crystals can be defined with certainty a precise meaning and probably an approximate correctness attaches to such results as these. The exact character of the chemical bond in most instances is not, however, actually known. It therefore hardly needs further emphasis that in these cases the assignment of atomic radii or even a clear definition of what is meant by the term atomic radius must await extensive increases in our knowledge of chemical combination.

The atomic dimensions which have just been discussed and which are of importance in determining interatomic distances are some sort of a measure of the maximum extension of atoms. If, however, the concentration of electrons about the atomic nucleus and the rate of falling off of this concentration are considered, a somewhat different kind of atomic radius is encountered. Comparisons between observed intensities of reflection and those calculated for various atomic models would yield information concerning these radii if enough were known of the laws of scattering of X-rays. Radii of this sort also result from attempts [13] to apply a method of calculating X-ray reflection data that has recently been proposed.[14] This procedure is based on the assumption that the distribution of the scattering elements (electrons) in any direction through a crystal can be obtained by treating all of the observed reflections together by the methods of Fourier analysis. If additional tests should show that its results have a real physical significance, a manner of interpretation will have been found which should prove of use not only to crystal analysis but to the problem of atomic dimensions.

In spite of the indefiniteness which is inherent in the problem of atomic dimensions at the present stage of its development, certain more or less empirical results concerning interatomic distances arise from accumulated diffraction data. Thus it is found [15] that among compounds having the same or similar types of atomic arrangement, interatomic distances are closely additive. The alkali halides constitute such a group of crystals. It has also been observed [16] that for most, but not for all, compounds exhibiting the ZnS, the ZnO, and the Cu_2O structures, pairs of crystals with equal electron sums have cells nearly identical in size. For sodium chloride-like crystals, however, corresponding atomic pairs are closer together the higher their valences. It has been concluded from this that the type of bond existing between atoms is the same in the NaCl, CsCl, CaF_2 and SnO_2 arrangements. This bond is supposed to differ from that found in the ZnO and ZnS-like structures.[17] Inasmuch as the former are probably ion compounds it is assumed that the atoms in all crystals like ZnO and ZnS are bound together by nonpolar electron-sharing unions. Arguments against this conclusion are readily found. Thus AgI and AgBr have very similar physical as well as chemical properties and it is hard to believe that the forces between their atoms are as different as is implied in the statement that the bond in AgBr is polar while that in AgI is nonpolar. An equally impressive objection to this conclusion is furnished by the fact that NH_4F has a ZnO arrangement [18] although the ammonium salts of the *less* electro-negative halogens are like either CsCl or NaCl.

It must be obvious from the foregoing illustrations that the factors determining atomic arrangement in crystals are not yet well understood. Nevertheless since regularities of one sort or another have already been observed and particularly since a practically limitless body of experimental material is awaiting study, this present inability to understand the structure of solids should not be a source of discouragement. It merely shows that the phenomena of the solid state are more complicated than had been expected.

More X-ray studies have thus far been carried out upon inorganic than upon organic compounds. This is due to the fact that many more inorganic compounds have simple structures with high crystal symmetry and thus are subject to more complete and detailed investigation by the limited methods now available. A few organic crystals, of which hexamethylene tetramine is the best example, have been adequately studied. The structure of this compound is a body-centered grouping of $C_6H_{12}N_4$ molecules. In these molecules the tetravalent character of the carbon and the trivalency of the nitrogen atoms are evident. The symmetry properties of the molecules probably existing in several other compounds have been established although the positions of their atoms are not satisfactorily known. Urea, pentaerythrytol, and ethane [19] are examples. These compounds all have relatively high symmetry. Nearly all organic crystals, however, have symmetries which are orthorhombic or lower. In spite of the very great limitations involved in studies of these low symmetry crystals, some X-ray measurements have been carried out upon a number of them. Both aliphatic and aromatic compounds have been examined and atomic arrangements have been suggested which are based upon the relatively meager data obtained in these investigations. For the development of these arrangements assumptions have been made concerning the shapes of the molecules involved. It is difficult properly to evaluate these results because too little is known of the extent to which the usual formulas are correct pictures of the atomic arrangements in organic molecules.

Certain results have already been obtained in crystal analysis which would not have been expected from the usual valence theory. For example the atomic arrangement in crystals [20] of B_2H_6 is apparently like that in C_2H_6 in spite of the fact that boron is trivalent and carbon tetravalent. At temperatures slightly above that of the room CBr_4 and CI_4 are cubic with atomic arrangements [21] undoubtedly like those prevailing in SnI_4 and [22] GeI_4. These structures are not simple groupings of RI_4 molecules as one might expect. Instead they seem to be face-centered distributions of more complex molecules of the composition $(RI_4)_2$. Crystals of $NaH(C_2H_3O_2)_2$ have also been studied in considerable detail. An atomic arrangement can be imagined for this compound which contains only three molecules and agrees with our usual chemical preconceptions. The observed data, however, require a unit which is eight times larger and thus contains 24 molecules of $NaH(C_2H_3O_2)_2$. In both of these latter instances the simple structures could have been deduced by ignoring some of the less conspicuous reflections. In the face of these inexplicable complications it is obvious that with organic crystals, as in other fields of crystal analysis, our earlier chemical ideas may easily prove faulty guides to atomic arrangement in crystals. X-ray measurements undoubtedly have much to offer to our knowledge of organic compounds but it seems probable that sure and sound progress will be better made by studying very thoroughly the simplest crystals and proceeding from them toward the more complicated.

The discussion thus far has been devoted to a consideration of results arising from a more or less complete knowledge of the atomic arrangement in crystals. X-ray diffraction data have also proved valuable in many directions even when the existing methods of study are incapable of deducing structures from these data. Of these uses the purely analytical applications are the most obvious and the least important. The atomic arrangement in a crystal is characterized by the type of structure and the size of the unit cell. All structurally isomorphous substances have, by definition, the same distribution of atoms but it only rarely happens that their units are of the same size. Hence most crystals can be distinguished from one another by the differences in the intensities or the positions of their X-ray diffraction effects.

It cannot be assumed, however, that chemically different and apparently non-isomorphous compounds will give rise to very different X-ray patterns. Thus it is found that the mineral sillimanite[23] ($Al_2O_3 \cdot SiO_2$) and the compound mullite ($3Al_2O_3 \cdot 2SiO_2$) yield effects which are nearly indistinguishable. Similarly all the strong lines[24] of Na_2CaSiO_4 and of $Na_4Ca(SiO_3)_3$ agree with one another in intensity and position. It will be evident from such observations that experience in X-ray procedures and some knowledge of their results are needed for analytical as for other purposes, if mistakes are to be avoided.

Powder photographic methods are the only ones which lend themselves readily to these analytical applications. Such photographs obtained from any but the simplest of solids contain many lines which, judged by spectrographic standards of accuracy, can be only approximately measured. For this reason diffraction photographs are poorly suited to the detection of small quantities of one crystalline species in the presence of large amounts of another kind of crystal. Experience has shown that X-ray methods of analysis are rarely of value in case microscopic procedures can be applied. They, therefore, find their chief usefulness among preparations which are either opaque to ordinary light or are composed of crystals too small to be seen clearly under the microscope.

An especially fruitful field of usefulness for X-ray methods lies in the examination of metallic alloys. These methods throw new light upon the solid solutions and the compounds customarily encountered in alloy systems. Some intermetallic compounds[25] like $SnMg_2$ or Mg_2Si have simple structures of the sort exhibited by strongly polar salts. Others such as occur in the system aluminum-copper[26] are very complex and exhibit phenomena which are not yet clearly understood.

Most solid solutions of one metal in another, like most solid solutions of isomorphous or potentially isomorphous salts, appear to consist of simple replacements of varying numbers of atoms in the crystal structure of the "solvent" by atoms of the "solute." Solutions of copper in gold, or iron in nickel, of NaCl in KCl or of AgCl in AgBr are instances of such simple isomorphous replacements. Some early observations have been supposed to show that, in accordance with previous predictions, the "solute" atoms distribute themselves regularly throughout the crystal when it is subjected to careful annealing. The best evidence now available does not, however, show this to be the case.

Besides these solid solutions involving only simple replacements, X-ray observations have pointed to the existence of other and more complicated types of solution. They have thus far been encountered chiefly among minerals. For example it has been found that albite ($NaAlSi_3O_8$) and anorthite ($CaAl_2Si_2O_8$) together with the whole series of feldspars of composition lying between them give rise to practically identical diffraction patterns. The solid solutions of acmite [$NaFe^{III}(SiO_3)_2$] and diopside [$CaMg(SiO_3)_2$] yield similar results. These phenomena are comprehensible only in case atomic replacements in such instances occur not singly and haphazardly, but in pairs.

In the field of mineralogy many apparent solutions are known between compounds which do not even have the same numbers of atoms in their unit cells. An example of this is furnished by anorthite and nephelite. Phase rule analysis, and optical examination, indicate that as much as thirty-five per cent of anorthite ($CaAl_2Si_2O_8$) dissolves in nephelite ($Na_2Al_2Si_2O_8$). X-ray examination[27] shows that the solid solutions have the same type of pattern as nephelite but that the size of the nephelite unit cell is not measurably changed by this insertion of two sodium atoms in place of one atom of calcium. Various explanations of such results suggest themselves but it is obvious that more data must be available before the true one is determined.

Solid solutions of the different kinds just discussed all partake more or less of the nature of atomic replacements. A different kind of phenomenon is furnished by steels and by such minerals as the augites. Spacing measurements combined with determinations of density indicate [28] that the carbon atoms in steel do not occupy symmetry positions in the structure but are present in some form of "interstitial" solid solution. Augites are minerals of the diopside type, $R''R_1''(SiO_3)_2$, which contain as much as ten per cent of Al_2O_3 and Fe_2O_3. Their X-ray patterns [29] are indistinguishable from those of the pure diopsides. From this and other properties it seems probable that the Al_2O_3 and Fe_2O_3 in them are also present in an "interstitial" solid solution. The "solutes" in these solutions consist essentially of a second phase distributed throughout the "solvent." Undoubtedly the particles of the "solute" phase are very minute but existing knowledge is not sufficient to tell whether they are of molecular dimensions or of a somewhat greater size. The solutions which are thus far recognized as "interstitial" have all been formed at elevated temperatures. It seems quite possible that at these higher temperatures where the cohesive forces within a crystal are weaker, many crystals can dissolve appreciable quantities of another and chemically different substance. The mechanism of this solution is perhaps similar to that occurring between liquids. As such a solid solution cools it would be expected that the second phase would become less soluble and would separate as molecules or as small aggregates of molecules distributed here and there throughout the structure. When X-ray measurements have been made at high temperatures they will undoubtedly add much to our knowledge of these questions and will greatly clarify the phase rule interpretations to be given to many physico-chemical studies of alloys and of the metal oxide systems important in geochemistry and allied fields.

X-ray diffraction methods have already been of value in studying the perfection and size of crystals and the effect of these factors on various physical properties. Efforts to establish the laws governing the intensity of X-ray reflections—to be used subsequently in advancing our knowledge of atomic arrangements—have provided many interesting facts concerning the perfection of single crystals. From these observations it has been found, for instance, that perfection of atomic arrangement is rarely maintained throughout a large volume of crystal. Most individual crystal specimens,[30] even those that appear clear and undistorted, behave as if they were "mosaics" of small perfect crystals lying more or less closely in parallel orientation. Relatively deformable crystals, like the alkali halides, seem to be always such mosaics. Others, like calcite, quartz and gypsum, are much better formed but the only very perfect crystals that have thus far been found are occasional specimens of the diamond.[31]

Both the intensities and the sharpness of X-ray diffractions depend upon the perfection of the reflecting crystal. The reflections from badly distorted crystals may be very broad and fuzzy. Broad diffractions of a similar appearance also arise from crystals of colloidal dimensions. It is therefore often difficult to know to which cause to ascribe diffuse X-ray effects. For instance it is hard to be certain whether the broad lines of α-iron that are obtained from martensitic steel are due to the extreme minuteness of the α-iron crystals or are produced through distortion of the α-iron by included or "dissolved" carbon. In spite of this ambiguity of interpretation X-ray studies of metals and alloys have emphasized the close relationships that exist between coarseness of crystallization and some of their physical characteristics. Thus the maximum of ductility and malleability is associated with unicrystalline or very coarsely crystalline metals; conversely the hardest and brittlest metals and alloys consist of interlocking aggregates of minute crystals. Experiments of this sort further show that one of the functions of a second finely distributed phase, like carbon in steel or promoters [32] in iron catalysts for ammonia synthesis is to inhibit the growth of larger crystals in such fine grained aggregates.

Early in the development of X-ray methods photographs were taken through metal foils which had been subjected to a variety of mechanical and heat treatments. Such photographs can be made to tell what sort of preferred crystal orientations are developed in the foils and to give measures of the rate of recrystallization in metals held at different temperatures. Recently steps [33] have been taken to make an amplification of this procedure of greater usefulness in the practical as well as the theoretical phases of metallurgy.

As crystals of colloidal dimensions become smaller and smaller they become increasingly incapable of producing perfect diffraction effects and the lines in their patterns therefore become broader and broader. The average size of these very minute crystals can thus be estimated. Such studies have already been made of the size of the metal crystals in colloidal gold and in several platinum metal preparations [34] used as catalysts.

X-ray methods should be of wide application in colloidal chemistry through their ability to determine whether a particular preparation is crystalline and to establish the relative or absolute sizes of the particles in such materials. Most inorganic colloids are probably microcrystalline. The metal sols of gold and silver are assuredly crystalline with the same structure that is exhibited by the macrocrystalline metal. The pattern of bauxite [$Al_2O_3H_2O$(?)] is obtained from alumina gels precipitated from hot solutions and from dialyzed alumina sols.[35] The hydrous oxide gel precipitated from solutions of alkali aluminates yields the pattern of hydrargillite ($Al_2O_33H_2O$). Gels of hydrous ferric oxide as well as those of silica always appear to be amorphous. Some precipitated metal sulfides seem amorphous but sols which can be made from them are crystalline. If an alumina gel is precipitated at so low a temperature that it shows no diffraction pattern with X-rays of the usual wave lengths, this gel will, on aging, grow crystals of a size sufficient to produce such a pattern. Studies of the rate of crystal growth under different conditions in the precipitates used for qualitative and quantitative analysis should be most helpful toward an understanding of the procedures employed in these branches of chemistry. They ought also to cast valuable light upon the mechanism of the transition from the amorphous to the crystalline state.

One of the most interesting and widely developed applications of X-ray diffraction methods has been the determination of crystal orientation in aggregates of different sorts. If diffraction effects are obtained from a single crystal their positions are determined by both the structure and the orientation of the crystal. If on the other hand photographs are prepared from aggregates in which the crystalline individuals are haphazardly distributed, the pattern naturally depends only on the structure. In many crystalline aggregates large numbers of the individual crystals are so arranged that they have one or more directions in common. The patterns from such materials which show preferential orientations are intermediate between those from single crystals and from powders. In the simple cases it has been possible to determine the character of this preferential orientation from analysis of these "fiber diagrams." Patterns of this sort have been obtained from worked metals[36] and from natural organic fibrous materials.

If, for example, wires are drawn from metals which, like molybdenum, tungsten or iron, have a body-centered cubic structure it is found that a dodecahedral [110] axis of the crystals forming the wire is parallel to the direction of drawing. For similar wires of such face-centered cubic metals as copper or platinum the cube [100] and the octahedral [111] axes have this preferred orientation. Related phenomena are observed from rolled foils of these metals. In the more complicated cases where there is more than one preferred orientation the relation between the several effects is a complicated one which depends in part upon the manner of working.

Many electro-deposited metal films show fibrous structures when their X-ray patterns [37] are investigated. Such fibrous deposits are favored by a low current density. They have been found to form more readily in some solutions than in others but all of the conditions determining their production have not yet been evaluated.

Platy substances, as well as those of a distinctly fibrous habit, yield crystal aggregates showing a preferred orientation. Graphite is such a conspicuously platy substance.

Cellulose [38] has been investigated in greater detail than other organic fibers. The diffraction patterns from several kinds of cellulose of both plant and animal origin appear to be practically identical. The direction of elongation of the cellulose fibers coincides in general with the same edge of the different cellulose crystals. The available data have been interpreted to mean that the unit cell is practically orthorhombic in symmetry and contains four molecules of $C_6H_{10}O_5$. In the absence of all crystallographic knowledge it is, however, hard to know how much weight can safely be put upon these conclusions. Experiment has also shown that mercerized cellulose [39] gives a pattern different from that of cellulose itself. Nearly all artificial silks [40] have the same pattern. If the process of mercerization is followed by means of X-ray examination it is found that the alterations in the X-ray patterns do not correspond to the changes in the physical properties of the cellulose. A plausible explanation of the behavior of cellulose has been based upon these and similar experiments. From this standpoint it is to be considered as a partially crystallized [41] material—crystals of cellulose imbedded in parallel orientation throughout a mass of amorphous compound. The additional experiments necessary to furnish a final answer to the problem of the nature of cellulose should be productive of results of great interest to the organic chemistry of complicated aliphatic compounds.

Up to this point discussion has been limited to a consideration of the diffraction effects obtained from crystalline solids. The X-ray patterns from matter in other than the crystalline state of aggregation have also been investigated to some extent. Experience has shown that except for the glasses nearly all solids are crystalline. It is probable that some precipitates of extremely insoluble compounds, such as metal sulfides, are essentially amorphous. Compounds which decompose at not too high a temperature with the loss of a gaseous component would be expected to become amorphous or to pass through an amorphous stage. The mineral kaolinite[42] is said thus to become amorphous when it is heated to the point at which it loses water. At higher temperatures this amorphous product is supposed to recrystallize into other compounds. No amorphous scattering[43] of X-rays was observed from the mineral dumortierite ($8Al_2O_3 \cdot 6SiO_2 \cdot B_2O_3 \cdot H_2O$) which had been heated to various temperatures within the range in which it gradually loses B_2O_3 and H_2O. Several oxides, of which Fe_2O_3, Sc_2O_3, ZrO_2 and Ta_2O_5 are typical, have apparently been prepared in an amorphous condition.[44] When these oxides are heated they phosphoresce strongly for a brief time. It has been observed that after this phosphorescence they yield crystalline diffraction patterns.

Most solids which are not crystalline appear on X-ray examination to be glassy. Glasses have many of the properties to be expected of liquids possessed of a very high viscosity. X-ray measurements confirm this identification of glasses as supercooled liquids to the extent of demonstrating that glasses and liquids both yield the same type of diffraction effects. These consist of one or more broad and not sharply defined lines. The character of the regularity which produces these bands has been the subject of much discussion. They have been ascribed to constant interatomic distances within the molecule, to constant distances between molecules, to chance aggregations of molecules similar to those which have been supposed to cause the scattering of ordinary light by liquids, and to the existence of crystalline arrangements of colloidal dimensions. Not enough is yet known to establish their true origin.

Qualitative measurements [45] have been made of the diffraction patterns obtained from a large number of glasses lying within the calcium metasilicate—sodium metasilicate—silica field. The bands from some of these glasses are very much narrower than others. It was found that the occurrence of these narrow band patterns and the positions of both the broad and the narrow bands are definitely functions of the chemical composition of the glass. An attempt was made to relate these glass patterns to the patterns of the crystalline phases found in this system. The results were not conclusive. An extension of this work to other glasses and systems of glasses together with more quantitative measurements than are photographically possible should add much to the slight knowledge now possessed concerning the fundamental nature of vitreous materials.

X-ray observations are also useful in studying the process of devitrification in glasses. Borax glass [46] gives a relatively complex pattern which may well be due to colloidal crystals of anhydrous borax. These glasses have been studied at various temperatures near to and below the point at which rapid devitrification takes place. It was hoped in this way to determine whether devitrification could be brought about through the gradual growth of such colloidal borax crystals. No corresponding sharpening of the glass bands could be observed. Devitrification invariably took place by the growth of large borax crystals upon the surface of little pieces of glass. The apparent existence of some kind of regularity of arrangement in liquids and glasses suggests that states of matter may perhaps exist which are intermediate between the completely amorphous and the completely crystalline condition. The physical and optical behavior of various long chain carbon compounds and the phenomena of liquid crystals have in the past led to this idea. Powder and Laue photographs of liquid crystals have shown no evidence of crystalline diffraction. The concentration of the liquid crystal phase in the preparations used was so small, however, that no effects should be expected. These experiments are therefore of little significance.

When long chain carbon compounds such as the fatty acids or various of their salts and esters are solidified upon a glass plate, the films thus produced give reflections having large spacings. These spacings vary in a regular fashion with the lengths of the molecular chains and it was at first assumed that the molecules of these compounds solidified with their lengths normal to the plane of the glass. The measured spacings were thus considered to represent the lengths of these molecules. Such a semi-crystalline state apparently corresponds to the "smectic" condition [47] that has often been discussed for compounds of this type. Closer examination [48] shows that these long chain films are truly crystalline and are possossed of low symmetry. The measured spacings cannot be the lengths of the molecules but are functions of these lengths which can only be evaluated when the crystallographic properties are known. Similar observations upon complex ketones [49] have formed the basis for inferences concerning the shape of the ketone group and for conclusions that in some of these compounds the chains are straight while in others they are doubled up. Such conclusions likewise are of uncertain worth as long as the crystallography of the crystals remains obscure.

From the existing experiments it must be concluded that though semi-crystalline states may be possible and perhaps are even probable for some complicated compounds, there is as yet no direct and unmistakable evidence for their existence.

In the preceding discussion it has been possible to do little more than mention directions in which X-ray diffraction analysis has so far been useful and to suggest developments it may be expected to make in the near future. It is hoped, however, that the treatment has been sufficient to indicate the way in which crystal analysis is at the moment growing into a more or less distinct branch of physical science, drawing its methods and material from various older branches of science and in return contributing information of use to them.

REFERENCES

1 For the sake of brevity references will be furnished only to those investigations which are not mentioned in R. W. G. Wyckoff, The Structure of Crystals (New York, 1924).
2 W. L. Bragg, R. W. James and C. H. Bosanquet, Phil. Mag. **41**, 309; **42**, 1 (1921); R. J. Havighurst, Proc. Nat. Acad. Sci. **12**, 380 (1926).
3 M. Born, Atomtheorie des festen Zustandes, 2te Auf. (Leipzig, 1923). For popular accounts see M. Born, The Constitution of Matter (London, 1923); P. P. Ewald, Kristalle u. Roentgenstrahlen (Berlin, 1923) Kap. 7.
4 W. L. Bragg, Proc. Roy. Soc. **A**, **106**, 346 (1924).
5 W. L. Bragg and S. Chapman, Proc. Roy. Soc. **A**, **106**, 369 (1924).
6 H. Ott, Zeit. f. Krist., **61**, 515 (1925); **62**, 201 (1925); Naturw., **13**, 644 (1925).
7 W. H. Bragg, J. Soc. Glass. Techn., **9**, 272 (1925); W. H. Bragg and R. E. Gibbs, Proc. Roy. Soc. **109A**, 405 (1925); R. W. G. Wyckoff, Science **62**, 496 (1925); Am. J. Sci., **11**, 101 (1926).
8 R. W. G. Wyckoff, Am. J. Sci., **9**, 448 (1925); Zeit. f. Krist., **62**, 189 (1925).
9 W. H. Bragg and G. T. Morgan, Proc. Roy. Soc., **104A**, 437 (1923).
10 Unpublished results by R. W. G. Wyckoff, T. P. McCutcheon and J. H. Müller.
11 J. Wasastjerna, Soc. Scient. Fenn. Comm. Phys.-Math. **I**, **37** (1923).
12 V. Goldschmidt, Vid. Akad. Skrifter I, Math.-Naturv. Kl. No. 2 (1926).
13 R. J. Havighurst, Proc. Nat. Acad. Sci., **11**, 502, 507 (1925).
14 W. Duane, Proc. Nat. Acad. Sci., **11**, 489 (1925).
15 W. L. Bragg, Phil. Mag. **40**, 169 (1920).
16 V. Goldschmidt, op. cit.
17 H. G. Grimm and A. Sommerfeld, Zeit. f. Physik, **36**, 36 (1926).
18 V. Goldschmidt, op. cit.
19 H. Mark and E. Pohland, Zeit. f. Krist., **62**, 103 (1925).
20 H. Mark and E. Pohland, op. cit.
21 The published crystal structures [H. Mark, Ber., **57**, 1820 (1924)] are apparently in error.
22 F. M. Jaeger, P. Terpstra and H. G. K. Westenbunk, Proc. Roy. Acad. Sci. Amster. **28**, 747 (1925).
23 R. W. G. Wyckoff in N. L. Bowen and J. W. Grieg, J. Am. Cer. Soc., **7**, 238 (1924); J. T. Norton, *ibid.*, **8**, 401 (1925); L. Navias and W. P. Davey, *ibid.*, **8**, 640 (1925); R. W. G. Wyckoff, J. W. Greig and N. L. Bowen, Am. J. Sci. **11**, 459 (1926).
24 R. W. G. Wyckoff and G. W. Morey, Am. J. Sci. in press.
25 L. Pauling, J. Am. Chem. Soc., **45**, 2777 (1923); A. Sacklawski, Ann. d. Physik, **77**, 241 (1925).
26 E. R. Jette, G. Phragmen and A. F. Westgren, Jour. Inst. Metals, **31**, 193 (1924).
27 Unpublished data by R. W. G. Wyckoff and N. L. Bowen.
28 A. Westgren and G. Phragmen, J. Iron Steel Inst., **109**, 159 (1924).
29 R. W. G. Wyckoff, H. E. Merwin and H. S. Washington, Am. J. Sci., **10**, 383, (1925).
30 See W. L. Bragg, C. G. Darwin and R. W. James, Phil. Mag. **1**, 897 (1926).
31 H. Mark, Naturw., **13**, 1042 (1925).
32 R. W. G. Wyckoff and E. D. Crittenden, J. Am. Chem. Soc., **47**, 2866 (1925).
33 G. L. Clark, E. W. Brugmann and S. D. Heath, J. Ind. Eng. Chem. **17**, 1142 (1925).
34 G. R. Levi and R. Haardt, Rend. Accad. Naz. Lincei, **3**, **91**, 215 (1926).
35 J. Böhm, Zeit, anorg. Chem., **149**, 203 (1925).
36 See for instance M. Polanyi, Zeit. f. Krist., **61**, 49 (1925); K. Weissenberg, *ibid.*, **61**, 59 (1925); H. Mark, *ibid.*, **61**, 75 (1925).
37 R. Glocker and M. Kaupp, Zeit. f. Physik, **24**, 121 (1924); G. L. Clark and P. K. Frölich, Zeit. Elektrochem., **31**, 655 (1925); R. M. Bozorth, Phys. Rev., **26**, 390 (1925).
38 See numerous papers from the Kaiser Wilhelm Institut für Faserstoffchemie by R. O. Herzog and collaborators.
39 J. R. Katz and H. Mark, Zeit. phys. Chem., **115**, 385 (1925); Cellulosechemie, **6**, 35 (1925); etc.
40 R. O. Herzog and H. W. Gonell, Kolloid.-Zeit., **35**, 201 (1924).
41 R. O. Herzog, Ber., **58**, 1254 (1925).

[42] W. H. Bragg, G. Shearer and J. W. Mellor, Trans. Cer. Soc., 22, 105 (1925).
[43] N. L. Bowen and R. W. G. Wyckoff, J. Wash. Acad. Sci., 16, 178 (1926).
[44] J. Böhm, Zeit. anorg. Chem., 149, 217 (1925).
[45] R. W. G. Wyckoff and G. W. Morey, Trans. Soc. Glass Techn., 9, 265 (1925).
[46] R. W. G. Wyckoff, unpublished observations.
[47] G. Friedel, Ann. de Phys., 23, 273 (1922).
[48] R. E. Gibbs, J. Chem. Soc. (London), 125, 2622 (1924); R. W. G. Wyckoff, F. L. Hunt and H. E. Merwin, Science, 61, 613 (1925).
[49] W. B. Saville and G. Shearer, J. Chem. Soc. (London), 127, 591 (1925).

CATALYSIS AND THE MECHANISM OF CHEMICAL REACTIONS

BY

HUGH S. TAYLOR

PROFESSOR OF PHYSICAL CHEMISTRY
PRINCETON UNIVERSITY

Lecture delivered at Columbia University in the special course on Contemporary Developments in Chemistry given in the Summer Session of 1926 on the occasion of the opening of the Chandler Chemical Laboratories

COLUMBIA UNIVERSITY PRESS
NEW YORK
1927

Printed from type. Published March, 1927

Printed in the United States of America

CATALYSIS AND THE MECHANISM OF CHEMICAL REACTIONS

By Hugh S. Taylor

Professor of Physical Chemistry, Princeton University, New Jersey

The study of reaction kinetics still provides the most fruitful method of approach to the problem of reaction mechanism. The results of such a study can now be supplemented, however, by a mass of additional data, involving the thermochemistry of the process and of possible intermediate steps, the influence of light on "order" of reaction and on velocity, the modification in nature and velocity which the utilization of catalysts entails. It is the object of the present lecture to discuss some of these cases in detail in the hope that some of the general laws underlying the promotion of chemical reaction may be elucidated and that the function of the catalytic agent may be more adequately understood.

The simplest type of chemical reaction and the one that appears to be most completely understood is the bimolecular thermal gaseous reaction. The classical example of this type is still the decomposition of hydrogen iodide into the elements or the synthesis of hydrogen iodide from its elements. The data of Bodenstein [1] are still the accepted data and, although there is evidence of surface action,[2] it is known that a considerable proportion of the observed rate occurs in the gaseous phase.

The velocity of such bimolecular thermal gaseous reactions in which two like molecules interact is given by an equation of the form

$$k = 2(\sqrt{2}\,\pi\sigma^2\bar{u}\,n^2)e^{-E/RT}$$

In this equation the first portion of the right-hand side expression is the classical kinetic theory expression (multiplied by 2, the number of molecules involved in the reaction), for the number of collisions per unit time between like molecules at the absolute temperature, T, n being the number present per cc., σ is the molecular diameter, $\bar{u}$ the root mean square velocity. The exponential expression takes account of the known fact that every collision does not lead to reaction and implies that only those collisions are effective in producing reaction in which the total energy of the colliding molecules exceeds a certain value E. The critical energies, E_1, for three well investigated cases of bimolecular reactions are given in the following table; with these values for E_1 the above expression gives good values for the velocity of these reactions over a range of temperatures.

TABLE I

Reaction		E_1	E_2	T
Thermal decomposition of	$2N_2O$	55,000	58,500	956
	$2HI$	43,900	44,000	760
	$2Cl_2O$	22,000	21,000	384

The absolute temperatures at which these several reactions proceed at the same rate under equivalent conditions is given in the third column of the table, whence it is apparent that the reaction requiring the higher energy increment also requires a higher temperature for equal reaction rate under equivalent conditions.[3] The same value is obtained, as the values under E_2 reveal, if the critical increment of energy of the reaction be calculated from the temperature coefficient of the reaction with the aid of the Arrhenius equation

$$\frac{d \log K}{dt} = \frac{E_2}{RT^2}.$$

The concordance between the data obtained by the two methods of deduction leads to the conclusion that the collision process is the activating process and that the kinetic energy of the molecules at the moment of collision is the activating energy.*

It can now readily be indicated why the thermal gaseous decomposition of hydrogen iodide is a bimolecular process instead of a dissociation of single molecules into atoms followed by rapid combination of the hydrogen and iodine atoms. From the thermal data,

$$\tfrac{1}{2}H_2 = H \qquad - 48.5 \text{ Kg. Cals.}$$

$$\tfrac{1}{2}I_2 = I \qquad - 18 \text{ Kg. Cals.}$$

$$HI = \tfrac{1}{2}H_2 + \tfrac{1}{2}I_2 - 1.5 \text{ Kg. Cals.}$$

* This conclusion has been discussed more rigorously by Hinshelwood and Thornton, Phil. Mag., 50, 1135 (1925); see also, Hinshelwood, The Kinetics of Chemical Change, Chapter III, Oxford University Press, 1926.

it follows, by summation that

$$HI = H + I - 68 \text{ Kg. Cals.}$$

The unimolecular decomposition of hydrogen iodide requires therefore an energy increment of at least 68 Kg. Cals. as contrasted with the 44 Kg. Cals. per two molecules of activating energy required in the bimolecular process. The kinetic theory teaches that the proportion of the molecules for which the kinetic energy exceeds E is given by the equation

$$\frac{n_E}{n} = e^{-E/RT}.$$

Hence the ratio of molecules of hydrogen iodide having an energy exceeding 22,000 calories to those having 68,000 calories for example at a temperature of 550° K. is

$$\frac{e^{\frac{-22000}{1100}}}{e^{\frac{-68000}{1100}}} \text{ or approximately } e^{40}.$$

It is thus obvious that the unimolecular process practically never occurs thermally under such conditions.

Photochemically it is easily possible for a hydrogen iodide molecule to acquire as much as 68,000 calories. This corresponds to a quantum of energy corresponding to a wave length in the violet portion of the spectrum. One would therefore expect, as Bodenstein actually found, that the photo-decomposition of hydrogen iodide was unimolecular, although it is possible that the dissociation into atoms is not the primary stage of the process but is preceded by an energy rich modification of the molecule. The precise mechanism subsequent to the absorption of an energy quantum is still under discussion * and is a question of great moment in the theory of reaction mechanism.

In addition to the homogeneous bimolecular thermal decomposition and the unimolecular photo-decomposition, hydrogen iodide can be catalytically decomposed at a variety of surfaces. At two such surfaces the reaction kinetics have been studied in detail. At the surface of gold the reaction rate is nearly independent of the pressure,[4] while at the surface of a platinum wire the reaction is approximately unimolecular.[5] The former reaction of zero order corresponds to the case discussed by Langmuir

* See, Stern and Volmer, Z. wiss. Phot., 19, 275 (1920); Franck, Trans. Farad. Soc., 21, 536 (1926). New work on the absorption spectrum of hydrogen iodide (Tingey and Gerke, J. Am. Chem. Soc., 48, 1838 (1926)), indicates that dissociation into atoms results from the absorption of light energy.

in his theory of catalytic action in which the surface is practically completely covered over the whole pressure range of the experiment at the given temperature. If this be true also over a given temperature range then the heat of activation obtained from velocity measurements over this range should be a true measure of the energy required to cause the adsorbed molecules to react. In this way Hinshelwood and Burk found that the zero order decomposition of hydrogen iodide on gold involved only 25,000 calories of activation heat as compared with the value of 44,000 calories for the thermal bimolecular process and the much higher value for a unimolecular homogeneous process. It is evident therefore that the association with the catalyst has, in this case, markedly lowered the heat of activation. It is not known whether the association between catalyst and reactant is through the iodine or the hydrogen of the hydrogen iodide. A knowledge of this would be illuminating.

In the unimolecular decomposition of the gas at a platinum surface an even lower heat of activation was found, namely 14,000 calories. It is possible to make a more detailed analysis of this case by reason of some recent work of Fryling of heats of adsorption of hydrogen on nickel and some unpublished data on platinum. Fryling's data [6] indicate that hydrogen may be adsorbed both as molecules and atoms, the latter revealing themselves by low values for the heat of adsorption at low partial pressures rising to a maximum and then falling away. Fryling attributes this curve to the superposition of the endothermic change

$$H_2 = 2H - 97.5 \text{ Kg. Cals.}$$

on the normal positive value of the adsorption heat of molecular hydrogen. Elsewhere I have analyzed these data [7] and have concluded that the adsorption of a hydrogen atom on a platinum atom may involve the evolution of as much as 50 Kg. Cals. On such a basis the decomposition of hydrogen iodide at a platinum surface would yield a process representable thermally by the equation

$$\underset{68}{Pt + HI} = \underset{50}{Pt(H)} + I - 18 \text{ Kg. Cals.}$$

We may now assume that the adsorbed hydrogen atom is struck by a hydrogen iodide molecule, whereupon the following reactions occur,

$$Pt(H) + HI = Pt(H_2) + I + x \text{ Kg. Cals.}$$
$$Pt(H_2) = Pt + H_2 + y \text{ Kg. Cals.}$$

The values of x and y are readily obtained when it is remembered that

$$I + I = I_2 + 36 \text{ Kg. Cals.}$$

and

$$H_2 + I_2 = 2HI + 3 \text{ Kg. Cals.}$$

whence, by summation

$$x + y - 18 + 36 = 3$$

or

$$x + y = -15.$$

Now, y is the heat of desorption of hydrogen molecules from platinum and has given measured values between —10 and —25 kg. cals., dependent on the portion of the platinum surface. It is evident therefore that the corresponding values of x must lie between —5 and +10 Kg. Cals. It follows, also, that, by the mechanism postulated, the unimolecular decomposition of hydrogen iodide has been changed from a single stage involving 68 kg. cals. to a series of stages no one of which requires more than 25 kg. cals. The energy of activation has probably been correspondingly reduced and the catalytic acceleration can very readily be understood.

We may now examine another interesting case, also studied by Hinshelwood and Burk,[8] the catalytic decomposition of ammonia at a platinum surface, in which our experimental data on heats of adsorption also lead to elucidation of their experimental observations. The homogeneous thermal decomposition of ammonia has never been observed. Very considerable energy increments would be required for the unimolecular process as indicated by the following thermal data, deduced on the assumption that the dissociation of nitrogen into atoms requires approximately 11.4 volts.[9]

$$N + N = N_2 + 263 \text{ Kg. Cals.}$$

$$6H = 3H_2 + 292.5 \text{ Kg. Cals.}$$

$$N_2 + 3H_2 = 2NH_3 + 22 \text{ Kg. Cals.}$$

whence

$$2N + 6H = 2NH_3 + 577.5 \text{ Kg. Cals.}$$

From these data the heat of breaking of a N-H linkage in ammonia may be assessed at $577.5/6 = 96$ kg. cals. No data are available as to the heat of activation of the bimolecular process, but the heat of activation must be high, as, even at 1000° C. the gas reaction is immeasurably slow (Cf. N_2O, page 2). Even in quartz vessels the decomposition is heterogeneous.

At a platinum surface, Hinshelwood and Burk found a unimolecular decomposition, inhibited by the hydrogen produced and having the abnormally high critical increment of 140 Kg. Cals. As with hydrogen iodide, we may assume the first step in the process to be the removal of a hydrogen atom from the ammonia and the adsorption of the hydrogen on the platinum surface. The thermal equation runs

$$\mathrm{Pt} + \begin{matrix}\mathrm{H}\\ \mathrm{H}\end{matrix}\!\!>\!\mathrm{N} - \underset{\substack{\| \\ 96\text{ Kg. Cals.}}}{\mathrm{H}} \rightarrow \underset{50}{\mathrm{Pt(H)}} + \begin{matrix}\mathrm{H}\\ \mathrm{H}\end{matrix}\!\!>\!\mathrm{N}- \; - 46 \text{ Kg. Cals.}$$

This process should cause a considerable reduction in the critical increment required to break a hydrogen atom from the molecule. By this process, however, the active platinum atoms become covered with adsorbed hydrogen atoms. In the hydrogen iodide case just considered these atoms were assumed to be removed by reaction with colliding hydrogen iodide molecules and the thermal data indicated this as quite possible. In the present case, removal of adsorbed hydrogen atoms with colliding ammonia is not possible. This conclusion can be reached on the basis of experimental evidence since Bonhoeffer and Boehm [10] have recently shown that no reaction occurs between free hydrogen atoms and ammonia. It is evident, therefore, that reaction between ammonia and the Pt (H) grouping, involving, as this latter does, a heat of formation of 50 Kg. Cals., will be still less likely to occur. The surface remains covered with hydrogen atoms except insofar as these are removed by evaporation or reaction with adjacent atoms to form molecules. On the assumed basis, the evaporation as atoms requires a critical increment of 50 Kg. Cals. This thermal analysis of the possible reactants adequately accounts for the observed inhibition by hydrogen.

The mechanism also accounts for the high apparent activation energy of the process. As Hinshelwood has pointed out,* the apparent heat of activation is related to the true heat of activation by the equation

$$E_{\text{app.}} = E_{\text{True}} + \lambda' - \lambda$$

where λ' is the heat of adsorption of the products and λ is the heat of adsorption of the reactants. On the mechanism postu-

* Kinetics of Chemical Change, pp. 176–178, Oxford University Press, 1926.

lated, E_{True} cannot be less than the 46 Kg. Cals. in the preceding equation nor λ' less than the 50 Kg. Cals. of heat of adsorption. λ is probably of the order of 10–20 Kg. Cals.[11] It is evident that $E_{app.}$ must therefore be of the order of 100 Kg. Cals.

It is evident therefore that data on the kinetics, adsorption and heats of adsorption of gases on catalyst surfaces is steadily providing a clearer picture of the mechanism of chemical reactions in such heterogeneous systems. The data of adsorption, of sensitivity to heat treatment, of poisoning and of promoters indicate, however, that the catalyst surface is not of uniform activity and that, therefore, it is the properties of the individual atoms and molecules of such surfaces that are of importance in such changes.[12]

Homogeneous Catalysis

The intensive development that has occurred recently in contact catalysis has not been paralleled by similar advances in homogeneous catalysis. In gaseous systems there is no good example of a homogeneous catalyzed reaction for which adequate kinetic data are available.

In general, the addition of foreign gases has little effect on bimolecular reactions. This is well shown by the recent experiments of Hinshelwood and his co-workers. Oxygen, nitrogen, chlorine and carbon dioxide do not alter the speed of chlorine monoxide decomposition, oxygen and nitrogen are without effect on nitrous oxide decomposition. In the former case it is probable that an intermediate chlorine oxide is formed which determines reaction rate, but there are no data which enable a quantitative formulation of its effect. Data on ozone decomposition are complicated by the effect of walls on the reaction.

Water vapor is the most popular example of a homogeneous catalyst for gaseous reactions. Here, also, however, there is no case known in which the kinetics of the influence of water vapor has been studied. There is a possibility that this influence may be studied kinetically in the combination of carbon monoxide and oxygen, since some new work by Benton[13] on quartz indicates that the experiments of Bodenstein and Ohlmer[14] showed too large an activity of such a surface. Benton finds that, even at 500° C., with quartz powder, the reaction is very slow and that it is fortuitous impurities that accelerate the reaction. It is possible that with a perfectly clean quartz surface the kinetics of the homogeneous reaction in presence and absence of water vapor might be studied. Dixon's experiments[15] on the combination of these gases in an explosion wave indicate that they are inert when completely dry but that the influence of water vapor increases steadily up to moderate partial pressures. It has always been suggested that this is probably due to a stoichiometric catalysis of the following type

$$2CO + 2H_2O = 2CO_2 + 2H_2$$
$$2H_2 + O_2 = 2H_2O.$$

However plausible this appears, it does not explain why water is necessary for the hydrogen-oxygen reaction. Is there any valid objection to considering the water vapor as a molecularly-dispersed contact catalyst on which the reactants are adsorbed and thereby activated?

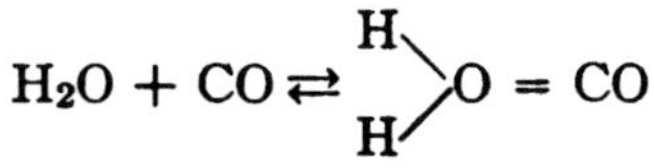

or

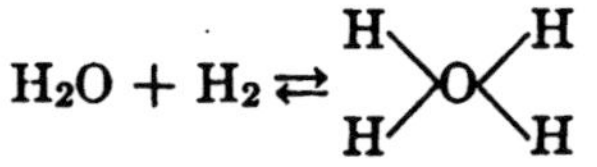

In both these cases the hydrogen atoms attached to the oxygens would be very liable to reaction with colliding oxygen molecules, a reaction familiar now from the study of the chemical properties of free hydrogen atoms.[16] Such a mechanism would account for the observed presence of hydrogen peroxide in such reactions.[17]

In homogeneous catalytic reactions in solution, the problem of mechanism is hampered by our lack of knowledge concerning the concentrations of reactants in such solutions. The complexity of the problem may be illustrated by myriad studies of catalytic processes of esterification, hydrolysis, hydration, dehydration, halogenation and the like. In spite of the immense number of the researches no completely satisfactory and comprehensive formulation of the catalytic effect has been made.

In acid catalysis of ester hydrolysis, the assumption of a hydrogen ion activity was adequate so long as only weak acids in presence or absence of their salts were considered. Extension of the studies to strong acids showed the abnormal variation of catalytic activity with either total acid concentration or hydrogen ion concentration as deduced from conductance measurements. Neutral salt action thereby attained a classical status in catalytic controversy. An activity of the undissociated acid molecule was among the assumptions made to meet the difficulties. Lapworth and his co-workers indicated that the active catalytic agent may be the non-hydrated hydrogen ion. The abnormal variations of acid catalytic activity with change in acid concentration, addition of neutral salts or solvents would in such case be ascribed to the influence of these several factors on the equilibria

$$H^{\cdot} + nH_2O \rightleftarrows H^{\cdot} \cdot nH_2O$$

where n may be a very variable quantity determined not only by the water concentration but also by the concentrations of the other constituents of the solutions. A beginning only has been made in the quantitative development of this idea.[18]

The solution of the problem of reaction velocity in its relation to catalyst concentration has been sought, also, in terms of the activities of the catalyst species. Thus, for example, in the investigations of Harned,[9] Akerlöf,[20] Lewis and his co-workers,[21] Brönsted and his students,[22] the attempt has been made to correlate the reaction velocities obtained with the data as to activity of the catalyst, deduced from studies of the electromotive behavior of acid-salt systems, freezing point, distribution experiments and the like. This attempt is actually parallel to that discussed in the preceding paragraph. The one attempts a thermodynamic formulation of the catalyst concentration; the other attempts a kinetic picture of the catalytic process. A more extended treatment of Brönsted's point of view will be given in succeeding paragraphs since it exhibits the applicability of the activity theory in a series of reactions for which no other formulation of the catalytic effect has been given. The theory of the subject is treated in Professor Brönsted's lecture in this series. Here we are concerned with the applications of the theory to neutral salt action.

The researches of Holmberg are of importance in the general problem of salt catalysis since marked variations in catalytic behavior of the salts are shown in the different examples. Furthermore, as Brönsted has recently shown,[23] these variations are in accord with the ideas developed by him towards a general activity theory of neutral salt action.

Holmberg[24] first studied the reaction between alkalis and dibromsuccinic acid, a reaction which stoichiometrically and kinetically proceeds according to the scheme:

$$\begin{matrix} CHBrCOO^{--} \\ | \\ CHBrCOO \end{matrix} + OH^- \rightarrow \begin{matrix} CH \cdot COO^{--} \\ \| \\ CBrCOO \end{matrix} + Br^- + H_2O.$$

The velocity constants of the process, at varying concentrations of the reactants and also in presence of salts, show a marked dependence on the total ionic concentration as is evident from the subjoined table for sodium as cation (Table II).

TABLE II

m_{SNa_2}	m_{NaOH}	m_{NaCl}	m_{NaNO_3}	$m_{Na_2SO_4}$	C_t	k
0.01222	0.01222	....	....	....	0.036	1.51
0.01727	0.01727	....	....	....	0.051	1.64
0.02473	0.02473	....	....	....	0.074	1.85
0.03109	0.03109	....	....	....	0.093	1.85
0.02154	0.02154	....	0.0313	....	0.096	2.08
0.02154	0.02154	....	....	0.0164	0.097	2.13
0.02473	0.02473	0.0354	....	....	0.110	2.11
0.03753	0.03753	....	....	....	0.113	2.07
0.04504	0.02252	....	....	....	0.113	2.19
0.02815	0.05630	....	....	....	0.113	2.19

The first and second columns record the concentrations of sodium dibromsuccinate and sodium hydroxide used, the third, fourth and fifth those of sodium chloride, nitrate and sulphate. In the sixth column is the total cation concentration and in the seventh the corresponding velocity constant.

Holmberg called this phenomenon, cation catalysis, since the velocity, with unchanged cation, appears to depend wholly on the total ion concentration, not upon the nature of the anions present. Holmberg showed, in accord with this, that variation of the cation, especially a change to calcium and barium salts, produced a profound change in velocity, an extraordinary acceleration with the divalent cations.

Brönsted shows that these results are quite in agreement with his activity theory of salts, applied to reaction velocity. He formulates the reaction measured by means of the equation

$$S^{--} + OH^{-} \rightarrow (S \cdot OH)^{---}$$

where S^{--} refers to the dibromsuccinic, OH^{-} to the hydroxyl ion and $(S \cdot OH)^{---}$ to a trivalent complex formed by the association of the two anions. Applying Brönsted's activity theory of reaction velocity the velocity should therefore be

$$v = k \cdot C_{S^{--}} \cdot C_{OH^{-}} \cdot \frac{f_2 f_1}{f_3}$$

where f_1, f_2 and f_3 refer to the activity coefficients of the several ions, subscripts indicating valencies of the ions. From Brönsted's solubility studies, he concludes that the activity factor varies in the ratio 1 : 1.8 for a change from 0.02 normal to 0.10 normal solutions. The velocity data for sodium salts in this range yield a value of 1.6. The independence of anion shown by the velocity measurements is also in agreement with Brönsted's theory, since the activity factor is determined in dilute solution largely by the valence and not by the nature of the ion. Furthermore, Brönsted's solubility studies have shown that, as regards the activity coefficient, at the same equivalent concentration, a divalent cation will influence a divalent anion more pronouncedly than will a univalent ion, and that this holds true to an even greater degree for a trivalent ion. Hence the enhanced effect of barium and calcium ions in Holmberg's measurements, which effect, however, is more pronounced than Brönsted would anticipate from solubility data.

The success of the Brönsted theory when applied to neutral salt catalysis is apparent when the second research of this series by Holmberg [25] is considered. In this case, the alkali decomposition of α-β-dibrompropionic acid and α-β-dibrombutyric acid was measured. In this case it was found that the ratio $\frac{k_{0.1}}{k_{0.02}}$, the ratio of the velocity constants in 0.1 and 0.02 normal solutions, had a value of 1.21 as compared with a value of 1.8 in the previous example. The similarity of the two sets of reactions would lead one to expect a similar salt effect. Brönsted indicates the difference between the two examples by pointing out that, whereas the case already considered involves a divalent halogenated anion, this second research involves a univalent anion, whence the reaction measured becomes

$$A^- + OH^- = (A \cdot OH)^{--}$$

and, as a consequence, the reaction velocity expression becomes

$$v = k \cdot C_{A^-} \cdot C_{OH^-} \cdot \frac{f_1^2}{f_2}.$$

For such a system, from activity theory, a value for $\frac{f_1^2}{f_2} = 1.22$ is theoretically calculable, in agreement with the observations recorded.

Similarly, for a later research of Holmberg,[26] the hydrolysis of acetyl glycollic acid

$$\begin{matrix} CH_2OAc \\ | \\ COO^- \end{matrix} + OH^- = \begin{matrix} CH_2OH \\ | \\ COO^- \end{matrix} + AcO^-$$

the reaction type is exactly analogous to the preceding case and so the same salt effect $\frac{k_{0.1}}{k_{0.02}} = 1.22$ should be, and is, found. The alkali hydrolysis of iodo-acetic acid also showed [27] a positive salt action as would be expected from the equation

$$CH_2ICOO^- + OH^- = CH_2OHCOO^- + I^-.$$

The conversion of iodo-acetic acid into sulpho-cyan-acetic acid [28] showed interesting results. In neutral solution the reaction showed the normal salt effect which would be expected from the equation

$$CH_2ICOO^- + CNS^- \rightarrow CH_2(CNS)COO^- + I^-.$$

The ratio $\frac{k_{0.1}}{k_{0.02}}$ was, in the mean, 1.29. Further, the ratio $\frac{k_{Ba^{++}}}{k_{Na^+}}$ was approximately 1.2. The reaction conforms to a velocity equation of the form

$$v = k \cdot C_{CH_2ICOO^-} \cdot C_{CNS^-} \cdot \frac{f_1^2}{f_2}.$$

In strong acid solution, on the contrary, where the dissociation of the iodo-acetic acid is practically suppressed, the reaction must be, primarily,

$$CH_2ICOOH + CNS^- = CH_2(CNS)COOH + I^-.$$

On Brönsted's activity theory this would give a velocity equation of the form

$$v = k \cdot C_{CH_2ICOOH} \cdot C_{CNS} \cdot f_0,$$

in other words the normal reaction process, Brönsted's Type O. Actually, and in complete agreement with this view, Holmberg found a velocity constant practically independent of acid and total ion concentration although this latter was varied sixfold and the former fourfold.

Hydrolysis of Lactones

Henry [29] and later Holmberg [30] showed that the hydrolysis of γ-lactones by alkalis occurs according to the reaction equation

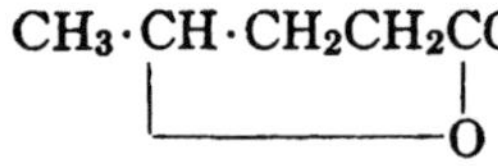

$$+ OH^- = CH_3 \cdot CHOH \cdot CH_2 \cdot CH^2COO^-.$$

It should, therefore, on Brönsted's theory show no neutral action and this was found by both the authors concerned.

The fidelity with which the Brönsted theory, as here indicated, interprets the varied manifestations of neutral salt effect in these several reactions, as contrasted with the complete inability of other theories to interpret so varied a behavior, merits for the theory the closest attention and consideration of the student of homogeneous catalytic action.

Rice [31] has shown that many such reactions show but small variations in temperature coefficient so that there is little to be learned from that line of attack. In certain instances, however, there seems to be some possibility of progress from this standpoint. Thus, Slater [32] showed that in the homogeneous catalytic substitution of chlorine in benzene using iodine chloride, stannic chloride and ferric chloride as catalysts the temperature coefficients were 1.05, 1.5 and 2.5 respectively. It is probable that some clues to mechanism can be obtained from such considerable variations in the coefficient.

The Inhibition of Homogeneous Reactions

It is possible also to learn something concerning the mechanism of chemical reactions from a study of the retardation of homogeneous reactions by inhibitors or "negative catalysts." Elsewhere,[33] I have called attention to the characteristics of these reactions and have suggested that the inhibitor acted in certain cases by reaction with the active forms of one or other reacting species to form molecular compounds.

Christiansen,[34] in a more penetrating analysis of the phenomenon, on the basis of the theory of bimolecular reactions set forth in the initial paragraphs of this paper, pointed out that the molecules of the reaction product, just after reaction, possess an available energy greatly in excess of the mean energy at the temperature considered. They contain not only the critical increment but also the energy equivalent to that liberated in the reaction. These "hot" molecules have sufficient energy to activate reactant molecules at the first encounter and, when these react, the resultants in their turn are again able to act as activators and so on. It is therefore possible that one elementary reaction will give rise to a whole series of such reactions and, so, Christiansen suggested that the inhibitors act by breaking the reaction chains set up in such processes by taking the energy from the "hot" molecules or reacting with them in some way or other.

On the theory of Christiansen, it should be characteristic of reactions subject to inhibition that the primary activation of a molecule of the reactants should lead to the formation of a reaction chain. As a reliable calculation of the number of such primary thermal activations in unit time is not possible in the present state of the general theory of reaction velocities, I suggested to Mr. Hans Bäckström a line of investigation which would lend considerable support to the Christiansen theory. If the thermal reactions are such as to lead to chains of reaction it would seem to be a plausible consequence that, in such cases, initiation of the primary reaction by the absorption of a light quantum should also cause a great number of molecules to react. With the aid of modern photochemical technique, this consequence could readily be tested experimentally.

Mr. Bäckström has carried out this study in the autoxidation reactions of benzaldehyde, chlorbenzaldehyde, oneanthaldehyde and of sodium sulphite solutions. In all cases it was found that a great number of molecules reacted for every quantum absorbed. These results will be published elsewhere in detail by Mr. Bäckström; it will suffice here to give one striking example.

In the oxidation of benzaldehyde by shaking with oxygen when illuminated with light of controlled intensity it was found that the oxygen absorption due to the illumination was 2.45 ccs. per minute. This corresponds to

$$\frac{2.45 \times 273 \times 6.06 \times 10^{23}}{22{,}412 \times 291 \times 60} = 1.04 \times 10^{18} \text{ molecules per second.}$$

On the least favorable assumption, that *all* the electrical energy used to produce the mercury arc light source, was active in bringing about reaction, the energy received from the light source by the aldehyde would have amounted to 3.42×10^4 ergs per second. If the further unfavorable assumption is made that all this energy was light of a frequency 3650 Å which is the first mercury line to be absorbed by benzaldehyde, the energy above indicated is equivalent to 6.36×10^{15} quanta per second. This corresponds to 163 molecules per quantum. This figure is of course a minimum, and there can be no doubt therefore that the oxidation of benzaldehyde is a chain reaction. A more reasonable calculation based on energy data and spectral distribution for other mercury lamps gives a ratio of 1860 molecules per quantum, still without correction for reflected and transmitted light. Figures for the oxidation velocity of oenanthaldehyde and sodium sulphite solutions under the same experimental conditions were of the same order so that there can be no doubt that these autoxidation reactions showing inhibition are chain-reactions. They are of great interest even apart from the problem of inhibition, in that they increase largely that class of reactions, of which the hydrogen-chlorine reaction is the prototype, in which there is an abnormally large deviation from Einstein's Law of Photochemical Equivalence.

It is very generally accepted that in such autoxidation reactions there is intermediate production of a peroxide, that, for example, in the case of benzaldehyde the stages in the process may be represented by the equations

$$C_6H_5CHO + O_2 = C_6H_5C\begin{cases}OH \\ O = O\end{cases}$$

and

$$C_6H_5CO_3H + C_6H_5CHO = 2C_6H_5COOH.$$

Mr. Bäckström has been able to prove definitely that the primary product of the photochemical reaction is the per-acid and that, *even in this primary process,* a large number of per-acid molecules are formed for every quantum falling on the system. Of all the oxygen absorbed under illumination at 0° C., he was able to titrate as per-acid as much as 94 per cent, and, by a calculation similar to that involved in a preceding paragraph, he concluded that 432 molecules of per-acid were formed for every quantum of absorbed light. There can be no doubt, therefore, that the photochemical formation of per-acid from aldehyde and oxygen is a chain reaction. The existence of a photochemical after-effect suggests also that the second stage in the process is also a chain reaction. It shows the phenomenon of inhibition as with the whole reaction process. Addition of anthracene slowed down the rate of reaction in the ratio of 1 : 2.4 in a single experiment in which the second stage of the process was studied by itself. The results indicate that the second reaction, in the case of benzaldehyde, initiates much shorter chains, on the average, than the primary reaction. That both reaction stages are marked chain reactions is ample cause for the well-known autocatalytic nature of the autoxidation of benzaldehyde.

As to the relative effect of inhibitors on the dark and the light reaction, evidence which may be regarded as supplementary evidence concerning the validity of the Christiansen theory, Bäckström comes to the following conclusions:

(1) The light reaction shows the phenomenon of inhibition to a very marked degree.

(2) An inhibitor for the light reaction invariably acts as an inhibitor also for the dark reaction. No exception has so far been found to this rule.

(3) In the case of benzaldehyde, the dark reaction is considerably more sensitive to inhibitors than the light reaction in the sense that a concentration of inhibitor which is insufficient appreciably to affect the light reaction may have a considerable effect on the dark reaction. There may be substances that act as inhibitors for the dark reaction without having any effect on the light reaction. In the case of sodium sulphite solutions, the two reactions show a close parallelism in their behavior towards in-

hibitors, whereas oenanthaldehyde occupies an intermediate position in this respect.

On Christiansen's theory the activation of aldehyde molecules should occur by collision with energy rich molecules of the reaction-product. This is analogous to the well known phenomenon of photosensitization, the activation of molecules by collision with molecules of a foreign substance that have been raised to a high energy level by the absorption of light quanta. It is therefore of interest that Bäckström has found that these autoxidation reactions can be photosensitized and greatly accelerated in the photo-oxidation by addition of such reagents as α-nitro-naphthalene, benzophenone, fluorene, benzil and phenanthrene-quinone.

The problem remains as to how the inhibitor breaks these well established chain reactions. Bäckström believes that the inhibitor is itself oxidized. The possibility of compound formation, first suggested by the writer, still remains as a possibility of chain-breaking. No decision is yet possible in this regard. In atomic chains of the hydrogen-chlorine type the walls of the vessel probably act as surfaces on which the atoms are adsorbed and their efficiency as chain producers eliminated. In liquid media this is less likely. It is in the highly specific influence of added agents that clues as to the exact mechanism of chain-breaking may be found.

REFERENCES

1 Z. physik. Chem., 29, 295 (1899).
2 H. A. Taylor, J. Phys. Chem., 28, 984 (1924).
3 Lewis, J. Chem. Soc., 113, 471 (1918); Hinshelwood and Hughes, J. Chem. Soc., 125, 1841 (1924).
4 Hinshelwood and Burk, J. Chem. Soc., 127, 1552 (1925).
5 *Ibid.*, 127, 2896 (1925).
6 J. Phys. Chem., 30, 818 (1926).
7 Proc. Roy. Soc., 113A, 77 (1926).
8 J. Chem. Soc., 127, 1105 (1925).
9 Birje and Sponer, Proc. Am. Phys. Soc., Phys. Rev., 27, (1926).
10 Z. physik. Chem., 119, 390 (1926).
11 By analogy with some unpublished work of Dr. W. A. Dew.
12 4th Report, Committee on Contact Catalysis, J. Phys. Chem., 30, 145 (1926).
13 J. Phys. Chem., 30, 1487 (1926).
14 Z. physik. Chem., 53, 166 (1905).
15 Phil. Trans. Roy. Soc., 175A, 617 (1884).
16 Taylor and Marshall, J. Phys. Chem., 29, 1140 (1925); Marshall, *Ibid.*, 30, 34 (1926).
17 See, Mellor, Treatise on Inorganic Chemistry, Vol. I.
18 Rice, J. Am. Chem. Soc., 45, 2808 (1923).
19 J. Am. Chem. Soc., 40, 1462 (1918); 44, 2194 (1922).
20 Z. physik. Chem., 98, 260 (1921).
21 J. Chem. Soc., 117, 1120 (1920); 121, 1613 (1922).
22 Z. physik. Chem., 102, 169 (1922).
23 Z. physik. Chem., 102, 169 (1922).
24 Z. physik. Chem., 79, 147 (1912).
25 Z. physik. Chem., 80, 573 (1912).
26 Z. physik. Chem., 84, 451 (1913).
27 Medd. K. Vetenskapsakad. Nobelinst. 5, 1919, Nr.11.
28 Z. physik. Chem., 97, 134 (1921).
29 Z. physik. Chem., 10, 96 (1892).
30 Z. physik. Chem., 80, 587 (1912).
31 loc. cit.
32 J. Chem. Soc., 83, 729 (1903).
33 J. Phys. Chem., 27, 322 (1923).
34 J. Phys. Chem., 28, 145 (1924).

CARBOHYDRATES

BY

SIR JAMES COLQUHOUN IRVINE

PRINCIPAL, THE UNIVERSITY OF ST. ANDREWS

Lectures delivered at Columbia University in the special course on Contemporary Developments in Chemistry given in the Summer Session of 1926 on the occasion of the opening of the Chandler Chemical Laboratories

COLUMBIA UNIVERSITY PRESS

NEW YORK

1927

Printed from type. Published March, 1927

Printed in the United States of America

CARBOHYDRATES

By Sir James C. Irvine

Lecture I

To lecture in Columbia University is not a new experience, and I have the happiest recollections of an occasion three years ago when, in this Hall, it was my privilege to speak to the chemists of New York. Bearing this in mind, I find much personal satisfaction in the thought that Columbia University has extended this further invitation to me, and has included the subjects in which I am specially interested as part of the program of the work of this session.

Confining myself, therefore, to the topic of carbohydrate chemistry we have to join forces in asking ourselves the questions: where do we stand, and whither are we going? They are large questions, and my chief difficulty will be to compress what I ought to say within the limits of my two lectures without involving you in a mass of minute detail. Yet detail cannot be altogether ignored, and although to answer our enquiries we must take an aeroplane view of the situation, it will be necessary, from time to time, to descend to earth and use the microscope. I may make the warrantable assumption that you are familiar with the general features of the landscape we have to survey and that you will understand the diffidence I have in directing your attention to features in that landscape which have been made or modified through the work of my Research School.

As a matter of convenience, I have divided the time at my disposal equally into two sections, the first to include Monosaccharides and Disaccharides, the other to embrace the more difficult problems of the Polysaccharides. Accepting this division, it devolves on us to-day to take up the consideration of the compounds which the text-books, with a lofty disregard for the subtleties involved, term "the simple sugars." My task will not be easy, for it will be necessary for me to demolish many of the ideas which, hastily formulated and hastily accepted, have crept into the standard literature on the subject.

Some preliminary observations of a general character may be permissible, for the investigation of sugars has become more than ever an exact and exacting study. No longer is it sufficient merely to isolate a new sugar from its natural sources, to ascribe to it a molecular formula and to ascertain what products it yields on hydrolysis by acids or enzymes. Investigation must proceed further and must penetrate the inner recesses of the molecules. It may be noted, too, that entirely new standards of accuracy and of purity have now been laid down for the examination of crystalline sugars, and this I consider to be an incidental but important development of the work of the American sugar chemists. To mention but one example, we find, in the researches of C. S. Hudson on the magnitude of the molecular rotations displayed by the α- and β- modifications of reducing sugars something more than brilliant speculation on the relationship between rotatory power and configuration. To secure such results and to attain the necessary accuracy, sugars must be purified to an extent never contemplated by our predecessors, and, in consequence, the whole plane of research on carbohydrates has been raised to a higher level with great advantage to the subject.

The chemist is confronted with problems entirely different from those of twenty years ago, for attention is now being focussed on the idea that, in order to understand even a monosaccharide, it is necessary to know the individual reactions of each part of the molecule, so that when we talk of the reactions of glucose we are referrring not so much to a series of properties with which glucose as a whole is endowed, but recognize the specific part played by each asymmetric system. This changed attitude of mind is, in itself, a sign of progress.

Let us take glucose as our type of a reducing monosaccharide and trace how this development has progressed. As you are aware, glucose was for long represented as a straight-chain hydroxy-aldehyde (Formula I) until, about thirty years ago, sufficient evidence accumulated to show that a cyclic structure (Formula II) afforded a better graphic representation of the properties of the sugar.

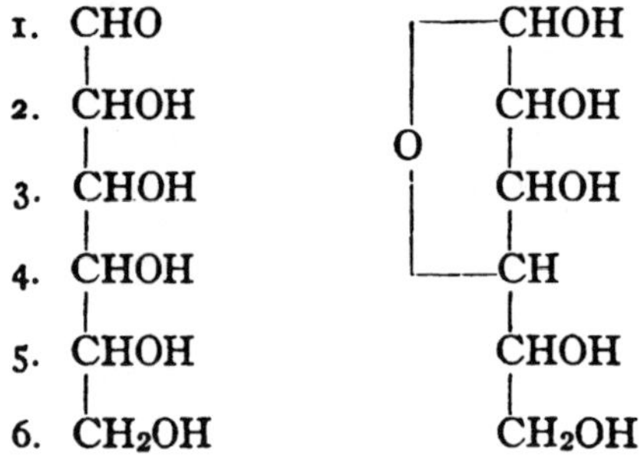

I II

I need not delay to give the reasons which led to this change, but may cite merely the phenomenon of mutarotation, and the formation of the artificial methylglucosides, as among the factors which made it necessary to allocate a hydroxyl, rather than an aldehydic, group to the terminal position. As formulated above it is possible to recognize that the hydroxyl groups in glucose fall into four distinct classes, these being:

(a) The terminal "reducing group" in the 1-position,
(b) Two secondary alcohol groups within the ring,
(c) One similar group external to the ring,
(d) The terminal primary group in the 6-position.

It will be observed that the above ring formula for glucose represents the compound as a butylene-oxide, and it may be stated that the choice of this particular ring was made on the grounds of collateral rather than of direct experimental evidence. But, if we admit that an oxygen-ring may be formed in glucose, it is conceivable that this ring may connect the terminal carbon atom with any other carbon atom in the chain giving the series of oxides shown in the following skeleton formulae:

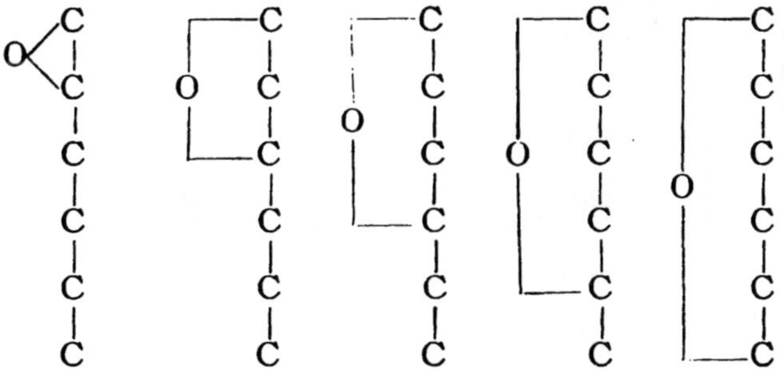

III IV V VI VII

The ring systems postulated above would vary greatly in stability, but, on theoretical grounds, it may be predicted that maximum stability will be attained in the systems indicated in formulae V and VI. Now, to the experienced worker in the sugars it had long been known that the butylene-oxide formula for glucose was not in all respects satisfactory, and an important advance, indicating the necessity for a less rigid view, was made twelve years ago, when the second isomeric form of tetramethyl glucose was isolated. The contrast between these isomeric sugars is remarkable, as shown in the appended scheme, which illustrates their origin and epitomizes their properties.

Source	Sugar	Properties
Glucose ⇄ α- or β-Methylglucoside →	Tetramethyl glucose	Crystalline Dextro-rotatory Stable
Glucose ⇄ γ-Methylglucoside →	Tetramethyl glucose	Liquid Laevo-rotatory Reactive

A detailed survey of their reactions shows that these methylated glucoses can differ only in respect of the position of the oxygen ring which, labile in the case of an unsubstituted sugar, is prevented from fluctuating from one position to another by the replacement of hydroxyl by methoxyl. This discovery constitutes an experimental proof that the oxygen ring in glucose does actually occupy different positions and the proof does not stand alone, as precisely parallel results are obtained with all reducing sugars.

The properties of these isomeric tetramethyl glucoses show that, most probably, they conform to the alternative formulae indexed as VIII and IX.

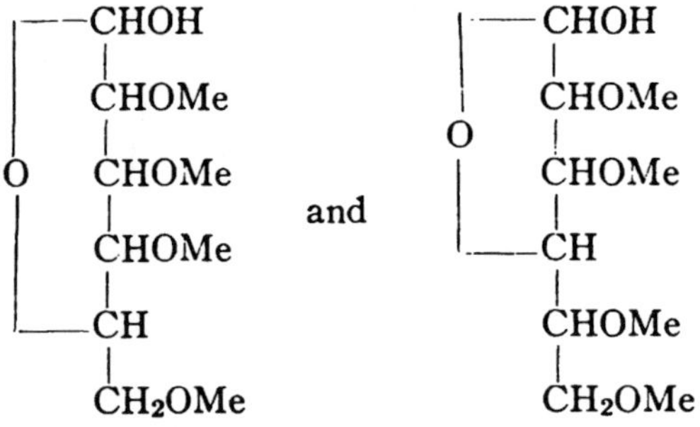

VIII IX

The allocation of these formulae is difficult but, if we accept Hirst's results on the oxidation of crystalline tetramethyl glucose, it would follow that this sugar corresponds to formula VIII, whilst the laevo-rotatory isomeride (termed "tetramethyl γ-glucose") would then be accommodated by formula IX. Here observe that this particular isomerism demands the use of a system of nomenclature in which the position of the substituting groups is indicated by numbers. Adopting this system the two methylated glucoses now under consideration would be termed respectively 2:3:4:6-tetramethyl glucose, and 2:3:5:6-tetramethyl glucose.

I have commenced with the question of the ring structure of glucose, as the subject has recently been prominent, and my opinion is frequently asked as to how I regard the claim that the constitution of glucose should be revised so as to represent the compound as an amylene-oxide. My answer is that there is no universally-applicable formula for glucose, and that the position of the ring is determined by the conditions under which the sugar is brought into reaction. The evidence submitted by Haworth in favor of the amylene-oxide formula is unconvincing, and leads to the mistaken opinion that this is the normal standard condition of the compound. Such a conclusion is entirely at variance with the structural formulae of such widely-distributed polysaccharides as cellulose and starch, and cannot be reconciled with the results Haworth himself obtained with cellobiose and maltose, if his observations are freed from a fundamental experimental error.

In my opinion we must discard the view that a reducing sugar has a fixed constitution, and should replace this conception by the more flexible idea that the molecule of these compounds adjusts itself according to the reagents or conditions to which it is exposed. It is, in consequence, necessary to ascertain the position of the ring in each derivative of a sugar under all experimental conditions employed, as otherwise endless confusion will inevitably result.

As an example we may consider the important reaction in which glucose, in common with other reducing sugars, combines with acetone. As is well known, the hexose condenses with two molecules of the ketone, but it has been proved that, in the case of glucose, the sugar transforms from the stable to the reactive variety before this reaction can proceed; in other words, the oxygen-ring alters as a necessary preliminary to condensation. A similar state of affairs is encountered in the case of fructose where only the stable variety reacts, and if a derivative of the unstable type, such as γ-methylfructoside, be employed, again the ring alters its position during the reaction.

Taking a broad view, and limiting ourselves to definite experimental evidence, we must recognize that glucose has already been shown to react in the following ways, in three of which different rings are present:

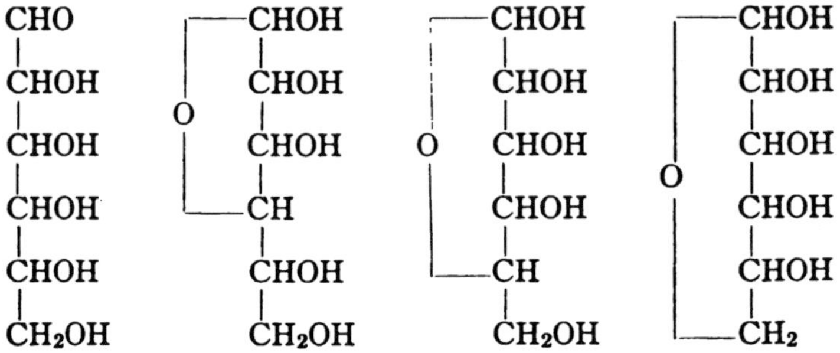

X XI XII XIII

As we have seen, two further structures are theoretically possible viz.: the 1 : 2 and the 1 : 3 oxygen linkage, but, so far as known, no derivatives of this type have been proved to exist. One point which emerges in this view of glucose structure is that the prevailing fashion of postulating oxygen rings must not blind us to the fact that glucose can play the part of an aldehyde.

Before leaving this subject, it is advisable to discuss the question of how the position of the oxygen ring in glucose is affected by the experimental conditions to which the sugar is subjected. Although perhaps premature to generalize too freely in this respect, it is possible to indicate how the ring is displaced by rise of temperature, if we consider the behavior of glucose when condensed with methyl alcohol under different conditions. At 20° the reaction yields "γ-methylglucoside" where the oxygen ring is presumably in the 1 : 4 position; at 100°, α-methylglucoside is produced, and here the ring couples the 1 : 5 positions, while if β-glucose be distilled in a high vacuum it is converted into β-glucosan which possesses two rings, one of them 1 : 5 and the other 1 : 6.

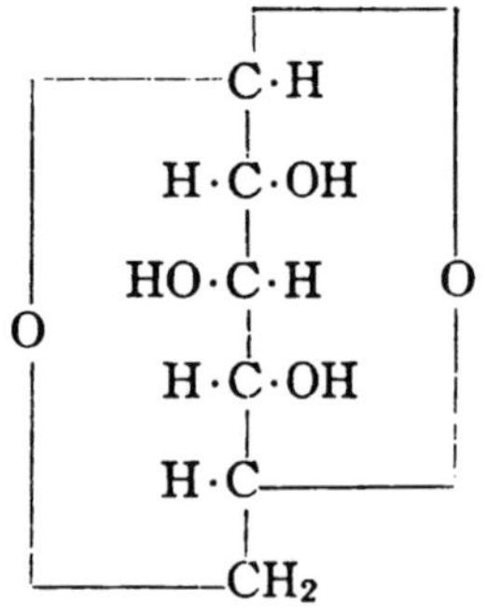

XIV

This is not the only glucose derivative containing two oxygen rings and, when confronted with such a case, it becomes a difficult matter to say which ring is to be regarded as part of the initial sugar-molecule and which is a genuine anhydro-ring. From what has been said, it will be evident that there are now good theoretical grounds for the belief that if we are to attain to the position of understanding the role played by the sugars in Nature, it will be necessary, in the future, to study the reactions of carbohydrates under the temperature conditions favorable to life, and to abstain from processes which are purely artificial.

So far, I have confined myself exclusively to one structural aspect of monosaccharide chemistry, and have treated the subject in so much detail that it is possible to review other developments in only the briefest possible manner. One general feature which may be discerned is that considerable novelty, and even daring, is now displayed in the choice of the reagents employed to act upon the simple sugars, and it is gratifying that, in the majority of cases, investigations on monosaccharides are carried out with a clearly-defined object in view. Among these enquiries may be mentioned the study of definite partly substituted sugars, that is to say, derivatives in which selected hydroxyl groups have been substituted, leaving others free for independent examination. As indicated in my opening statement, this type of work is of exceeding value, and already is proving of advantage in the interpretation of such obscure physiological processes as the action of glucose in combating the effect of an over-dose of insulin.

Further lines of enquiry in which rapid progress has been made, include the masterly researches conducted within a short distance of this room on the baffling and perplexing nitrogen derivatives of sugars, surely one of the most important sections of carbohydrate chemistry. Then, there is the search for the quantitative relationships between optical rotatory power and configuration, which has engrossed the attention of the physical chemist since the earliest days of van't Hoff's doctrine, but led to no satisfactory conclusion so long as the quest was confined to compounds of simple type. Nevertheless, when the complex examples furnished by the sugar group were examined, the discerning genius of Hudson brought to light a series of generalizations which now enable us to apply rotatory power as a test of structure.

There remains one other type of problem which, after a period of little progress, now shows signs of a revival. I refer to the synthesis of monosaccharides, and more particularly to the attempts being made to study the consecutive reactions in which formaldehyde, when supplied with light energy, undergoes autocondensation. This represents a field of sugar chemistry which has remained too long uncultivated, for the mind continually travels back to the fundamental question of the origin of sugars. Many problems which lie at the very foundation of sugar chemistry are still unsolved: What is the first sugar molecule formed in the plant, and from what simpler compounds does it arise? To what cause does the molecule owe its optical activity, and why is the glucose configuration favored above all others? What is the chemical process which operates so easily in Nature, and is so difficult in the laboratory, whereby glucose and fructose are interchangeable? How are the hexose molecules dehydrated and condensed with each other? These questions are but part of the great problem which lies to be solved before we are in a position to interpret the reactions of the living plant—and how far we are from providing the answers!

If now we turn to consider the chemistry of Disaccharides, we find again that, in recent times, the greatest advances have been made in purely structural studies, and that the analytical, rather than the synthetical, method has been prominent. Now, in approaching such a problem as the constitution of a disaccharide, the essential factors to be established are:

(1) What monosaccharide residues are present in the compound sugar,

(2) Which of these constituents is responsible for the reducing action, if any, of the disaccharide,

(3) In what stereochemical form (α- or β-) does the reducing component exist,

(4) In what stereochemical form (α- or β-) has the non-reducing component become united to the reducing sugar residue,

(5) Through which hydroxyl groups are the two monosaccharide molecules condensed together,

(6) What oxygen rings are present in the monosaccharide residues.

The answers to the first four questions are fairly easily found by enzyme hydrolysis, either of the free sugars or of derivatives, but the most direct and, practically speaking, the only method of solving the remaining questions depends on the use of methylated sugars. The principle and working processes involved in this particular method of determining structure are well known and need not be repeated, beyond stating that, if a disaccharide be fully methylated and the product hydrolyzed, two alkylated hexoses will be formed. Determination of the number and position of the alkyl groups in these scission sugars supplies the evidence required to formulate the parent disaccharide, as the positions through which the hexoses were united are protected from methylation and are therefore represented by hydroxyl groups in the hydrolytic products. The process, which was initiated and developed by the St. Andrews School, has been widely applied, and has been used by Haworth in ascribing to disaccharides the structural formulae which, up to the present, have found general acceptance. Two years ago it was my belief that a stage had been reached when the constitutional chemistry of disaccharides might be considered to be on a secure foundation, and that the time was ripe to turn to research on the synthetical polymerization processes leading to the formation of polysaccharides.

This view, expressed in an address to the Royal Institution in London, is no longer tenable; we are only at the beginning instead of at the end of knowing how the monosaccharides are attached to each other in the disaccharides. It may at once be stated that this disquieting situation is not attributable in any way to the suggestion that the oxygen-ring in glucose should be represented as amylene-oxide rather than butylene-oxide in type. The fact that the entire structural classification of the disaccharides has broken down is due, not to a difference in interpretation of experimental results, but to mistakes of fact. To make this clear, and to attempt to save sugar chemistry from confusion, it is necessary to summarize the position.

As already explained, the application of the methylation method of solving the constitution of disaccharides involves the identification of the methylated monosaccharides formed on hydrolysis. In practice it is found that a limited number of methylated glucoses function as "reference compounds" in research of this description, and, for the moment, consideration may be limited to two examples. These are isomeric forms of trimethyl glucose and, tentatively, although not finally (as will be evident later), they may be formulated as containing an amylene-oxide ring:

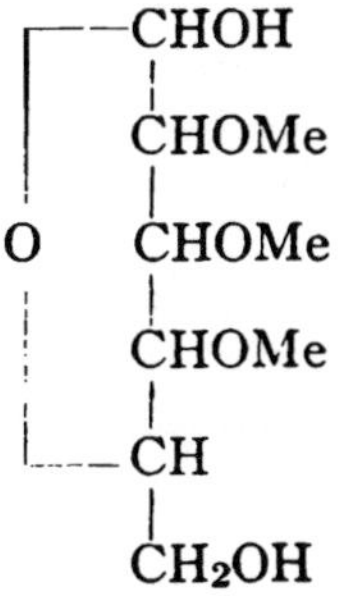

: 3 : 4-trimethyl glucose. Liquid.

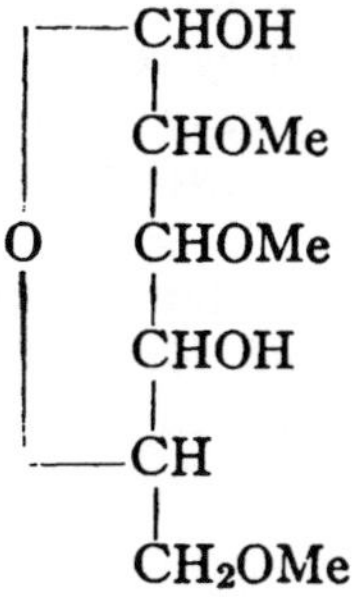

2 : 3 : 6-trimethyl glucose. Crystalline.

If then, we obtain the liquid variety of trimethyl glucose from a disaccharide we conclude that one glucose molecule has been attached to the terminal —CH_2OH group of the other; but, consideration will show that if the crystalline trimethyl glucose is produced, the result is not decisive, as the oxygen-ring in this sugar can couple either the 1 : 4 or the 1 : 5 positions, so that two interconvertible forms of 2 : 3 : 6-trimethyl glucose can exist. These two isomeric forms of 2 : 3 : 6-trimethyl glucose may now be formulated, as it will be necessary to make repeated reference to them.

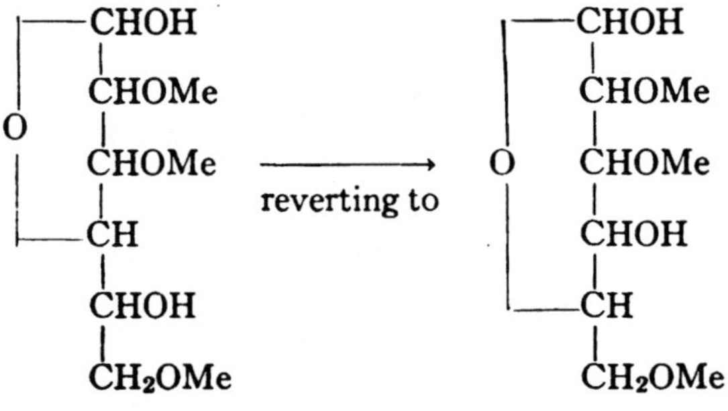

Unstable. Stable.

This isomeric change has been verified experimentally, and it follows that all constitutional formulae for disaccharides, founded on the formation of this particular reference sugar, are invested with uncertainty. *Obviously, the same 2: 3: 6–trimethyl glucose would be formed irrespective of whether the constituent monosaccharides were joined through the 4th or 5th positions, as the crystalline sugar actually isolated is always the stable form.*

The point is a somewhat elusive one, and to build up a structural scheme for the disaccharides on a faulty logical basis is no more than unfortunate, but, as already indicated, the whole question has been further complicated by mistakes in facts. This may be illustrated by reference to the sugar-maltose, an example which, as events have proved, is critical and affects the whole scheme. The disaccharide, when methylated and hydrolyzed, was *said* to give:

(1) Crystalline tetramethyl glucose together with

(2) The liquid form of trimethyl glucose, and, from this result, the constitution of maltose was claimed to be:

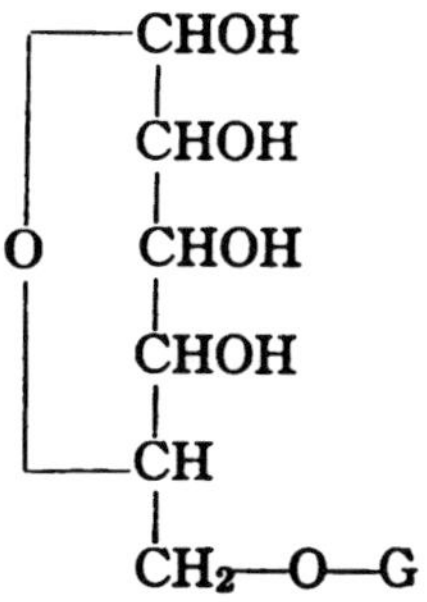

where G = the second glucose residue

This view which was again expressed by Haworth so recently as January of this year has been refuted by the results of an investigation on the structure of starch, and on duplicating Haworth's work, it has transpired that the liquid specimen of trimethyl glucose he isolated from maltose was, in reality, an impure preparation of the crystalline 2:3:6-isomeride. This mistake, which is repeated again in parallel work on raffinose, has been far-reaching, and, taken in conjunction with the failure to recognize that the formation of 2:3:6-trimethyl glucose is not diagnostic, renders impossible the acceptance of Haworth's views. Inspection of the following list of di-hexoses, to most of which a structural formula has been assigned, will indicate how fundamentally these errors permeate the work. To facilitate reference the sugars showing a close relationship have been bracketed in pairs.

{ 1. Maltose
{ 2. Isomaltose

{ 3. Melibiose
{ 4. Lactose

{ 5. Cellobiose
{ 6. Isocellobiose

{ 7. Gentiobiose
{ 8. Fischer's "Isomaltose"

As the constitution ascribed to maltose by Haworth is incorrect, that of melibiose, which he deduced by analogy, must also be discarded. Further, as maltose and cellobiose give the same methylated glucoses his constitutional distinction between these sugars vanishes. Including the *iso*-forms of maltose and cellobiose there are now too many disaccharides competing for one and the same formula, and we are forced to the conclusion that, in the case of one pair, the glucose components are joined through position 4 while, in the remaining pair, the union is through position 5. This is represented below:

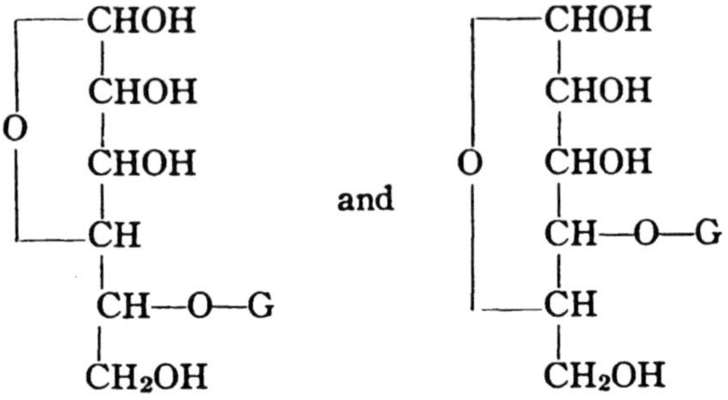

(G = Non-reducing glucose residue)

Glucose 5-glucoside. Glucose 4-glucoside.

The formula for lactose likewise becomes conjectural as two alternatives are possible, and thus the error regarding the formula of maltose affects the complete series of disaccharides (with the possible exception of gentiobiose), and extends to the trisaccharide series. The constitution at present accepted for raffinose, for example, contains the melibiose residue now shown to be incorrect and, in addition, represents a sucrose residue to which an inaccurate formula is applied. The methylation method of solving structure is discriminative and accurate, but its success depends on the correct identification of methylated sugars and this demands care and experience. Amidst so much uncertainty, one definite conclusion remains, viz.: that no universal oxygen-ring (amylene-oxide or otherwise) can be adopted for glucose.

We must now recognize that, as each hydroxyl group of glucose can unite with both the α and β forms of another hexose, two disaccharides are derivable from each hydroxyl position. We can therefore say that, so far as the diglucoses are concerned, cellobiose and *iso*-cellobiose form one such pair while maltose and *iso*-maltose form the other. Four different sugars have therefore to be allocated between two formulae, so that in place of being able to ascribe a definite constitution to each compound, we are left in a position of uncertainty. The only disaccharide formula which seems to have been established is that of gentiobiose, a sugar now regarded as identical with Fischer's synthetic *iso*-maltose.

The most disconcerting feature of the situation is that, so long as we cannot discriminate structurally between maltose and cellobiose, the chemistry of starch and of cellulose must be retarded, but, on the other hand, there is every reason to hope that these polysaccharides will themselves supply the evidence required to formulate in structural terms the disaccharides into which they are convertible. Progress is already far advanced in this direction. It will be generally acepted as a rational view that the essential structure of starch is repeated in maltose, and that the union of the glucose residues in the polysaccharide is in the same position as in the disaccharide. Within the past two years Pringsheim has put forward evidence leading to the conclusion that both starch and glycogen are built up of γ-glucose residues, and it is clear that this can be accommodated only by the allocation of the glucose 5-glucoside structure to maltose. The results obtained in the methylation of starch are consistent with this view, which may be accepted for the time being, and accordingly the structure glucose 4-glucoside should be reserved for cellobiose. The combined evidence of many related researches has found further support in a recent paper by Zemplén who has degraded cellobiose to a stage where the product ceases to form an osazone and thereby he comes to the opinion that cellobiose is glucose 4-glucoside.

To sum up the situation, we can, so far as present knowledge takes us, allocate the disaccharides under discussion to the following formulae:

{ Gentiobiose
Fischer's *Iso*-maltose

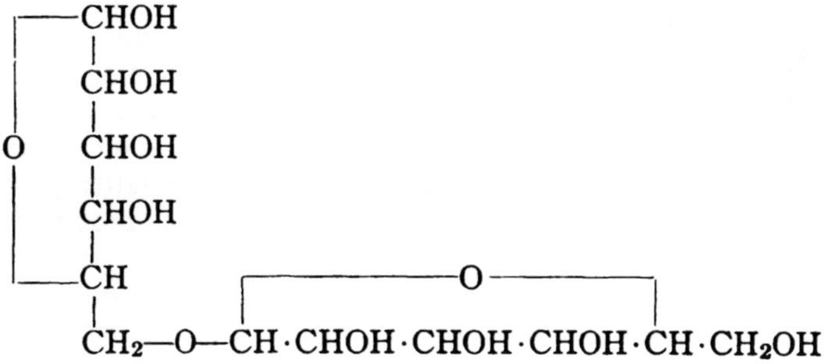

{ Maltose
Iso-maltose

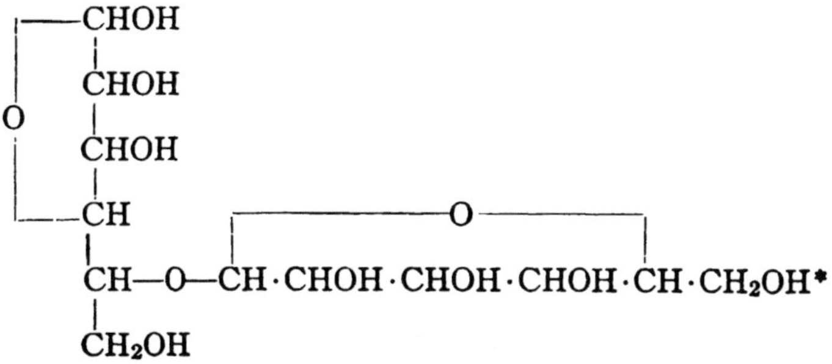

* Melibiose may be represented also by the above formula, where * = galactose.

{ Cellobiose
{ *Iso*-cellobiose

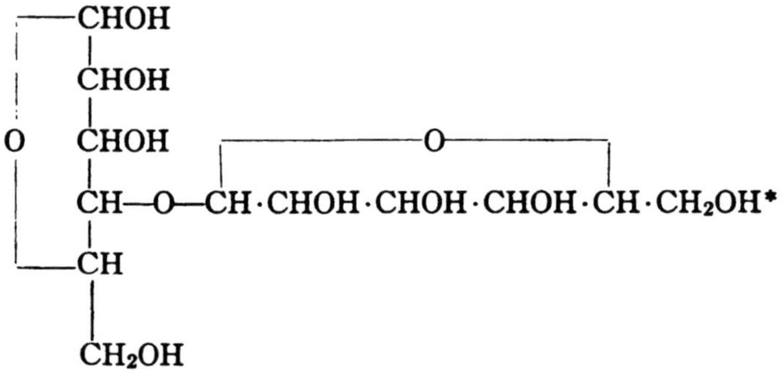

* Lactose may be represented also by the above formula, where * = galactose.

There remains one other disaccharide of which the constitutional formula must be corrected as it has been adopted on the basis of evidence which I have always regarded as insecure. I refer to sucrose, the first complex sugar to which the methylation process was applied by Purdie and myself in the early days of our collaboration. An octamethyl sucrose was duly obtained and this, when hydrolyzed, yielded the crystalline 2:3:4:6-tetramethyl glucose together with a dextro-rotatory liquid consisting of tetramethyl fructose. At a later stage the latter product was recognized as tetramethyl γ-fructose so that, provided the constitution of this sugar could be determined, the structure of sucrose would immediately follow. Haworth, who investigated this section of the sucrose problem, claimed, on the evidence of oxidation, that tetramethyl γ-fructose is an amylene-oxide and therefore possesses the formula now given:

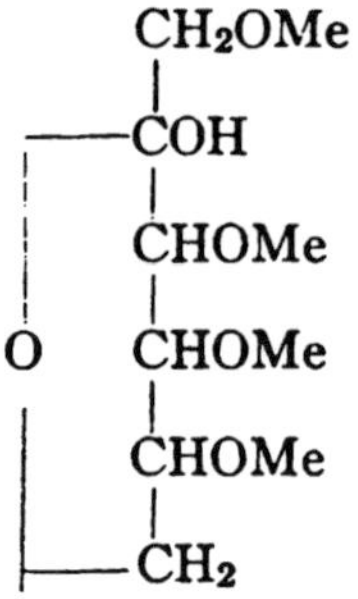

In consequence, he formulated sucrose as under:

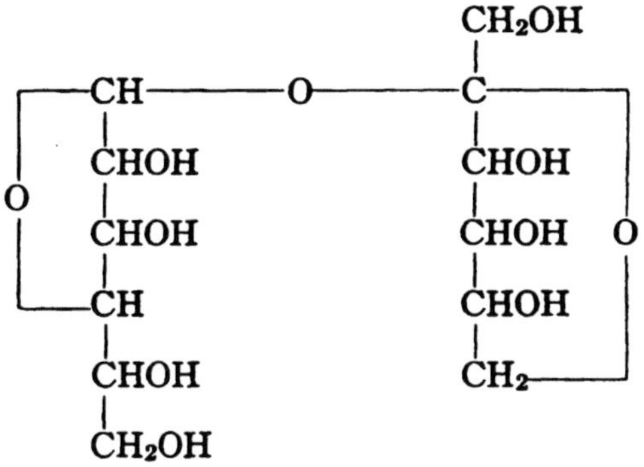

A few months ago he transferred the oxygen-ring of the glucose fraction from the 1:4 to the 1:5 position, a change which is justified only in view of Hirst's oxidation of tetramethyl glucose, but greater importance must be attached, for many reasons, to the fructose component of sucrose. For a considerable time it has been evident that Haworth's formula is incorrect and that his representation of γ-fructose is at variance with the properties of the ketose. This doubt became a certainty when it was recognized that Haworth's interpretation of the results he obtained on oxidizing tetramethyl γ-fructose involved a serious theoretical mistake, it being overlooked that while degradation of *d*-fructose gives members of the *d*-arabinose series, the latter are the optical antipodes of the same compounds obtained from natural arabinose. It is therefore not surprising that when McOwan repeated Haworth's oxidation of tetramethyl γ-fructose he failed to confirm the results.

I need not elaborate the point, and indeed on an occasion such as this, I would prefer to avoid any references so destructive but, in the interests of sugar chemistry, it is necessary to make the situation clear. The constitution of sucrose is not established, and the fresh evidence now available through McOwan's work may be interpreted as leading to either a propylene-oxide or a butylene-oxide structure for the γ-fructose residue. If, however, we consider the results of collateral researches, particularly those dealing with the acetone derivatives of fructose, there is strong evidence in favor of the larger ring, so that, meanwhile, sucrose may be formulated as:

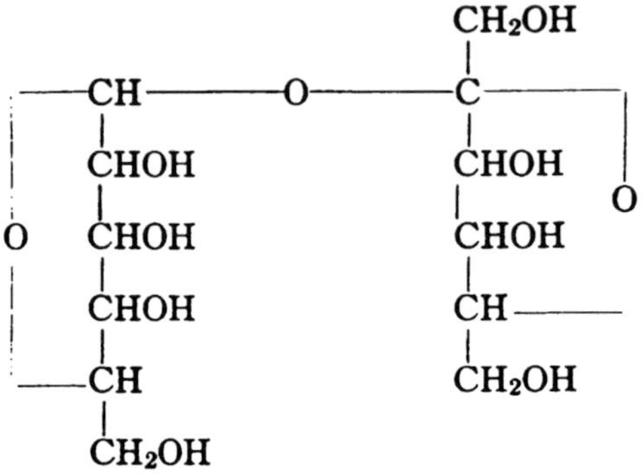

It is now possible to develop a formula for raffinose in agreement with experimental facts. As we have seen Haworth's structure for the trisaccharide is incorrect, being founded on a formula for melibiose now discredited and involving, therefore, a repetition of the experimental mistake which led him to false conclusions in the case of maltose. Raffinose may be represented as:

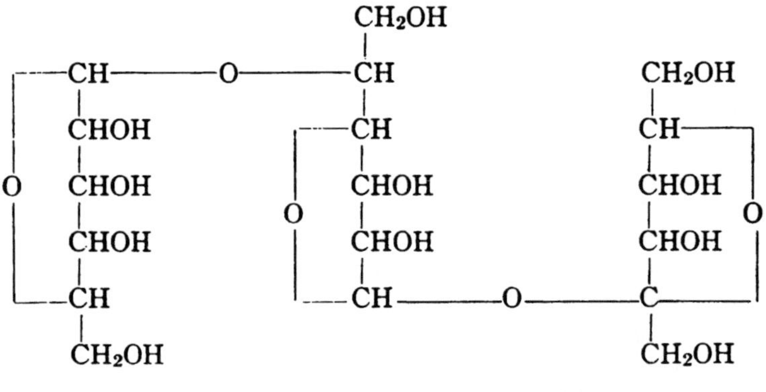

Galactose Residue Glucose Residue Fructose Residue

Melibiose Residue Sucrose Residue

In closing this section of our review, I can but repeat cautious words I used some years ago in speaking on the same subject before the Chemical Society of London. There should be no finality in our views of sugar structure, as each set of suggestions represents merely a stage in knowledge. Theories serve their purpose only for a time, and only so far as they direct our way and lead ultimately to fundamental truth. The pursuit of structural sugar chemistry is, in fact, like the pursuit of golf, as summed up by Tom Morris, the philosopher of the links, "The game is always fighting the player—and the player never wins."

Lecture II

In completing our discussion of carbohydrates, I propose to devote this lecture to a review of the constitutional aspects of polysaccharide chemistry. I invite your attention, therefore, to problems which are theoretically related closely to those presented by the simple sugars, yet nevertheless seem infinitely remote to the experimenter working at the bench.

In place of definite crystalline compounds to which the orthodox processes of the chemist are applicable, the investigator has to deal with amorphous non-volatile substances, soluble only in the sense of giving colloidal systems, and displaying the contradictory properties of being, at one and the same time, remarkably stable and highly reactive. In addition to the physical complications to which I have referred, all the difficulties inherent in carbohydrate research are present, and the combination places severe restrictions on the choice of reactions applicable to polysaccharides and capable of yielding results which can be interpreted in terms of structure. Quite apart from difficulties, it is disconcerting to an organic chemist, who, like myself, has been trained in the exact school of stereochemical research, to conduct investigations with material which may not be strictly homogeneous and, in most cases, is actually a mixture. If we ask the question, what are the criteria of a "pure" cellulose (or starch) no very satisfactory answer can be given, and, when investigating polysaccharides, it is, in consequence, necessary to bear constantly in mind that we are not studying the reactions of chemical individuals, but only unfolding the average properties displayed by a group of closely-related compounds.

What I have said is an expression of my own experience, gained at the bench and in the study, as a result of extending my earlier researches on mono- and disaccharides to the investigation of polysaccharides. I approached these complex compounds with precisely the same object in view as in studying the simple sugars—a desire to ascertain the inner structure of the molecules—and this induces me to make a statement which I trust no one will misunderstand. It is merely to point out that the technical applications of polysaccharides far exceed those of the sugars proper, and, particularly in the case of cellulose and starch, a voluminous scattered literature has come into existence. In these publications much work is described which is of excellent scientific quality, but there is also a mass of work which is empirical, imperfect and even contradictory. It is frequently the case that investigations on polysaccharides are directed towards some technical objective, and fundamental chemistry finds little place in the scheme. Fortunately there has always been a small band of workers who, recognizing that it is best to begin at the beginning, have approached the problems of polysaccharides by the avenue of molecular sugar chemistry. To attain the ultimate goal, the services of both types of worker are, no doubt, necessary but, for many years to come, it will be difficult to reconcile views taken from such widely-divergent angles.

Following the policy laid down in the first lecture, I shall endeavor to sketch for you the present position and future prospects of polysaccharide research, limiting myself largely to questions of molecular structure, and, for the most part, to investigations with which I have been personally connected.

If we now consider the two outstanding polysaccharides related to glucose it is a difficult matter to say whether cellulose or starch is the more important, and the more worthy of first attention. But as a matter of convenience I begin with cellulose which appears to display the greater molecular simplicity. As a result of conventional choice, which research has proved to be well-founded, cotton cellulose is selected as a standard variety of cellulose, and the reactions hereafter described have been obtained with this particular variety. From the chemical point of view the substance is $(C_6H_{10}O_5)_x$ and it has long been known that hydrolysis converts it into a syrup which was claimed, although not universally admitted, to be glucose. Again, it was recognized that, calculating on the basis of a C_6 chain, three hydroxyl groups are present in cellulose and, here, until recently, what may be termed discriminative information leading to structure ended.

The first problems to engage our attention were fundamental, the conversion of cellulose into the corresponding hexose, the identification of this sugar and the determination of its amount. It appeared safer not to make any assumptions regarding these factors which were supplied by preparing cellulose triacetate under conditions in which the reaction proceeds quantitatively and without depolymerization. Thereafter, the product was heated with methyl alcohol containing hydrogen chloride, a process which effected depolymerization and hydrolysis to the parent sugar; this, in turn, formed methylglucoside through reaction with the solvent and was thus protected from decomposition. In this way, it was shown by the best of evidence, viz.: the production of pure crystalline glucose, that cellulose is convertible into the hexose in a yield of 98 per cent of the theoretical amount, calculating on the basis of the equation

$$(C_6H_{10}O_5)_n \rightarrow n\ C_6H_{12}O_6$$

The logical development of this first series of investigations demanded in the next place enquiry as to the positions of the glucose molecule which are concerned in the formation of $C_6H_{10}O_5$ from $C_6H_{12}O_6$ and, for precisely the same reasons as those discussed in the previous lecture, it is evident that the information cannot be contributed either by the hydrolysis of cellulose alone or of its acyl derivatives. Once again, the method of methylation proved effective, but, in this case, owing to solubility factors it was necessary to employ the methyl sulphate reaction exclusively. The process was laborious, as the methoxyl content of the product increased only slowly, but frequent repetition of the reaction gave trimethyl cellulose as a friable powder which still preserved the fibrous nature of the original material. Hydrolysis of this cellulose ether, under conditions designed to avoid experimental loss, gave 2 : 3 : 6-trimethyl glucose as the sole product, and consideration will show that this result, in itself, greatly simplifies the constitutional problem. Let me remind you that two alternative views may be held regarding the nature of the cellulose molecule, the first being that the polysaccharide consists of molecules of large dimensions built up, on the glucoside model, by the union of many glucose residues. Alternatively, it is conceivable that cellulose may consist of simple molecules highly polymerized. The first alternative which was formerly advocated by Hess, and is represented below, must be discarded.

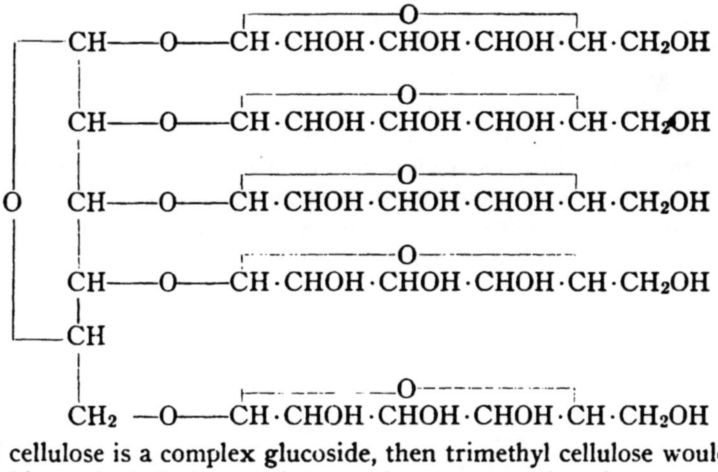

If cellulose is a complex glucoside, then trimethyl cellulose would yield, on hydrolysis, a mixture of sugars ranging from tetramethyl glucose to glucose itself. Such is not the case, and, in view of the result actually obtained, we are forced to the conclusion that the molecular unit of cellulose is an anhydride of glucose, or of a simple polyglucose. When it is remembered, however, that cellulose is convertible into the disaccharide cellobiose, it is at once clear that the number of hexose residues present in the molecule must be at least two, and this permits us to ascribe the simplest possible formula to the cellulose molecule. As we have seen, cellobiose is glucose 4-glucoside, and as, moreover, cellulose is completely convertible into 2:3:6-trimethyl glucose it follows that the anhydro-cellobiose unit postulated above must be constituted as:

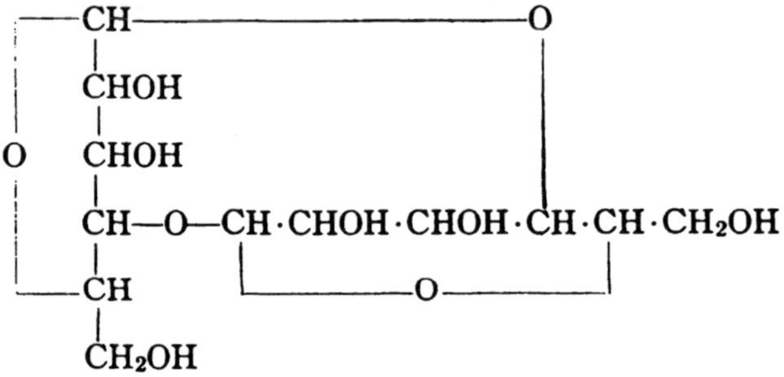

It is conceivable that this hypothetical "cellosan" is actually one of the molecules which form cellulose by polymerization, but many reasons exist leading to the belief that other anhydrosaccharides play a greater part in this respect. Among these reasons may be mentioned only one, the fact that the yield of cellobiose obtainable from cellulose falls far short of the amount demanded by the above formula. The discrepancy is, in fact, so great that it is necessary to consider the addition of a third glucose residue in the molecule, and here it may be remarked that two possibilities exist, as the additional glucose residue may become attached either symmetrically or unsymmetrically. Only the former alternative, however, need be considered at this stage as, to discuss the evidence leading to the exclusion of the unsymmetrical structure would involve us in considerable detail. A symmetrical anhydro-trisaccharide which would fulfill all the requirements, so far as chemical reactions are concerned, would be represented as:

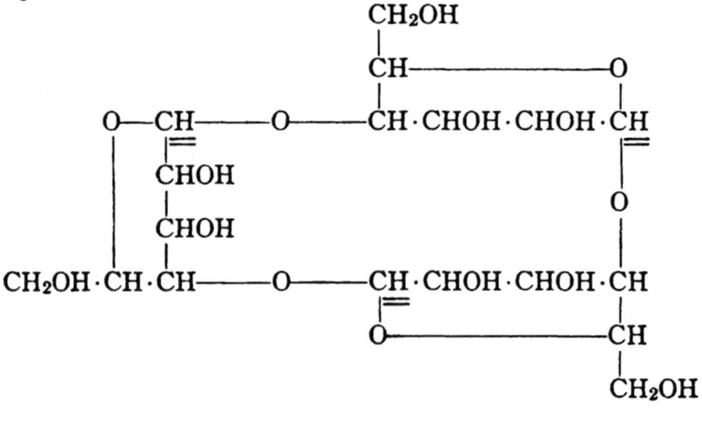

This is by no means the limiting case, as additional glucose residues may be added in similar manner, so as to give the further series:

A. An anhydro-tetrasaccharide,
B. An anhydro-pentasaccharide,
C. An anhydro-hexasaccharide, etc.

These types may be distinguished as odd and even according to the number of glucose residues present, and, as the methylation process cannot discriminate between them, it is necessary to fall back upon supplementary evidence, in order to gain an idea as to which type is present in cellulose. All the examples containing an even number of C_6 chains are open to the same criticism: that they fail to account for the yield of cellobiose obtained from cellulose, although it must be admitted that this reaction is unsuited for exact quantitative application. On the other hand, purely physical investigation, such as X-ray spectrographic analysis, favors the idea that the cellulose molecule contains four or six hexose chains, and although it is far from my intention to minimize the value of this evidence, there remains the special importance attached to the actual isolation of a depolymerized cellulose containing three glucose residues. Cellulose has been degraded, by means of controlled acetolysis, to a stage immediately preceding the formation of cellobiose acetate, the successive reactions being represented diagrammatically as under:

Cellulose
↓
Cellulose Acetates
↓ — *A.*
Cellulose Dextrin Acetates
↓
X
↓
Cellobiose Octaacetate — *B.*
↓
Glucose Pentaacetate

In the stages indexed as *A* some hydrolysis takes place, but the essential change is depolymerization, while *B* represents reactions which are mainly hydrolytic. The stage indexed as *X* thus includes the group of compounds formed when depolymerization is maximum and hydrolysis minimum, and it is here that the unpolymerized cellulose molecule is most likely to be found. The examination of this mixture has revealed that the essential compound present is actually an anhydro-trisaccharide, and the yields obtained indicate that this molecule forms at least one-third of the cellulose aggregate.

Taking everything into account, the most prudent view which I can recommend is to regard cellulose as a polymeride, towards the formation of which different anhydro-saccharides may contribute. Amongst these, is the anhydro-triglucose already isolated, but the corresponding anhydro-di-, tetra-, and hexaglucoses should also be included, there being positive evidence for their retention, and only indirect evidence for their rejection.

This opens out an overwhelming prospect of further research and leads me to consider a closely-related field of carbohydrate chemistry in which I am sure lies one path to future progress. I refer to the phenomenon of polymerization as applied to anhydro-sugars, and here we must recognize the lead given by Pictet, when he described the first experiments in which β-glucosan was converted from the monomeric to the polymeric form, by fusion in presence of catalysts. The conditions of this process are certainly far removed from those operative in Nature, but even admitting the comparatively drastic conditions of the reaction, the result is a decided and, indeed, the first step towards the synthetical study of the polysaccharides. This becomes apparent when we consider the constitution of these "polyglucosans," which appear to fall into two classes of distinctly different type, as unfolded by means of the methylation process. Both odd and even numbers of glucose residues combine together giving the series: di-, tri-, tetra-, and hexa-glucosans, but this progressive increase in molecular complexity is not polymerization in the ordinary sense, so much as condensation. For reasons with which I need not now detain you, it has been shown that the initial step is the conversion of one molecule of glucosan into glucose, a process which is followed by a species of glucoside-formation and further condensation.

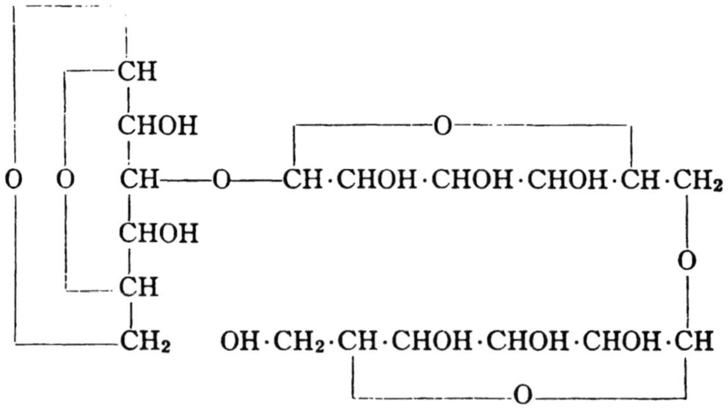

An extension of this research has shown that similar changes ensue when glucose is dehydrated under milder conditions, and the prospect is not altogether excluded that, ultimately, conditions will be found under which glucose can be dehydrated and polymerized with a facility and in a manner comparable with the economical synthetical processes which surround us in Nature.

After this digression, we may resume consideration of the polysaccharides related to glucose and turn our attention to the case of starch. Here the problems involved are, in many respects, more difficult than those presented by cotton cellulose, and we are at once confronted with the fact that not only do the starches derived from different sources vary profoundly in important respects, but starch from one and the same source is not homogeneous. It is possible, however, to arrive at constitutional factors which are common to all varieties of starch and, from now onwards, I shall attach distinct limitations to the expression "starch," leaving out of account, for the time being, that there are individual components in the starch grain, and also inorganic or nitrogenous constituents associated with the polysaccharide, confining myself to the main constituents which have been shown to possess the same molecular structure. It is evident that the diagnostic properties of starch are summarized in the statement that, empirically, the compound contains three hydroxyl groups for every C_6 residue, and further that by the action of selected enzymes it may be converted either into maltose or iso-maltose. As the structure of maltose is known, it now remains to consider the constitution of iso-maltose, and this is provided by the results obtained in methylating starch, a process which presents unusual experimental difficulties owing to the tendency of the reaction to become arrested at definite stages. The preparation of a trimethyl starch is, indeed, a task demanding the most laborious operations, yet there is consolation in the fact that the obstacles to be overcome throw significant light on the nature of the starch molecule. According to the conditions employed it is possible to identify three consecutive products, viz.:

A. Dimethyl Starch	Methoxyl = 30 per cent. This corresponds with the replacement of six hydroxyl groups out of a total of nine.
B. Methylated Starch	Methoxyl = 37 per cent. Seven groups out of nine methylated.
C. Trimethyl Starch	Methoxyl = 43.7 per cent. All the hydroxyl groups methylated.

If we consider the behavior of compound C, it is found that hydrolysis gives 2:3:6-trimethyl glucose and no other sugar, a result which at once proves that the hydroxyl groups are uniformly distributed in starch, each C_6 chain having three such groups in attachment. It also follows that maltose and *iso*-maltose are structurally identical so far as the position of attachment of the glucose residues is concerned, the one sugar being glucose 5: α-glucoside, and the other glucose 5: β-glucoside. Using the corrected formula for maltose, the simplest structure for the starch molecule would be that of a "maltosan" constituted as under:

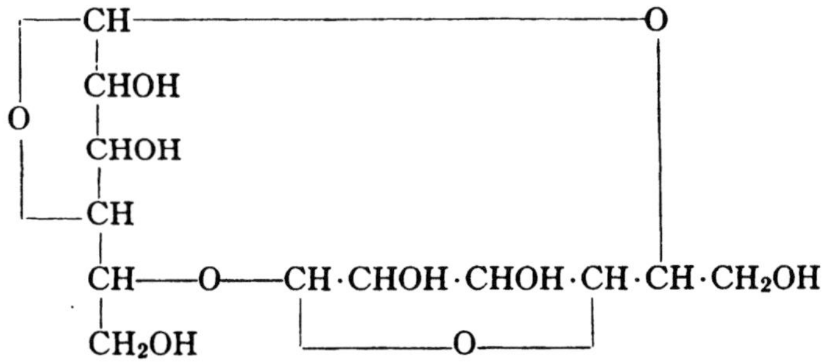

This represents a double anhydride formed by the union of two γ-glucose molecules and, on first inspection, the structure is satisfactory. But it is inconsistent with the fact that starch under one set of hydrolytic conditions is converted into maltose and gives *iso*-maltose when the conditions are changed, nor does it account for the formation of a definite methylated starch containing 37 per cent of methoxyl. For these reasons, we carry our speculations a stage further and include a third glucose residue. Here again we are justified in limiting discussion to the symmetrical system now formulated:

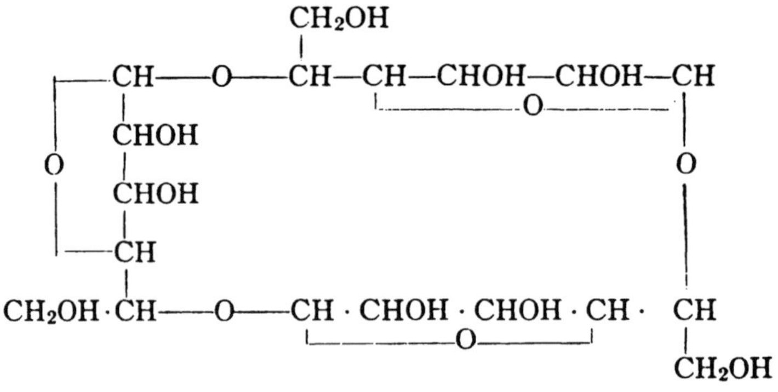

The methylation results demand that, as a minimum, nine hydroxyl groups are present in the starch molecule and this condition is satisfied by the anhydro-trisaccharide formula which, however, does not altogether provide for the alternative hydrolysis of starch to give exclusively maltose or *iso*-maltose without the formation of glucose. This objection is removed, and the methylation results equally well accommodated if the ring be doubled in such a manner that two anhydro-trisaccharide residues are coupled together, and, in consequence, it will be seen that the evidence accumulated by purely chemical methods and by biochemical methods coincide in the same formula. To simplify description, this "hexahexosan" formula may be represented diagrammatically:

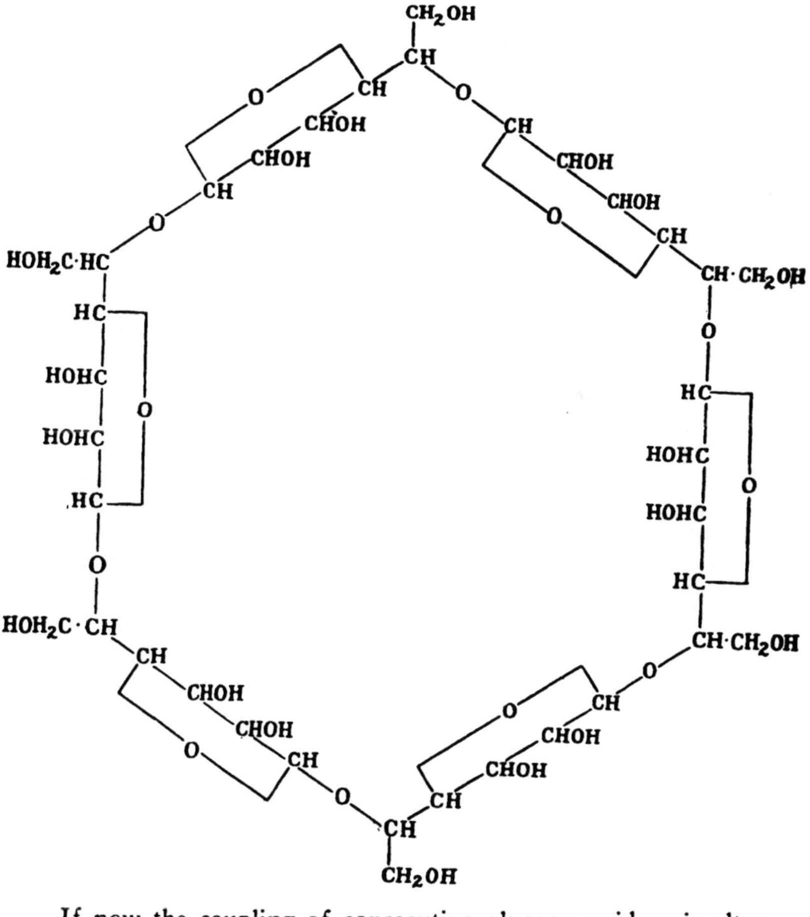

If now the coupling of consecutive glucose residues is alternatively α and β, then it is at once clear that either maltose or *iso*-maltose could result from starch on hydrolysis. This extension from a C_{18} to a C_{36} ring is attributable to Ling, in whose hands the bio-chemical study of starch has made striking progress, and although hesitation may be felt in accepting the idea of an unbridged ring of such dimensions, the first consideration is to accommodate the facts within the formula.

Evidence is gradually accumulating to show that varieties of starch which differ chiefly in respect of solubility owe their dissimilar properties to the different degree of polymerization which one and the same molecular unit has undergone.

This brings me to the vexed question of the distinction between starch and glycogen. In this subject the present position is still confusing, and the information I have to add does not go far to clear up the situation. Glycogen, on methylation, behaves in very much the same manner as potato starch, and in the course of the substitution, similarly loses the non-carbohydrate constituents. As we would expect, methylated glycogen is also convertible on hydrolysis into 2:3:6-trimethyl glucose, so that, empirically, the results are parallel. One difference observable is that, throughout, the specific rotations of the methylated glycogens are uniformly higher than those of methylated starches, showing the same composition, and precisely the same result emerges when the acetates of starch and of glycogen are prepared. This may be explained either on the ground of different polymerization or in the stereochemical configuration (α or β) of adjacent glucose residues, but speculation must not be allowed to out-run experiment, and here I leave the subject without further reference to much new data which still remains uncoordinated.

One other important disaccharide calls for attention and, in closing, I may devote a few minutes to the interesting and instructive question of the constitution of inulin. Here we have an example of a polysaccharide built up exclusively of fructose residues, and occupying in consequence, a unique position among natural products. Readily hydrolyzed, it yields the ordinary laevo-rotatory form of fructose but, as in the case of sucrose, this result cannot be accepted as evidence that the variety of the ketose present in the parent compound is identical with that which is actually isolated. Once more we have an example of γ-fructose being liberated on hydrolysis and reverting to the stable form during the process of isolation. That such is the case was evident for the first time when inulin was converted into trimethyl inulin, from which trimethyl γ-fructose was produced on hydrolysis:

Inulin (laevo) ⟶ Fructose (laevo)
↓
Dimethyl inulin ⟶ Dimethyl γ-fructose (dextro)
↓
Trimethyl inulin (laevo) ⟶ Trimethyl γ-fructose (dextro)

Again we recognize how the methylation process can reveal intimate details of structure which other methods fail to identify. The basis of the inulin molecule can be deduced from a study of the complicated reactions of dimethyl, trimethyl and tetramethyl γ-fructose. The oxidation of the two latter sugars shows decisively that γ-fructose cannot have the amylene-oxide structure attributed to this sugar by Haworth and incorporated by him in his formula for sucrose. It also emerges that in trimethyl γ-fructose both the primary alcohol groups are substituted and the same holds true for dimethyl γ-fructose as is evident from the fact that the sugar is incapable of reverting to the stable type. A survey of the combined evidence leads to the conclusion that the molecule of inulin contains the following grouping:

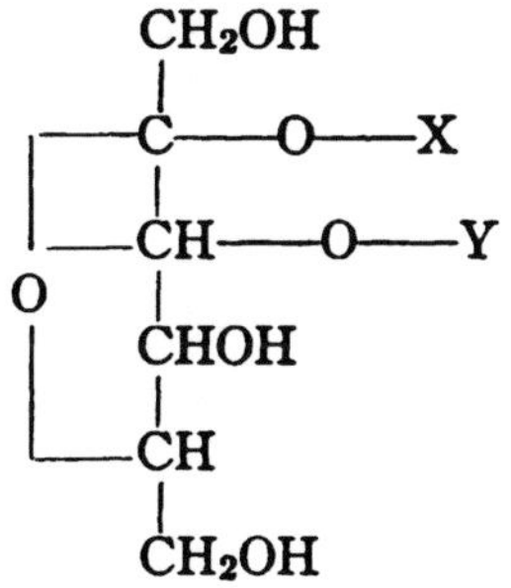

it being possible to repeat the same structural unit at X and Y. There is lacking, however, any evidence that inulin gives a difructose on hydrolysis and, in this respect, the compound differs from starch or cellulose, so that theoretically the simplest view is that the polysaccharide is a polymeride of anhydro-γ-fructose:

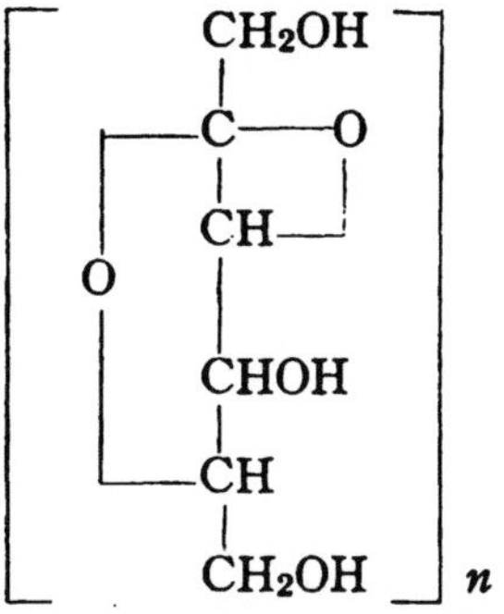

If I have not exhausted the subject of polysaccharides, I fear I have come to the end of your patience. It is inevitable in closing that one's mind should go back to those whose labor has carried knowledge even a little way into the darkness of the unknown and has enabled us to obtain this fleeting glance into a country so largely unexplored. For the moment I think gratefully of those loyal co-workers who have given me zealous ungrudging help, of the research schools scattered over the world where similar activities are at work and the same ideal in view, and at last there comes the reflection of what we owe to those who have gone before us. To Emil Fischer, the master of carbohydrate chemistry and to the long line of investigators stretching back to Lavoisier we pay now our tribute of admiration and gratitude.

OXIDATIVE CATALYSIS

BY

EDWARD C. KENDALL

RESEARCH CHEMIST OF THE MAYO CLINIC AND
PROFESSOR OF BIOCHEMISTRY, UNIVERSITY OF MINNESOTA

Lecture delivered at Columbia University in the special course on Contemporary Developments in Chemistry given in the Summer Session of 1926 on the occasion of the opening of the Chandler Chemical Laboratories

COLUMBIA UNIVERSITY PRESS

NEW YORK

1927

Printed from type. Published March, 1927

Printed in the United States of America

OXIDATIVE CATALYSIS IN THE BODY

EDWARD C. KENDALL

Section of Biochemistry

THE MAYO FOUNDATION, ROCHESTER, MINNESOTA

Calorimetric relationships concerning the combustion of organic material in the animal organism have been determined with ever increasing refinements, and to-day definite laws exist concerning the heat value of food substances.[5, 9, 33] The manner in which metabolites are oxidized in the animal organism still remains a problem. The processes are apparently influenced by certain agents which appear to be catalytic in nature and which come under the general class of enzymes. Oxidation-reduction reactions occurring in the animal organism have been studied in great detail, but the exact mode of action of any of the catalytic agents affecting these reactions has not yet been stated in terms of chemistry.

Agents Affecting Oxidation

Through observations in clinical medicine it was first suggested that the basal metabolic rate is influenced by changes in the activity of some of the ductless glands. Undoubtedly the clinicians that were familiar with the disease now called exophthalmic goiter were impressed with the voracious appetite of this group of patients, but it required careful determinations of metabolic rate before these clinical impressions could be expressed in quantitative terms.

Magnus Levy,[35, 40] in 1895 established the fact that patients with an over-active thyroid may have a basal metabolic rate which is higher than normal, and also that thyroid-deficient patients have much lowered basal metabolic rates.

Another of the ductless glands concerning which clinicians and physiologists have made many prophesies is the adrenal.[41] Surgeons and clinicians have often remarked that they were convinced that the adrenal gland was deeply involved in metabolic processes.

Pharmacological investigation of epinephrin has shown that it influences blood pressure, the distribution of the blood, the fatigability of muscle, the clotting time of the blood, the response to nerve stimulation and the concentration of blood sugar. Cannon has incorporated these investigations into a theory for the action of epinephrin in which epinephrin is called into function in times of emergency and stress. More recently Boothy, Aub, Marine, Baumann and others have shown that epinephrin influences the basal metabolic rate. Boothy and Sandiford [10] were the first to suggest that the calorigenic action is the probable mechanism for the physiological activity of this compound.

The clinical study and metabolic investigation of diabetes mellitus indicated that the pancreas was involved in the combustion of sugar. The isolation of an active constituent from the pancreas has resulted in definite proof that sugar metabolism in the animal organism is largely controlled through the chemical reactions of the substance known as insulin.[19, 34]

A more direct chemical approach to the problem of oxidation in the animal organism resulted in the identification of the sulphydryl grouping by Heffter, Thunberg, Meyerhof, and Hopkins. After a vigorous attack upon this problem, Hopkins [24] was successful in isolating a sulfur compound which he identified as a dipeptide of glutamic acid and cysteine, and named it glutathione. This remarkable compound can be shown to effect the oxidation of animal tissue even after it has been boiled, dried, and ground to a fine power.

The four substances, thyroxin, epinephrin, insulin and glutathione have been shown to markedly influence the rate of oxidative processes. The time during which they can be shown to function and the quantitative relationship between the weight of the substance used and the effect produced indicate that they function as catalysts. Indeed, the quantitative response following the administration of thyroxin, which can be expressed in a mathematical formula, is of such a nature as to form a typical example of a catalyst. The other substances cannot be studied in as satisfactory a manner since they function over a shorter period of time, but evidence suggesting catalytic action will be given in this paper concerning one other of this group of compounds, glutathione. That these substances are physiologically active and determine the rate of energy production in the body, is an established fact. That they function as catalysts and pass through a definite cycle of reactions when they enter into metabolic processes is the hypothesis which will be presented and developed in this paper.

Oxidation and Reduction of Derivatives of Ox-indole Propionic Acid

In 1914, the writer isolated thyroxin from the thyroid gland.[27] From a chemical study of thyroxin and its derivatives, it seemed probable that the structural formula best representing the configuration of the molecule was a tri-iodo derivative of ox-indole propionic acid.[30] A discussion of the chemical configuration of thyroxin cannot be included in this paper, but the recent work of Harington has conclusively shown that it is not a derivative of ox-indole propionic acid.

In 1919, an investigation was undertaken with the object of synthesizing compounds related to thyroxin in the hope that the chemical properties of these substances would help to explain how thyroxin functions in the animal organism. This work has resulted in the synthesis of 2-oxo-dihydro-indole-3-propionic acid and its hydro derivatives.[30] Many halogen derivatives of these compounds have now been prepared. At an early stage of the work it was shown that an unusual chemical grouping could be made in these compounds. This reaction involved the loss of two atoms of hydrogen from the various derivatives which were found susceptible to this type of oxidation. It was subsequently shown that in a slightly alkaline solution molecular oxygen would bring about this reaction and finally that these derivatives could be oxidized with dibromo indophenol and other oxidizing dyes, in an aqueous solution at a pH of 7.4.

These results suggested the determination of the oxidation-reduction potentials of this series of compounds. This work which was carried out with J. M. Ort showed that 2-oxo-dihydro-indole-3-propionic acid reacted in an unusual manner towards oxidizing agents.

The work of Biilmann, La Mer, Conant,[15, 16, 17] Clark and others,[14] has resulted in an electrochemical formulation of reversible and irreversible oxidation and reduction of organic compounds. Excellent reviews of this subject have recently been published by Clark [12] and Conant.[15] Clark has shown that with substances which are reversibly oxidized the oxidation-reduction potential is determined by the ratio of the reduced to the oxidized form of the compound. By an equation involving this ratio, and the concentration of hydrogen ions it is possible to predict the oxidation-reduction potential of any mixture of the two forms of any compound after certain constants have been established. Conant and others have investigated irreversible reactions. A typical example of irreversible reduction is the change of an azo compound through the hydrazo to the simidine derivative, and an example of irreversible oxidation is the effect of oxidizing agents on the aminonaphthols resulting in the formation of a quinone and the liberation of ammonia.

Investigation of 2-oxo-dihydro-indole-3-propionic acid showed that it is irreversibly oxidized. The result of oxidation is the formation of a bond the exact position of which has not been determined. This oxidized form of the lactam can be easily reduced with the formation of the starting material through the action of hydriodic acid.

It was found that neither the fully reduced nor oxidized forms affect the platinum electrode, but in the presence of dibromo indophenol an addition-product is apparently formed resulting

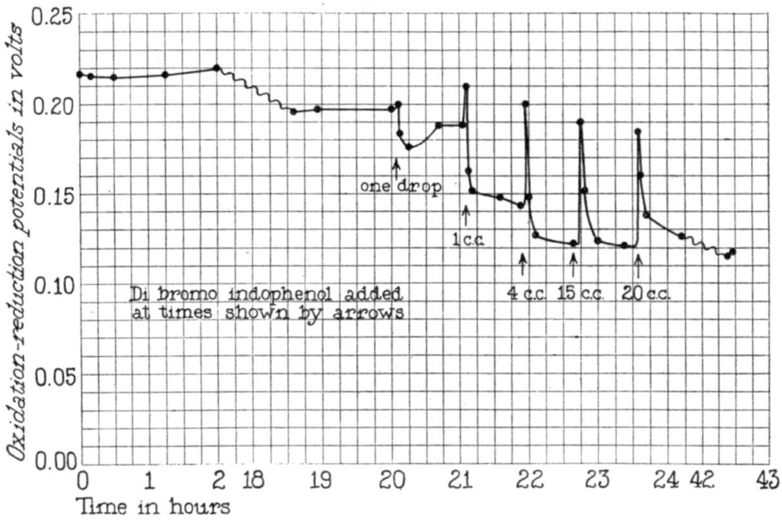

Fig. 1.—Oxidation of di-iodo-2-oxo-dihydro-indole-3-propionic acid showing the increase in reducing potential after addition of dibromo indophenol.

in oxidation of the indole derivative and reduction of the indophenol. The resulting potential of the solution up to a certain maximum which cannot be exceeded, is dependent upon the concentration of the indole derivative and the amount of dibromo indophenol added (Fig. 1).

The result of the investigation of the oxidation of 2-oxo-dihydro-indole-3-propionic acid may be summarized in the statements that the essential chemical property of the molecule which characterizes its reaction and permits it to interact with oxidizing substances is its ability to form addition-compounds probably through its atom of nitrogen, and that no condition has been found which permits the reduced and oxidized forms to exist in equilibrium with each other.

Oxidation-reduction Potentials of Cysteine and Glutathione

In 1923, Dixon and Quastel investigated the oxidation-reduction properties of cysteine and glutathione. They showed that neither cystine nor oxidized glutathione affects the platinum electrode, and that the oxidation-reduction potential of the solution depends upon the concentration of cysteine or reduced glutathione. The fact that the reduced and oxidized forms of these sulfur compounds can be in solution not in equilibrium with each other, suggested a study of the chemical relationship between these compounds containing sulfur and the indole derivatives already outlined. A re-investigation of the oxidation-reduction potentials of cysteine and glutathione was therefore undertaken. During this work which was carried out with F. F. Nord the results of Dixon and Quastel were confirmed.[28] Cysteine does and cystine does not affect the platinum electrode. Cysteine is readily oxidized by dibromo indophenol, methylene blue, molecular oxygen, and hydrogen dioxide. Investigation of the velocity of oxidation by dibromo indophenol or methylene blue showed that the curve representing the rate of oxidation of cysteine and glutathione is similar to the curve representing the velocity of oxidation of 2-oxo-dihydro-indole-3-propionic acid. Cysteine is irreversibly oxidized at a pH of 7.4 and the oxidized form does not affect the potential.

The Influence of Oxygen on Cysteine

The results outlined in the preceding paragraph were obtained in a solution from which oxygen was completely excluded. In such a solution cysteine is not oxidized by indigo carmine. When a solution containing cysteine and indigo carmine is permitted to come in contact with a small amount of oxygen and the solution remains unagitated for a few hours, the indigo carmine is reduced. The reduction is first apparent through the formation of a layer of the completely reduced dye, about 1 cm. below the surface of the solution, which slowly extends from this point to the bottom of the flask. After this reaction has taken place more indigo carmine can be reduced and the amount which can eventually be reduced by this solution is dependent upon the amount of cysteine present. The reduction of the indigo carmine is apparently brought about by an addition-product formed between cysteine and oxygen. In the presence of this compound indigo carmine is easily reduced by cysteine and the speed of reduction is dependent upon the concentration of the oxygen addition-product. The last mentioned compound appears to react as a catalyst bringing about an interaction between the indigo carmine and the —SH group. After it had been shown that cysteine is capable of reducing indigo carmine in the presence of an oxygen addition-product, the influence of this compound on cystine was investigated. Cystine cannot oxidize reduced indigo at a pH of 7.4, but if the oxygen addition-product of cysteine is formed in the solution, crystine readily oxidizes reduced indigo.

Reversible Oxidation-reduction System of Cysteine-Cystine

In the presence of the oxygen addition-product cysteine and cystine are brought into equilibrium with each other, and al-

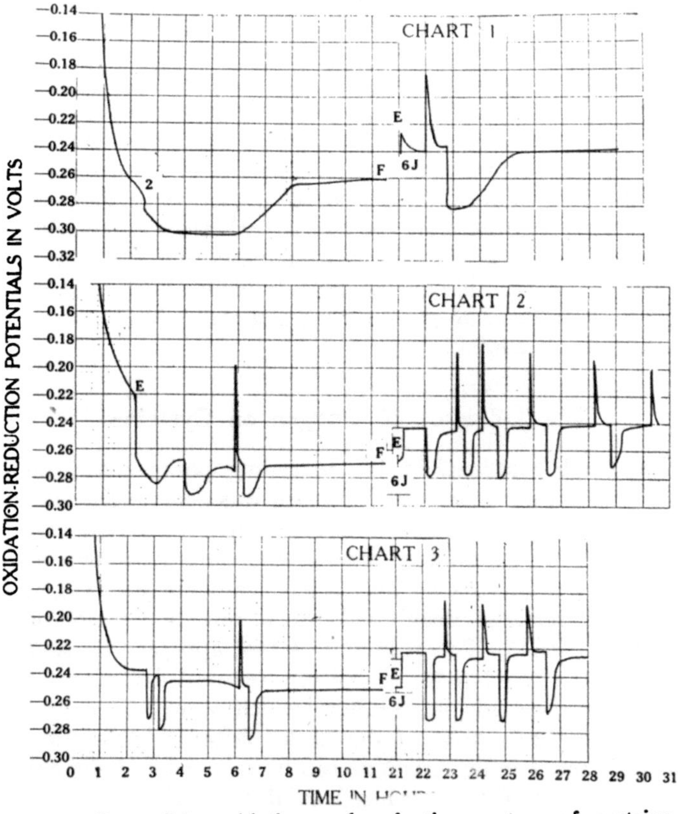

Fig. 2.—Reversible oxidation and reduction system of cysteine and cystine in the presence of an oxygen addition-product. The three charts show three different equilibrium points. The drop in potential followed by the quick recovery was produced by the addition of indigo carmine and the increase in potential with slow recovery was brought about by the addition of reduced indigo to the solution, 28.

though no reversible system can be made with cysteine and cystine alone, a reversible system is formed by these two substances in the presence of the oxygen addition-product, which is a third component of the system. It was subsequently shown that hydrogen dioxide in the presence of indigo carmine will form an addition-product which functions in a manner entirely similar to the compound formed by acting on cysteine with molecular oxygen. Sodium disulfide moreover forms a sulfur addition-product having similar properties.

In the presence of the oxygen or sulfur addition-product, mixtures of cysteine and cystine give definite oxidation-reduction potentials which are determined by the ratio of cysteine to cystine, and the equilibrium point is maintained over many hours without change. If to such a solution an oxidizing agent is added, the potential is immediately changed depending upon the amount of oxidizing agent used. After reaching a minimum the potential returns to its former value and the velocity of this reaction depends upon the concentration of the third component present and the potential of the solution before addition of the oxidizing agent. If after the equilibrium point has again been reached a suitable reducing agent, such as reduced indigo, is added the potential of the solution becomes more reducing depending on the amount of the reducing agent used. After the maximum departure from the equilibrium point has been reached, the potential again returns to the value shown before the addition of the reducing agent. The time required for the return to the equilibrium point of these solutions depends upon three factors: the amount of oxidizing or reducing substance added, the potential at the equilibrium point, and the concentration of the oxygen addition-product (Fig. 2).

Chemical Properties of the Oxygen Addition-product

These results show that although cysteine and cystine cannot form a reversible system, the presence of a third component makes available the reducing energy of the —SH grouping, and the oxidizing energy of the —SS grouping; the properties of the oxygen addition-product can be summarized by the statement that it behaves in a manner similar to compounds which do form a reversible system, that is, the third component itself is capable of existing in oxidized and reduced forms. The oxidized and reduced forms are always in equilibrium with each other, but in addition to this they are also in equilibrium with cysteine and cystine. Under these conditions when any suitable oxidizing or reducing substance is added to the solution the concentration of the oxidized and reduced forms of the third component are altered until either cysteine or cystine has reacted with the added substance, and the equilibrium point of the solution which is established by the concentration of cysteine to cystine has been reached. This reversible system, therefore, is somewhat similar to solutions of buffers for the regulation of the hydrogen ion concentration; however, no buffer solution is known in which the mechanism has been assumed to consist of three components and which works in such a manner that a change from the first to the second component necessitates interaction with the third component. This mechanism appears to offer the simplest explanation of the reversible system of cystine in the presence of the oxygen addition-product.

Glutathione has been shown to be entirely similar to cysteine and cystine except that it is capable of forming and retaining the oxygen addition-product with much greater ease.

The essential chemical property of this series of compounds is their ability to form addition-products with oxygen, sulfur and other compounds. The utilization of the energy of the —SH and the —SS groupings depends upon the formation of essential addition-products between the compounds to be oxidized and reduced and the source of energy.

The interpretation of this reaction must involve a change in the valency of the versatile sulfur atom. The essential difference in the chemical properties of cysteine and cystine and those of the oxygen addition-product consists in the great ease with which the last mentioned compound can change its state of oxidation. The sulfur atom as it exists in cysteine and cystine is not readily acted upon by mild oxidizing and reducing agents, and the change in its state of oxidation therefore is brought about only through relatively drastic treatment. In the presence of the third component, that is the oxygen (or sulfur) addition-compound, the mildest oxidizing and reducing agents are capable of affecting cysteine and cystine, but this influence is brought about presumably through the change in valency in the sulfur atom of the third component as a primary step.

Reversible Oxidation-reduction of Compounds Containing Nitrogen

After it had been shown that 2-oxo-dihydro-indole-3-propionic acid is irreversibly oxidized and that the probable mechanism of the oxidation was through the primary formation of an addition-product, experiments were tried with certain other substances containing nitrogen to see if this is a general reaction having wider application. The first substance tried was adrenalone. Adrenalone will not oxidize reduced indigo, but it is easily oxidized by dibromo indophenol, naphtho dichloro indophenol, and indigo carmine. When a solution of adrenalone which has been oxidized is treated with reduced indigo it is found that the oxidized adrenalone is capable of acting as an oxidizing agent, resulting in the oxidation of an equivalent amount of reduced indigo.

There can be but little doubt that oxidation of adrenalone involves the two hydroxyl groups of the benzene ring. In order to determine the ease of oxidation of this grouping closely related compounds were examined. It was then shown that neither pyrocatechol nor chloro aceto pyrocatechol can be oxidized by dibromo indophenol and the other oxidizing dyes mentioned at pH 7.4. When the chlorine atom in chloro aceto pyrocatechol is substituted by the methyl amine group the molecule becomes highly sensitive to oxidation; that this is through the primary formation of addition-products between the methyl amine group and the oxidizing agent was clearly indicated by the almost specific action of certain oxidizing agents. The response of this molecule to oxidation, however, shows that it is not of a reversible nature similar to that of indigo carmine, dibromo indophenol, or methylene blue, and it is not irreversible as is the oxidation of 2-oxo-dihydro-indole-3-propionic acid, but it does closely resemble the oxidation of cysteine.

Intensity of Oxidation Required for Adrenalone

In order to study the oxidation of adrenalone more fully and to compare it with ox-indole propionic acid, cysteine and glutathione, oxidation-reduction potentials were determined in connection with the oxidation of this compound. During this work, which was carried out with E. J. Witzemann, several unexpected and puzzling reactions were obtained which were difficult to reconcile with our present concept of variations in potentials accompanying oxidation and reduction. Conant has pointed out that substances which are irreversibly reduced or oxidized require an "apparent reduction or oxidation potential." Adrenalone doubtless requires a certain intensity of oxidation before it is affected, but this intensity cannot be directly related to the oxidation-reduction potential of the solution. This is shown by the following results. Adrenalone is oxidized by dibromo indophenol, naphtho dichloro indophenol, and indigo carmine. The potential of the solution after equilibrium has been reached, depends in each case upon the oxidizing dye used. With dibromo indophenol it may be +0.010 v., with naphtho dichloro indophenol —0.060 v. and with indigo carmine, —0.220 v.; but if one or more molecular equivalents of hydrogen dioxide are first added to the solution the indigo carmine may be reduced by adrenalone at a potential of +0.24 v. In the two solutions last mentioned the indigo carmine is in the reduced form, but the potentials of the solutions differ by more than 0.4 v.

These reactions could be readily explained if adrenalone were oxidized by all oxidizing agents that can produce a potential in the solution equal to or greater than —0.220 v. This, however, is not the case. At a pH of 7.4 adrenalone is not oxidized by hydrogen dioxide nor by molecular oxygen although both of these oxidizing agents produce a potential of +0.30 v. or somewhat more, in the solution. When hydrogen dioxide is added to a solution of adrenalone containing indigo carmine in the reduced form, it does not oxidize the reduced indigo carmine. There is no appreciable change in color, but the oxidizing potential of the solution is greatly changed. Furthermore, the hydrogen dioxide remains in the solution in an active form, since it will oxidize reduced indigo quantitatively.

These results show that there is no apparent oxidation potential of adrenalone which is comparable with that of the azo dyes investigated by Biilmann and by Conant, and they also emphasize another factor of great importance: the configuration of the oxidizing dye.

This effect is clearly brought out by the influence of hydrogen dioxide and molecular oxygen on adrenalone. Hydrogen dioxide and molecular oxygen will not oxidize adrenalone although they produce within the solution a high oxidizing potential. Hydrogen dioxide will not oxidize reduced dibromo indophenol or reduced indigo carmine at an appreciable rate in such a solution. If, however, blood serum is added to the solution the catalase [39] of the serum will liberate oxygen from the hydrogen dioxide, and then molecular oxygen will rapidly oxidize indigo carmine, and will slowly act on dibromo indophenol. Under these conditions all excess hydrogen dioxide is destroyed and the liberated oxygen is completely removed from the electrode chamber by passing oxygen-free nitrogen through the solution. It is then found that the amount of reduced indigo that can be oxidized by the solution is proportional to the amount of adrenalone which has been oxidized.

Compound-Formation Between Adrenalone and the Oxidizing Agent

With experiments of this nature it was observed that molecular oxygen itself does not oxidize adrenalone at a pH of 7.4. However, in the presence of 5 per cent of its molecular equivalent of indigo carmine, adrenalone is rapidly and completely oxidized with molecular oxygen. The interpretation given to these results is that an addition-product is formed between adrenalone and indigo carmine. The formation of this addition-product is not dependent upon the oxidation potential of the solution, and it is for this reason that adrenalone does not appear to require an apparent oxidation potential which must be produced by the oxidizing compound. After the indigo carmine has oxidized its molecular equivalent of adrenalone, molecular oxygen reoxidizes the reduced indigo carmine, which then in turn oxidizes another equivalent of adrenalone; in short, indigo carmine acts as a catalyst inducing molecular oxygen to oxidize adrenalone.

Dibromo indophenol is reduced by adrenalone, but in these solutions reduced dibromo indophenol is not oxidized at appreciable velocity with molecular oxygen; for this reason dibromo indophenol does not catalyse the oxidation of adrenalone with molecular oxygen as rapidly as does indigo carmine. Further evidence indicating that compound-formation is essential for the oxidation of adrenalone is furnished by methylene blue. Methylene blue creates within the electrode chamber a greater intensity of oxidation than is obtained with indigo carmine, and Clark has shown that it stands between dibromo indophenol and indigo carmine in the scale of oxidizing intensity;[13] however, methylene blue oxidizes adrenalone at a rate much slower than does indigo carmine. These results may possibly be further related to the general configuration of the molecule and the oxidizing groups of the respective compounds. With dibromo indophenol and indigo carmine the active part of the molecule is the quinone or carbonyl group. In methylene blue the active part of the molecule is associated with a change in the degree of oxidation of one atom of nitrogen. It therefore seems probable that compound-formation between adrenalone and those dyes which easily oxidize it is made possible by interaction between the methyl amine group of adrenalone and the oxidizing group of the dye. This addition-product is unstable and breaks down with the formation of the reduced form of the dye and oxidation of the adrenalone. This interpretation of the reaction is in harmony with that given for the oxidation of 2-oxo-dihydro-indole-3-propionic acid which is brought about by dibromo indophenol, but not by hydrogen dioxide.[29] Ox-indole propionic acid and adrenalone can apparently react with dibromo indophenol with the formation of an addition-product. In the indole derivative two atoms of hydrogen are removed from the molecule with the formation of a bond. The formation of this bond is an irreversible reaction at a pH 7.4. With adrenalone two atoms of hydrogen are removed and the molecule is left in a form which is capable of acting as an oxidizing agent with suitable reducing agents such as reduced indigo. This, therefore, is a reversible reaction at a pH of 7.4.

Although the oxidized form of adrenalone has not been isolated in crystalline form, it seems highly probable that the oxidation involves the hydroxyl groups in the benzene ring. However, oxidation of ortho-hydroxyl groups would result in the formation of an ortho-quinone and these compounds are so unstable in the presence of water that the oxidation would probably result in an irreversible reaction if a simple ortho-quinone grouping were formed. This suggests that the methylamine group is not only the essential part of the molecule permitting the formation of an addition-product between adrenalone and the oxidizing dye, but that it is also the essential part of the molecule stabilizing the oxidized form of this compound. The ortho-quinone group could form an addition-product with the methylamine group similar to the addition-product between adrenalone and the oxidizing dye. Such an addition-product should be capable of oxidizing other substances. Oxidized adrenalone will oxidize cysteine and reduced indigo, and it will apparently oxidize epinephrin.

The results concerning the oxidation of this series of compounds are at first sight paradoxical and impossible. Indigo carmine will not oxidize cysteine but it will oxidize adrenalone and the oxidized adrenalone will then oxidize cysteine. There are two possible explanations for this. One is that cysteine can form an addition-product with oxidized adrenalone, and it is through this agency that the sulphydryl group is oxidized to the disulfide form. The other possibility is that after adrenalone has been oxidized with indigo carmine there is an intra-molecular rearrangement resulting in the formation of a group which has a higher oxidizing intensity than indigo carmine. The latter explanation appears to be the more reasonable from our present knowledge of the reaction, but more data will have to be obtained in order to prove which is the correct explanation.

Ox-indole propionic acid and its halogen derivatives, cysteine, glutathione, adrenalone and its derivatives form a series of compounds which possess similar but unusual chemical properties which indicate that these substances enter into reactions involving oxidation and reduction because of the chemical properties of certain groups in the molecule, which are capable of forming addition-products with those compounds which induce oxidation or reduction.

Ox-indole propionic acid can be oxidized with mild oxidizing agents. This substance injected into the animal organism produces no visible effect. If, however, an anhydride form is prepared and this compound is injected a drop in blood pressure and stimulation of the respiratory center are observed. The increase in the physiological activity of this molecule brought about by loss of a molecule of water suggests an increase in the intensity of activity of some grouping in the molecule which probably involves the nitrogen atom. By a slight alteration, the molecule acquires sufficient intensity to permit it to enter into chemical reactions influencing physiological processes.

When adrenalone is oxidized with indigo carmine it is capable of oxidizing cysteine. As a working hypothesis this is interpreted as a change in the adrenalone molecule after oxidation, involving the production of groupings having a higher intensity of oxidation than indigo carmine itself. These two experimental facts, combined with the evidence that apparently only those compounds which can form addition-products with this group of substances are capable of interacting with them, form the basis of an hypothesis concerning the functioning of the four substances, thyroxin, epinephrin, glutathione and insulin, which appear to act as catalysts in the animal organism. The probable chemical reactions which constitute the cycle of changes through which some of these four substances pass when they function as catalysts, will be discussed, and for purposes of logical sequence epinephrin will be considered first.

Epinephrin

It has been shown that adrenalone influences basal metabolic rate.[26] Whether the oxidation and reduction of adrenalone in the body is the entire explanation for the effect of this compound on basal metabolic rate, cannot be stated at this time, but it seems probable that this reaction occurring in the tissues would furnish a poising agent which would affect the intensity of oxidation. Moreover, it can be shown that adrenalone has the ability to react with certain substances through the probable formation of addition-products and thereby to affect oxidation independent of the intensity of oxidation as measured by the oxidation potential. The selective action of this compound may be involved in its influence on the basal metabolic rate.

When epinephrin is added to a solution of oxidized adrenalone, the solution loses in large part its oxidizing power; this is interpreted as reduction of the oxidized form of adrenalone, and it is of significance because epinephrin is not directly oxidized by indigo carmine, although adrenalone is oxidized by this reagent. Adrenalone does not influence the stability of oxidized adrenalone, and for this reason the oxidizing power of the solution can be easily measured through the addition of suitable reducing agents, such as reduced indigo.

When the reaction of epinephrin toward oxidizing agents was investigated it was shown that epinephrin reacts with dibromo indophenol, but it was impossible to prove the presence of an oxidized form of epinephrin through its action on reduced indigo. This is evidence that adrenalone was not obtained in the oxidation of epinephrin. In the experiments with epinephrin there was no other substance present in the solution with which epinephrin in the oxidized form could react, but in the tissues of the animal organism it is highly probable that the oxidized form of epinephrin, which may act as a cyclic catalyst but which has not yet been obtained in vitro, can enter into reaction with other substances, and its specific influence on certain groupings together with its poising action on oxidizing intensity are suggested as its probable mode of action.

Thyroxin

Investigation of the chemical properties of thyroxin showed that it contains a carboxyl, an amine and an hydroxyl group and that 65 per cent of the weight of the molecule is iodine.[30] It was suggested that the activity of this substance is primarily due to its amine group and that the iodine is not reversibly broken off and added to the molecule when the compound functions as a catalyst, but that the iodine modifies the activity of the reactive groups, and that the molecule as a whole is responsible for the specific chemical reactions induced by this substance. When, therefore, in the preparation of 2-oxo-dihydro-indole-3-propionic acid and its hydro derivatives, compounds were prepared containing a carboxyl, an amine and an hydroxyl group which possessed chemical properties closely simulating those of thyroxin, and furthermore when it was shown that these substances could be oxidized with mild oxidizing agents including molecular oxygen, it seemed probable that a possible explanation for the functioning of thyroxin lay in its alternate oxidation and reduction, and that the presence of the molecule in the tissues influenced the chemical reactions involving oxidation and reduction. The inability of the acetyl of thyroxin to influence the rate of metabolism and the fact that thyroxin under suitable conditions is sensitive to oxidation were the first indications pointing to a possible mechanism for the chemical explanation of the functioning of thyroxin.

The recent work of Harington with the identification of the organic radical to which the iodine is attached in thyroxin shows that the hypothesis suggested for the functioning of thyroxin, in which a bond is formed in the indole nucleus cannot be the correct explanation for the activity of thyroxin; but certain fundamental facts, independently established, are highly suggestive.

Boothby and Baldes [7] have recently shown that the rate at which thyroxin influences basal metabolism following its injection can be expressed as a compound exponential curve. This is in harmony with the fact that thyroxin as injected is physiologically inert, and the assumption that some alteration occurs which permits thyroxin to influence basal metabolic rate. Clinical observations by Plummer, and basal metabolism determinations by Boothby and Sandiford [8] have shown that hyperthyroidism does not influence the combustion of any one particular class of metabolites, but that it modifies the rate of metabolism of proteins, carbohydrates and fats depending on the composition of the food intake. The fact that thyroxin can function for five weeks after a single injection is evidence for the stability of the molecule, and analysis of the urine for iodine has shown that the iodine is not excreted and presumably not broken off from the thyroxin molecule while the latter is functioning.

The chemical reactions involved during the functioning of thyroxin in the animal tissues are now being actively investigated and until quantitative evidence is obtained any hypothesis becomes merely speculation. No evidence can be given at this time concerning the cycle of changes in the thyroxin molecule and a discussion of the catalytic effect of this compound which has been so thoroughly investigated in respect to its influence on basal metabolism will be reserved for a future publication.

Glutathione

After it had been shown that cysteine and glutathione can be made to form a reversible system in the presence of an oxygen or sulfur addition-product, a question of great interest arose concerning the application of these reactions to the thermostable oxidation-reduction system of proteins described by Hopkins and Dixon. It was readily shown that there is nothing in boiled yeast which in the absence of oxygen can reduce indigo carmine, and that muscle and liver which have been boiled cannot reduce more than traces of indigo carmine. Methylene blue is reduced by these preparations and it therefore appeared probable that the reduction of methylene blue is brought about through the formation of —SH groupings which can react directly with this dye. In the absence of the oxygen addition-product, however, the —SH groups which are present cannot reduce indigo carmine. Hopkins and Dixon pointed out that there was a quantitative relationship between the amount of protein and the amount of methylene blue which could be reduced. They also stated that they have been unable to find any substance not of animal origin capable of reducing cystine or oxidized glutathione at a pH of 7.4. In a later publication, Hopkins [23] states that the active agent which reduces the —SS groupings is the fixed —SH grouping in the protein. In addition to the activity of the fixed —SH group, Hopkins has also shown that carbon dioxide is a product of the reaction, and this finding indicates a more involved process than merely the direct reduction of the —SS grouping by the fixed —SH group.

After it had been shown that cystine and oxidized glutathione are readily acted upon by suitable reducing substances in the presence of an oxygen addition-product it seemed probable that an oxygen addition-product may also be the essential intermediate in the reaction between the fixed —SH group of protein and the —SS grouping in oxidized glutathione. However, in this reaction, the oxygen addition-product is not made by the addition of hydrogen dioxide or molecular oxygen on cysteine but it is more probably an oxygen grouping similar to the quinone or carbonyl form of combination. Such an addition-product between a quinone or carbonyl group and the fixed —SH group in protein could react with the —SS grouping of cystine or oxidized glutathione in an irreversible manner resulting in decomposition of the organic compound containing the carbonyl group with subsequent liberation of carbon dioxide, the reduction of the —SS grouping to —SH, and the oxidation of the fixed —SH group in protein to the —SS form. It has been impossible to make a muscle preparation which did not slightly reduce methylene blue in the presence of glutathione. These experiments indicate that the reduction of methylene blue is a test for the production of —SH groups and that the reduction of indigo carmine can only be brought about through the formation of the oxygen addition-product or its chemical equivalent, which is the essential intermediate in the formation of any reversible oxidation-reduction system of cysteine or glutathione.

The results with the muscle preparation confirm the experiments with cysteine and glutathione indicating that the mechanism by which these substances function is through the formation of addition-products either with oxygen, hydrogen dioxide, sodium disulfide or possibly chemical groupings such as the carbonyl or quinone.

Insulin

Recent work with insulin has shown conclusively that the utilization of glucose is increased by the presence of insulin in the animal organism.[19] Although it has been impossible to account for all the sugar injected in experimental animals, either as unchanged, burned, or stored as glycogen, this failure may be due to experimental conditions. Of the many possible effects of insulin on glucose which have been suggested only two will be mentioned. The first is that insulin is necessary for the oxidation of glucose and the second is that insulin is involved in the conversion of glucose into fat. The identification of the —SH group in insulin by Abel is highly suggestive that this substance may exert its power in a manner somewhat similar to cysteine and glutathione.[40a] Its reaction with glucose may be explained by the configuration of the molecule which may be of such a nature that the specific grouping occurring in glucose is so modified that the —SH or —SS group in insulin is capable of interacting with the glucose molecule. The evidence for the specific reactivity of some of the compounds described in this paper, and of the ability of the —SH group to form addition-compounds, is suggestive that this may be the possible explanation for the action of insulin.

Since cysteine and glutathione can form reversible oxidation-reduction systems in the presence of an oxygen addition-product, it is possible that insulin would provide the mechanism for both oxidation or reduction of glucose, depending upon the conditions present in the tissues; oxidation of the glucose leading to carbon dioxide or reduction of the glucose to fat. The reversible system of glutathione responding to both oxidizing and reducing substances is evidence that glucose could be acted on by a system containing sulfur compounds, if in addition to this the glucose molecule could be modified[25a, 40a] sufficiently to be affected by oxidizing or reducing agents which are brought into the sphere of reaction by the —SH or —SS groupings.

Discussion

The four compounds, thyroxin, epinephrin, glutathione and insulin have been shown to exert marked influence on the metabolic activity of the animal organism. This fact has been established independently of the chemical nature or the cycle of the changes occurring in the substances when they function. The weight of the substances required to bring about marked changes in the nutrition and rate of the production of energy in the body, in relation to the mass of material transformed owing to their presence, is evidence that they function as catalysts.

Through the investigation of oxidation reduction properties of some derivatives of 2-oxo-dihydro-indole-3-propionic acid, and through the study of the oxidation of cysteine, glutathione and adrenalone, it is possible to show that these compounds are acted on by certain oxidizing agents, which have carbonyl or quinone groups present in the molecule. The response to oxidations shows that these compounds do not form directly reversible oxidation-reduction systems, and that the oxidized form is not in equilibrium with the reduced form. The essential feature of the oxidation of all of these substances appears to be the formation of an addition-product between the compound oxidized and the oxidizing agent. It can also be shown that the oxidation of these substances is not directly related to the oxidizing intensity, but is dependent upon the chemical groups in the oxidizing agent.

The results with the derivatives of ox-indole propionic acid suggest that the oxidation is dependent upon a change in the valence of nitrogen. This change permits the formation of the addition-product. The derivatives of adrenalone appear to be oxidized by a mechanism entirely similar, in which the primary step is the formation of an addition-compound with the oxidizing agent, which is brought about by a change in the valence of the nitrogen.

Quantitative evidence concerning the oxidation-reduction of cysteine and glutathione shows that the reactivity of the reversible system is probably due to a change in the valence of sulfur, which is sensitized, by the formation of an oxygen addition-product, to slight alterations in the oxidizing or reducing power of the solution, due to the presence of other substances. Our knowledge of the physiological action of thyroxin, epinephrin, glutathione and insulin appear to justify the working hypothesis that these substances affect oxidation in the animal organism through change in the molecule, which are brought about by the formation of addition-products between these compounds and metabolites, and that the reactivity of these substances can be shown to be due to the change in valence of sulfur or nitrogen.

The configuration of the respective molecules in these four compounds is such that the change in valence in the sulfur or nitrogen permits oxidation of certain groupings in the molecule which can then enter into processes of oxidation and reduction, bringing about alternations in the rate at which the animal organism can produce energy.

BIBLIOGRAPHY

1 Abel, J. J., Crystalline insulin, Proc. Nat. Acad. Sc., 1926, xii, 132–136.

2 Aub, J. C., Bright, E. M., and Forman, J., The metabolic effect of adrenalectomy upon the urethanized cat, Am. Jour. Physiol., 1922, lxi, 349–368.

3 Aub, J. C., Bright, E. M. and Uridil, J., Studies upon the mechanism of the increased metabolism of hyperthyroidism, Am. Jour. Physiol., 1922, lxi, 300–349.

4 Aub, J. C., Forman, J. and Bright, E. M., The effect of adrenalectomy upon the total metabolism of the cat, Am. Jour. Physiol., 1922, lxi, 326–348.

5 Benedict, F. G., The measurement and significance of basal metabolism; Lectures on nutrition (Mayo Foundation Lectures, 1924–1925), Philadelphia, W. B. Saunders Co.

6 Biilmann, E. and Blom, J. H., Electrometric studies on azo-and hydrazo-compounds, Jour. Chem. Soc., 1924, cxxv, 1719–1731.

7 Boothby, W. M., and Baldes, E. J., Some quantitative relationships of thyroxine calculated from its calorigenic action, Jour. Pharmacol. and Exper. Therap., 1925, xxv, 139.· (Proc. Am. Soc. Pharmacol. and Exper. Therap.)

8 Boothby, W. M., and Sandiford, I., The total and the nitrogenous metabolism in exophthalmic goiter, Jour. Am. Med. Ass., 1923, lxxxi, 795–800.

9 Boothby, W. M. and Sandiford, I., Basal metabolism, Physiol. Rev., 1924, iv, 69–161

10 Boothby, W. M. and Sandiford, I., The calorigenic action of adrenaline, Am. Jour. Physiol., 1923, lxvi, 93–126.

11 Cannon, W. B., Bodily changes in pain, hunger, fear and rage, New York, D. Appleton & Co., 1915.

12 Clark, W. M., Recent studies on reversible oxidation-reduction in organic systems, Chem. Rev., 1925, ii, 127–178.

13 Clark, W. M., Cohen, B. and Gibbs, H. D., Studies on oxidation-reduction, VIII Methylene blue, Pub. Health Rep., 1925, xl, 1131–1201.

14 Clark, W. M., and Collaborators, Studies on oxidation-reduction I to X, Pub. Health Rep's., 1923, xxxviii, 443–455; 666–683; 933–958; 1669–1718; 1924, xxxix, 381–414; 804–823; 1925, xl, 649–659; 1131–1201; Supplement to Pub. Health Rep's. No. 54, 61 pp.; and No. 55, 34 pp.

15 Conant, J. B., The electrochemical formulation of the irreversible reduction and oxidation of organic compounds, Chem. Rev., 1926, iii, 1–40.

16 Conant, J. B., and Lutz, R. E., An electrochemical method of studying irreversible organic reductions, Jour. Am. Chem. Soc., 1923, xlv, 1047–1060.

17 Conant, J. B., and Lutz, R. E., The irreversible reduction of organic compounds, I; The relation between apparent reduction potential and hydrogen-ion concentration, Jour. Am. Chem. Soc., 1924, xlvi, 1254–1267.

18 Dixon M. and Quastel, J. H., A new type of oxidation-reduction system, I, Cysteine and glutathione, Jour. Chem. Soc., 1923, cxxiii, 2943–2953.

19 Grevenstuk; A. and Laquer, E., Insulin: Seine Darstellung, physiologische und pharmakologische Wirkung, mit besonderer Berücksichtigung seiner Wertbestimmung (Eichung). Ergebn. d. Physiol., 1925, xxiii, II, 1–267.

20 Harington, C. R., Chemistry of thyroxine, I, Isolation of thyroxine from the thyroid gland, Biochem. Jour.., 1926, xx, 293–299.

21 Harington, C. R., Chemistry of thyroxine, II, Constitution and synthesis of desiodothyroxine, Biochem. Jour., 1926, xx, 300–313.

22 Heffter, F., Ueber die Wirkung des Schwefel auf Eiweiss, Hofmeister's Beitr., 1904, v. 213.

23 Hopkins, F. G., Glutathione; Its influence in the oxidation of fats and proteins, Biochem. Jour., 1925, xix, 787–819.

24 Hopkins, F. G., On an autoxidizable constituent of the cell, Biochem. Jour., 1921, xv, 286–305.

25 Hopkins, F. G. and Dixon, M., On glutathione; II, A thermostable oxidation-reduction system, Jour. Biol. Chem., 1922, liv, 526–563.

25a Irvine, J. C., The constitution of polysaccharides, Chem. Rev., 1925, i, 41–71.

26 Kendall, E. C., Influence of the thyroid gland on oxidation in the animal organism, Jour. Indust. and Engin. Chem., 1925, xvii, 525–534.

27 Kendall, E. C., Isolation of the iodine compound which occurs in the thyroid, Jour. Biol. Chem., 1919, xxxix, 125–147.

28 Kendall, E. C., and Nord, F. F., Reversible oxidation-reduction systems of cysteine-cystine and reduced and oxidized glutathione, Jour. Biol. Chem., 1926, lxix, 295–337.

29 Kendall, E. C., and Ort, J. M., The oxidation-reduction potentials of 2-oxo-dihydro-indole-3-propionic acid and some of its halogen derivatives, Jour. Biol. Chem., 1926, lxviii, 611–630.

30 Kendall, E. C., and Osterberg, A. E., The chemical identification of thyroxin, Jour, Biol. Chem., 1919, xl, 265–334.

31 Kendall, E. C., Osterberg, A. E., and McKenzie, B. F., The preparation of 2-oxo-dihydro-and 2-oxo-hexahydro-indole-3-propionic acid and some of their halogen derivatives; Studies on thyroid activity, V, Jour, Am. Chem. Soc., 1926, xlviii, 1384–1401.

32 La Mer, V. K., and Rideal, E. K., The influence of hydrogen ion concentration on the auto-oxidation of hydroquinone, Jour. Am. Chem. Soc., 1924, xlvi, 223–231.

33 Lusk, Graham, The elements of the science of nutrition, 3rd Ed. Philadelphia, W. B. Saunders, 1917.

34 Macleod, J. J. R., Insulin. Physiol. Abstracts, 1924, iv., 21–68..

35 Magnus-Levy, A., Ueber den respiratorischen Gaswechsel unter dem Einfluss der Thyreoidea sowie unter verschiedenen pathologischen Zuständen, Berl. klin, Wchnschr., 1895, xxxii, 650–652.

36 Marine, D., and Baumann, E. J., Influence of glands with internal secretions on the respiratory exchange, III; Effect of suprarenal insufficiency (by removal) in thyroidectomized rabbits, Am. Jour. Physiol., 1922, lix, 353–368.
37 Idem. V and VI., Further data on the effect of suprarenal insufficiency (by removal) in rabbits, Jour. Metabol. Research, 1922, i, 777–802; ii, 1–18.
38 Meyerhof, O., Chemical dynamics of life phenomena, Philadelphia, Lippincott, 1924.
39 Morgulis, S., Die Katalase, Ergebn. d. Physiol., 1924–1925, xxiii, I, 308–367.
40 Müller, F., Beiträge zur Kenntniss der Basedow'schen Krankheit, Deutsch. Arch. f. klin. Med., 1893, li, 335.
40a Oppenheimer, C., Die Fermente, 1926, xi, 1705.
41 Sajous, Chas., E. de M., Thermogenesis as the fundamental function of the adrenal medulla and cortex, Endocrinology, 1925, ix, 441–455.
42 Thunberg, T., Die biologische Bedeutung der Sulfhydrylgruppe, Ergebn. d. Physiol., 1911, xi, 328–344.

IMMUNOLOGY AS A BRANCH OF CHEMISTRY

BY

H. GIDEON WELLS

PROFESSOR OF PATHOLOGY, UNIVERSITY OF CHICAGO, AND DIRECTOR OF MEDICAL RESEARCH, OTHO S. A. SPRAGUE MEMORIAL INSTITUTE

Lecture delivered at Columbia University in the special course on Contemporary Developments in Chemistry given in the Summer Session of 1926 on the occasion of the opening of the Chandler Chemical Laboratories

COLUMBIA UNIVERSITY PRESS

NEW YORK

1927

Printed from type. Published March, 1927

Printed in the United States of America

IMMUNOLOGY AS A BRANCH OF CHEMISTRY

By H. Gideon Wells

From the Otho S. A. Sprague Memorial Institute and the Department of Pathology, University of Chicago

Immunity may be appropriately called the chemical warfare of existence. Man, like all other living creatures, goes through life beset constantly by enemies seeking to drag him down. The larger enemies that attack man by physical means have for the most part been destroyed, but there still remain vastly more numerous and more dangerous invisible enemies to attack him by chemical means. We all harbor in our mouths, throats and intestines and on our skins, countless bacteria that are capable, once they get a proper opportunity, of attacking us and killing by the poisons that they produce. Against them we constantly war, destroying them by our own poisons or they destroy us. Compared to the chemical attacks against us, the physical attacks of our enemies and our own physical defenses, are insignificant.

The chemical processes of defense are usually referred to as the reactions of immunity, whether they accomplish a complete immunity or not, and the science of immunology is properly a subdivision of biological chemistry, although up to the present time most of its disciples have been bacteriologists or pathologists. But it is to be noted that the chief advances in knowledge have come through the work of chemists, such as Pasteur, Ehrlich, and Bordet. It seems safe to predict that the advances in the future are also to be made largely by chemists, either alone or in association with the immunologists. And what can be a more important field for the chemist than the problems of the methods by which life and health are preserved?

In the brief time at my disposal in one lecture period, the most I can attempt to do is to point out the extent to which the problems of immunity are problems of chemistry, and the nature of some of the advances that have been made and especially of those that need to be made. In this attempt I am handicapped by the fact that immunology, in its ignorance of the exact nature of the problems that it is studying, has been compelled to devise an arbitrary nomenclature to designate processes and agents of unknown character that cannot be more accurately labeled until we know more about them. Speaking to chemists, I shall endeavor to avoid these makeshift appellations as far as possible, and to discuss the subject on an honest chemical basis.

To begin with, let us admit that at the present time we do not know the nature of the chemical forces, or the materials to which are attached the forces that enable us to ward off the chemical attacks of bacterial and some other poisons. We merely recognize them by their manifestations and measure them by their results. The chemist will at once see that in this respect the protective agencies are on quite the same footing as the enzymes. We may never yet have isolated an enzyme, we may not know just what an enzyme is, but we know much about what enzymes do, and we make great practical use of them. Once we were in the same position concerning electricity, with which our familiarity and practical application far outran our understanding. Our knowledge concerning the nature of the processes and agencies of immunity have largely come through the application of the minds and methods of chemists. Our utilization of these agencies has been largely developed by bacteriologists, pathologists, and physicians, desperately seeking in a pioneer way to secure help in their urgent struggle to prevent and relieve human suffering. It is not strange that with so acute a necessity the practical application has received far more attention than the theoretical investigation, but we have now reached a stage where we realize that we must understand more before we can accomplish more.

Still, in the course of many studies by many means we have learned some things about the fundamentals of our immunology. First and foremost is the fact that these reactions of immunity are not simply means of defense against bacteria, but that they are fundamenal biological reactions. They are exhibited not only against bacteria or other parasites, but against any foreign protein that makes its way through the protective epithelial layer that covers and lines the entire body.

The Specificity of the Proteins

Our proteins are very specific things. At least some of the proteins of every species are demonstrably different from the proteins of every other species—indeed probably there are some differences in at least some of the proteins of every individual. How could a dog distinguish his master by scent alone from all other individuals unless there were some chemical differences in the odors, and how can these odors be different unless the protoplasm of the cells that manufacture these odors is different from that of the cells of some other dog's master? Presumably on the same basis of difference in chemical make up, one individual is injured by hens' eggs, another by cows' milk, another by horse dandruff, although the great majority of us do not suffer from contact with such foreign proteins.

Evolution seems to be largely the evolution of new combinations of the score of amino acids which make up the protein molecule. In all living cells we find these colloidal aggregations of amino acids that we call proteins, and no matter whether we examine the proteins of simple single celled creatures, like yeast and bacteria, or complex organisms like cabbages or tree toads or statesmen, we find the same limited variety of amino acids, combined in various proportions but quite the same in character. The other known components of cells, such as the inorganic salts, carbohydrates, and fatty compounds, are few in number, and pretty much the same in the cells of all different species. Only the proteins exhibit the variability that makes possible the infinite variety of living creatures. Abderhalden has accommodatingly calculated that 20 different amino acids could form at least 2,432,902,008,176,640,000 different compounds, and this without including the enormously greater number that might be made by varying the proportion of the different amino acids in a single protein.

The living body takes great pains to see to it that the specificity of its proteins is maintained, otherwise individuality would be lost. For this purpose we have the elaborate process of digestion, whereby the many different sorts of proteins are disintegrated into their building stones, the amino acids, and so the individuality of the food proteins is lost before their absorption. The body can then select such combinations of these amino acids as it needs to build up into its own structures, while it burns up the rest to furnish energy. It is by this process that pretty much the same sort of an individual can be produced from cows' milk, seal meat, bananas, codfish or corn.

Immunity and Proteins

Under ordinary normal conditions, no intact foreign proteins probably ever get past the epithelial barriers and into the body, but if such an abnormal thing does happen the body has means to meet the situation. As this situation is exceptional, the means of defense is limited unless something happens to stimulate special development of the defense. This defense is accomplished by what we call the reactions of immunity since these are chiefly evoked against the agents of disease. This is because the foreign proteins that are most likely to invade the body are contained in the parasites that have invasive power, but in principle the reaction is quite the same whether the foreign protein enters the blood and tissues in the form of typhoid bacilli, or snake venom, or horse serum injected with a hypodermic syringe. We have several different methods of recognizing the changed reactivity of the body, to which we give various names, but there is much doubt whether there are several different reactions. Perhaps there is but one sort of reaction which we recognize by several different methods. The point is that the reaction seems to be one which tends to alter the foreign protein so that it is no longer harmful, and, for the chemist the item of greatest interest is that the reactions of immunity seem to be concerned exclusively with colloids and to accomplish their effects by altering the colloidal state of the foreign proteins.

Zinsser has suggested that the reason the immunological reactions are exhibited against colloids rather than against crystalloids is that the latter can enter the body cells and be destroyed therein; the colloids, not diffusing into the cells, must be destroyed outside the cells by agents secreted for this purpose by the cells. These secreted agents we find present in the blood serum and recognize as the reacting agents of immunity to which we give various names. This idea is well supported by the fact that if we digest a protein into smaller complexes its power to stimulate the reactions of immunity is lost, but if we resynthesize these cleavage products into a collodial protein-like molecule again, this product is capable of inciting immunity reactions.

Perhaps you have thought of a contradiction to this statement, that immunity may be incited only against foreign colloids, remembering the case of the morphine fiend or the hardened tippler. To be sure, these people do exhibit less susceptibility to their pet drugs than a novice, but such increased tolerance seems to be something quite different from the immunity observed against bacteria and other foreign proteins. It depends rather upon an altered tolerance of the tissues and organs to the drug, and is not accompanied by or dependent on the presence in the blood of specific agents that neutralize the poison, as is the case with such a poison as that of diphtheria, for example.

While some investigators have believed that lipoids may also serve to incite immunity reactions, this has not been established to the point of general acceptance. The lipoids, and perhaps many other substances, may, however, combine with proteins, and these compound proteins may thus have new properties that are manifested by immunity reactions. In this way we explain the reactions that may be produced by such simple non-colloidal chemical substances as iodine, iodoform, salicylic acid, salvarsan and so on. Landsteiner and others have added much to our knowledge of the essential principles of immunity by producing artificial compounds of proteins attached to simpler chemical radicals and studying their behavior in immunological reactions. This field of research offers great promise for future development by chemists and immunologists.

The Manifestation of Immunity

Here, in order to make clear what is to follow, I must explain briefly the character of the immunological reactions with which we are most familiar. This may be done by citing a few specific examples. For example, if we take the blood serum of a man suffering with or recently recovered from typhoid fever, or the serum of an animal immunized with typhoid bacilli and add to it a suspension of typhoid bacilli, these will lose their motility and become agglutinated. This is called the agglutination reaction and is used as a test for typhoid fever, the Widal test. If to this serum we add a solution of typhoid bacilli, or of proteins obtained from typhoid bacilli, a precipitate will be obtained, this constituting the precipitin reaction, which is used in medico-legal work to identify the character of blood stains and other protein materials. If living typhoid bacilli are added to the immune serum they will be killed, that is, the serum is bactericidal. With immune serum for some other bacteria, or for red corpuscles, the cells are not only killed but dissolved, which process is called bacteriolysis or cytolysis. This bactericidal or lytic reaction depends upon the joint action of two constituents in the serum, one, formed during immunization, is called the amboceptor because it unites the cell to the lytic agent, which is called complement, since it complements the amboceptor action. The union of these agents is also much used in the form of the complement fixation reaction, which is the basis of the well-known Wassermann test for syphilis. Perhaps related to the complement fixation reaction is the anaphylaxis reaction, one of the most spectacular of the immunological processes. This is demonstrated by injecting into an animal a minute amount of a foreign protein, and repeating this injection after ten days or more. It is then found that the second dose of protein is very toxic, so that a few milligrams may be an immediately fatal dose, although in an animal not thus sensitized by a preliminary injection the same protein may not be appreciably toxic even in enormous doses. As little as 0.000,000,05 gm. of egg albumin may suffice to sensitize a guinea pig to this harmless foreign protein, and 1 or 2 milligrams of egg albumin given two weeks later will be enough to kill the sensitized animal. A single drop of horse serum has been known to kill a man sensitized to horse proteins, although 100 cc. may be well tolerated by a normal man.

And then there is the more familiar antitoxin, which in some unknown way neutralizes the poisonous effect of a very small number of poisons secreted by certain bacteria, also a number of the venoms, and a few toxic vegetable proteins. There are several other reactions, less commonly used, which it is not necessary to discuss at this time, especially since it is by no means certain that all these different reactions may not depend on a single sort of antibody, the different immunological reactions merely being different methods for demonstrating the presence of the same antibody.

The Sensitiveness of Immunity Reactions

The chemist cannot fail to be impressed by the wonderful delicacy of the immunity reactions and their striking specificity. They are far more delicate than the methods of the biochemist. For example, the biuret reaction will detect proteins in a dilution of one in one thousand or so, the precipitin reaction will detect one part in ten thousand or even in one hundred thousand; the complement fixation reaction detects one part in a million, while by the anaphylaxis reaction less than one millionth of a gram of protein may be detected. But the superiority of the immunological reaction over ordinary chemical tests is much greater than is suggested by these figures, for the chemical tests merely tell that some sort of protein is present, while the serum reactions tell us just what one of all the infinitely great number of possible proteins is present in our solutions. This fact depends upon the uncanny specificity of the reactions. Not only can we tell, in medico-legal work, that a given blood stain was made by human blood rather than by the blood of a dog or a cow, but we can get evidence on biological relationships. For example, we can show by these reactions that the blood of the chimpanzee and orang utan is more closely related to human blood than to the blood of the common monkeys. Indeed the bloods of anthropoids are more closely related to human blood, according to immunological tests, than the bloods of horses and donkeys are related to each other, although these two animals are so closely related to one another that they can hybridize.

The immunological reactions will also pick out a single protein present in minute amounts in a mass of other proteins. For example, in the lymph coming from the thyroid gland can be identified the minute quantities of thyroid protein present therein.

Some of the manifestations of this immunological specificity are so remarkable as to stretch our credulity, were they not attested by scientists who control their observations carefully. Especially are such reactions seen in man himself in his manifestations of hypersensitivity or anaphylaxis, most familiar in the form of hay fever, asthma and hives. These are commonly reactions to foreign proteins, such as pollens and food proteins, but sometimes proteins from other sources. The person sensitive to ragweed pollen may not react to rye pollen, the victim of guinea pig asthma may ride horseback with impunity while the subject of horse asthma may play with guinea pigs. We even read of a woman reacting to her husband's dandruff while not affected by the dandruff of her neighbor's husband, and of a person sensitive to hen's eggs who could eat the meat of roosters but not the meat of hens!

The Application of Immunity Reactions to the Study of Proteins

This remarkable specificity of these reactions seems to depend on the corresponding specificity of the proteins. In fact, it is probably justifiable to look upon immunology as merely one field of protein chemistry. The reactions of immunity offer great aid to the protein chemist, for with the use of but minute quantities of material many things can be learned about a sample of protein that would require much material and weeks of labor, and many things are disclosed by immunological methods that our chemical methods are incompetent to reveal at all. For example, the immunologist can tell from what species of animal a minute quantity of protein is derived, and he can detect in a mixture of proteins just what these are, no matter what the proportions may be. If a trace of protein is present in the urine, he can tell readily whether it is human protein or a foreign protein coming from the food or introduced fraudulently. In a sample of meal or flour he can detect with little labor the presence of other meals or flours introduced for adulteration.

In research work the immunological identification of proteins is invaluable. To illustrate by a specific case, a biological chemist obtained from milk proteins a fraction soluble in alcohol. Before that time, alcohol-soluble proteins had been found chiefly in grains and not at all in animal materials, so he suspected that his material might be merely a cleavage product of some milk protein or a protein made alcohol-soluble by some admixture. Study of a small fraction of a gram of this material by an immunologist soon revealed, with very little waste of time, that this material was a true protein and not a cleavage product; also that it was different from any of the proteins that had previously been found either in cow's milk or in the blood of cows. With this information to go on, the chemist was justified in taking the time and material necessary to isolate this alcohol-soluble protein in large enough quantities to study chemically and to describe its chemical composition and properties, thus establishing the existence of an overlooked component in so much-studied a material as milk.

For another illustration of the way in which problems of protein chemistry may be aided by immunological methods, let me cite an experience I have had in studying the question of the solubility of coagulated proteins. Egg albumin coagulation, which occurs so readily and completely, is generally looked upon as a complete irreversible process, and for all practical purposes it is. But some observations have made me doubt this, for I have found that 5 cc. of a suspension of thoroughly washed, heat-coagulated egg albumin which has stood for some months in salt solution, may contain enough redissolved egg albumin to sensitize occasionally, but not always, a guinea pig so that it will give an anaphylactic reaction with egg albumin. As the minimum sensitizing dose of egg albumin is not far from 0.000,000,05 gm., it is evident that the solubility of coagulated egg albumin is extremely low. Experiments with the delicate complement fixation test establish similar values for the amount of free egg albumin dissolved in water in which coagulated egg albumin has been suspended for several weeks or months.

These figures give a striking illustration of the delicacy of the immunological methods and their value in studying certain problems in protein chemistry. In no other way could such minute amounts of protein be detected in a solution. Furthermore, the immunological tests not only disclose the presence of this minute amount of protein in solution, but specificity tests establish also that it is the same protein that was coagulated by heat which has been redissolved in its original form and not merely a product of hydrolysis of the coagulated protein. Chemical tests on relatively large amounts of such fluids might at the best possibly disclose that amino compounds, perhaps even proteins or their cleavage products, are present in the solution, but they could not possibly establish the character of the dissolved molecules as intact molecules of egg albumin, a feat easy of accomplishment by the immunological tests.

I may also cite some other personal experiences in using immunological methods in studying chemical relationships of the proteins. For example, Dr. D. Breeze Jones of Washington obtained from the seeds of squashes and cantaloups two crystallizable proteins. These were found to be identical in composition, as far as our crude chemical methods for the study of protein composition now permit us to say. Examination of the crystals by crystallographic methods also showed identical form and angles. Immunological tests showed these two proteins from different seeds also to be immunologically identical. These observations, with many others, serve to show that immunological relationships between proteins depend upon chemical similarity or identity.

The same chemist has been studying a number of globulins isolated from different beans, and I have been studying their immunological relationships. Among other things, Dr. Jones found a chemical similarity between the globulins of the mung and adzuki beans, and I found them to be immunologically closely related or identical. This observation has particular interest because a government botanist had independently concluded that these two beans are so closely related to each other, and so different from the rest of the beans, that he recommended that they should be separated from the rest to form a subvariety of *Phasealus*.

Still more striking are the experiences that Dr. Julian Lewis and I have had with some of the alcohol-soluble proteins of cereals. Proteins soluble in relatively strong alcohol (70 to 80 per cent), called prolamines because rich in prolines, are not common, being found chiefly in the cereal grains, gliadin of wheat and zein of corn being perhaps the best known. A number of others have been isolated and studied by Gortner and Hoffman of the University of Minnesota. They were obtained from wheat, durum, emmer, einkorn, oats, barley, spelt, corn, kafir, teosinte and sorghum seeds. The chemical properties, as known from such data as are contained in the literature and those obtained by Gortner and Hoffman, indicate that these proteins may be divided into a "wheat group,'" which includes the proteins isolated from the genus *Triticum*, and a "corn group," including those isolated from maize, teosinte, kafir and sorghum. The genetic behavior and nuclear structure of these cereals have been extensively studied by plant breeders, especially with the wheat group, and these studies seem to reveal interesting evolutionary relationships. It has been shown that einkorn, apparently the ancestor of modern wheat, is characterized by having seven chromosomes, that emmer and durum, presumably intermediate stages, have fourteen chromosomes, and wheat and spelt have twenty-one chromosomes, as if their evolution had been accomplished by polymerization. The fertility or total or partial sterility of inter species is said to be in accordance with these chromosome numbers. In the "corn group" it has been found that teosinte will hybridize with ordinary corn, indicating a genetic relationship.

Because of these interesting genetic relationships, as well as on account of the known chemical similarities and differences between the proteins of the various members of these two groups, their immunological relationships are of particular significance, and hence we have carried out studies by means of four different immunological methods.

Our immunological results agree perfectly with the conclusions drawn from the chemical data of Gortner and Hoffman, who found that the chemical properties of these alcohol-soluble proteins indicate the existence of a "wheat group" and a "corn group" of prolamines, the members of each group being chemically similar to each other. Our results also showed that these alcohol-soluble proteins from cereal grains fall into these two classes. Gliadin from wheat and rye and the prolamines from durum, einkorn, emmer, and spelt, of the genus *Triticum*, seem to be very closely related, apparently identical, according to all our tests. The alcohol-soluble proteins from common maize, from kafir corn, and from teosinte, were found to be closely related to one another, but not to the alcohol-soluble proteins of the *Triticum* group.

Accepting the morphological evidence that the modern wheat with its 21 chromosomes has developed from the simple seven chromosome einkorn, perhaps through a fourteen chromosome stage as illustrated by durum and emmer, we find that through these evolutionary stages one of the most abundant and characteristic proteins of the seeds has retained a constant chemical composition. This would seem to go much farther than the observation of Friedenthal that the blood of a Siberian mammoth preserved in ice perhaps 25,000 years, shows close immunological relationship to the blood of modern elephants. It is perhaps more related to the observation of Landsteiner that the bloods of man, chimpanzee and orang utan contain common agglutinins for red corpuscles not present in the blood of monkeys, which agglutinins presumably have existed in the blood of the anthropoids and man since these species split off from the common stock from which they and the monkeys arose.

The Nature of the Reactive Agents

Although we are sure enough that almost all foreign proteins are able to stimulate the processes of defense, and that probably nothing else can do so, we do not know what the reactive agents are that manifest the immunological processes. All we know is that they are found in the blood serum of the immunized animal more abundantly than in normal animals, although greater or less activity may be shown in the blood of animals that have not been immunized. Despite innumerable attempts, none of these immune antibodies, as they are called, has ever been isolated. Like most enzymes, the antibodies are found so intimately associated with the proteins that there is strong reason to suspect that they are proteins, or that the antibody effect is accomplished by some special radical or some special configuration of the protein molecule.

In favor of the view that the antibody is associated definitely with certain proteins, rather than that it is simply an extraneous substance adherent to proteins isolated from the blood, is the behavior of diphtheria antitoxin, the most studied of the antibodies. When antitoxin is concentrated for therapeutic purposes, it is found that the active antibody is recovered in one small part of the entire protein fraction of the serum, the water-soluble globulin (pseudoglobulin). Despite the preliminary removal of the water-insoluble globulin, the antitoxic activity remains with and comes out of solution with the pseudoglobulin, and does not appear in the albumin fraction left after removal of the pseudoglobulin, or a modification of ordinary pseudoglobulin by rearrangement in or addition to the molecule, is unknown. No one has ever been able to separate the antibody from the protein by dialysis, electrolysis or other means, but that does not prove that antitoxin is a protein. On the other hand, occasionally some investigator concentrates antitoxin or some other antibody and secures an active solution which seems to contain no protein at all, or at least so little that it does not give the ordinary chemical tests for proteins. But in view of the great delicacy of the immunological tests as compared with the chemical tests for proteins, these scanty items of evidence are not convincing as to the non-protein nature of the antibodies. It is possible that the antibody character resides in relatively small groups or radicals which can function only when attached to some colloidal complex of large molecular dimensions.

There is a striking similarity to be noted between antibodies, toxins, enzymes and insulin. Each of them acts in quantities that are so infinitesimally small that it is not possible to say whether the proteins which may usually be found in the solutions are themselves the active agent or merely contaminations. Especially similar are the antibodies and the insulin, since each seems not to be the active agent itself, but merely one that accelerates or makes possible the action of other catalysts—that is, they are catalysts of the second order. Thus, the amboceptor serves to activate the complement; the insulin serves to make possible the burning of sugar by the tissues. Neither of itself accomplishes the final reactions. The amount of material necessary to accomplish this in the case of insulin is similar to the extremely minute figures obtained in quantitative studies of toxins, etc.

The Colloidal Chemistry of Immunity

In any event, the reactions accomplished by the antibodies are distinctly of the character of reactions with which we are familiar in the field of colloidal chemistry, as was first pointed out by Bordet, when he compared the well-known reaction of agglutination of bacteria by immune serum with the precipitation of colloidal suspensions in rivers when they meet the salt water of the ocean; that is, the precipitation of colloids by electrolytes. Bordet made the important observation that *agglutination does not occur if both the bacterial suspension and the agglutinating serum are dialyzed free from salts before mixing;* but if, to such mixtures, a small amount of common salt is added, agglutination and precipitation of the bacteria occur at once. This observation brought the phenomenon of bacterial agglutination into close relation with the precipitation of colloids by electrolytes. Bacteria are held in suspension by virtue of the fact that the mutual repulsion by the like negative charges of their cells overcomes the forces of attraction, presumably surface tension. Agglutination therefore occurs whenever the similar electric charges are decreased to amounts such that they will no longer overcome the pull of surface tension. Or conversely, surface tension may be increased until it overcomes the dispersion effect of the similar charges. In all physico-chemical respects the behavior of bacteria and agglutinin resembles the behavior of colloidal mixtures which form an electrically amphoteric colloidal suspension, so that the ions of electrolytes or the electric currents, by discharging them unequally, cause precipitation.

It is interesting to note that agglutination exhibits the "zone reaction" familiar in colloidal chemistry, that is, an excess of either of the reagents will inhibit the reaction. For example, an agglutinating serum which gives a strong reaction at a dilution of one to five hundred may show little or no agglutination at a dilution of one to ten. The same is true of all the reactions of this sort. For example, a precipitating serum which gives strong precipitin reactions with the specific antigen when diluted 1–10,000 or even much more, may show no precipitation whatever in concentrations of 1–100 or greater. This corresponds with the well-known facts concerning colloidal reactions; *e.g.*, when two colloids of opposite electrical charges are mixed in suitable proportions they may be precipitated, but an excess of either one may partially or entirely prevent the precipitation or redissolve a precipitate that has been formed. Another typical phenomenon of colloidal chemistry is also shown in the "Danysz effect," which is the term applied to the fact that if an excess of bacteria is added to a given quantity of agglutinating serum a little at a time, there will remain more unagglutinated bacteria in the suspension than if they had all been added at once. Such a behavior is identical with that seen in colloidal reactions; *e.g.*, if a piece of blotting paper is torn into pieces which are dropped into a dye solution one at a time, they will take up more of the dye than will the same amount of blotting paper added in one piece.

Since the precipitin reaction, like the agglutination reaction, occurs between the colloids of the serum and the colloids used for immunization, it also resembles the reaction seen when colloids of different charges mutually precipitate one another. For example, since the bacteria in a suspension in salt solution, like the colloidal gold particles of a gold solution, carry negative electrical charges, it is likely that their agglutination or precipitation are similar chemical reactions, that the changes are governed by well-known laws obtaining in the precipitation of a colloid bearing an electrical charge of one sign by the addition of a colloid with an opposite electrical charge, and that the precipitate contains both the precipitating and the precipitated colloids. The precipitin reaction is analogous, at least to a certain extent, with these colloidal reactions, but differs from them in the remarkable specificity of the immunity reaction. Indeed, the many attempts that have been made to explain the immunological reactions according to known analogies in physical chemistry have all collapsed in the face of this outstanding phenomenon of specificity. Nothing like it is known in physical chemistry—the nearest thing to it is the specificity of certain enzymes, but many enzymes are not specific, *e.g.*, pepsin and trypsin will digest almost all sorts of proteins. The problem of specificity offers a brilliant chance for attack by the chemist.

The most we can now say about these agglutination and precipitation reactions is that they are characterized by the reduction of the degree of dispersion of colloids, either in suspension or in solution, so that the aggregated particles become visible, form coarse flocculi and precipitate. The maintenance of suspension of colloidal particles probably depends upon the repellent forces of like electrical charges on the particles which oppose coalescence, and the Brownian movement which opposes gravitation; whereas the aggregation of colloidal articles depends on the cohesive forces which draw the particles together. Therefore, the existence of a suspension, and equally the agglutination of its particles (whether these are molecules, micellae or larger masses), depends upon the relative strength of these opposing forces. The agglutination and precipitation reactions produced by a specific immune serum acting upon a specific protein, are closely related to the corresponding reactions in colloidal chemistry. In many essential respects these immunological reactions are similar to, and perhaps identical with, the precipitation of colloids under the influence of other colloids of different electrical charges, or by electrolytes.

As yet we cannot explain exactly how these colloidal reactions of immunity are brought about, and their striking specificity does not support the idea that the simple colloidal reactions alone are all that is involved. The specificity of the reactions suggests some more definitely chemical reaction, presumably superimposed on the colloidal reactions which undoubtedly take place. Until we know more than we now do concerning the colloidal reactions and their relations to chemical reactions in general, and until the underlying principles of solution are better understood, the details of these remarkably delicate and specific immunological reactions cannot well be explained.

Numerous suggestions have been made concerning the processes involved. For example, that agglutinins attack some protective colloid about the bacteria or other cells which are to be agglutinated, whereby they become like unprotected inorganic suspensions which are precipitated by salts and other electrolytes. Or that the antibody alters the protein so that it is more capable of combining with or adsorbing salts, which then reduce the colloidal dispersion. Or that the antibody increases the cohesive force which brings the particles together and overcomes the repellent effect of the like electrical charges of the particles.

The complement-fixation reaction likewise seems to be a typical colloid-chemical reaction, dependent upon the physical state of the colloids and the *pH* and electrolyte content of the fluid in which the colloids are dispersed. The complement function is manifested only when the colloidal particles resulting from union of antigen and antibody are of a certain size and when electrolytes are present, and in some reactions artificially constructed colloidal emulsions may be made to serve either as antigen or as complement. Furthermore, the fact that the adsorbent precipitate responsible for the complement fixation reaction forms more rapidly in the ice chest than in the incubator is in favor of its being physical rather than chemical in the usual sense.

In any event, it seems evident that the problems of immunity are to a large extent problems of colloidal chemistry, of great interest and significance to the chemist working in this field.

The Neutralization of Toxins by Antitoxins

This is probably the best known of all the immunity reactions because of its practical value in the treatment and prevention of diphtheria and tetanus. The value of antitoxin in these two diseases lies in the fact that diphtheria bacilli and tetanus bacilli produce their harmful effects by secreting a soluble poison of high potency, against which a specific antitoxin can be prepared. Unfortunately for the success of medical practice, but a few of all the disease-producing bacteria owe their effects to poisons of this type, the true *toxins*. Most of the harmful bacteria liberate poisonous products only when they die and disintegrate, and against these cellular poisons (*endotoxins*) no antitoxins can be produced, for some reason as yet unknown. The organism responsible for one sort of food poisoning, botulism, is one of the few other important bacteria for which an efficient antitoxin can be prepared, and antitoxins can be obtained for snake venoms and for the poison causing some of the symptoms in scarlet fever.

We know nothing more about the nature of antitoxins and how they neutralize toxins than we do about the rest of the immunity reactions. All we can say is that the toxin-antitoxin reaction is characterized by the fact that here a specific chemical poison is neutralized by a specific antibody, without the destruction of either, for neutral toxin-antitoxin mixtures may be dissociated by various means into the active components. Ehrlich thought that the reaction was like that of the neutralization of a strong acid by a strong base. To fit this theory to the many observed facts it became necessary to imagine the toxin as composed of many components of varying degrees of toxicity and with varying affinities for antitoxin, and the theory finally became so complex as to collapse under its own weight.

Arrhenius and Madsen studied the physical chemistry of toxin-antitoxin neutralization and advanced the hypothesis that it resembled more the reaction between weak acids and bases, in which equilibrium occurs according to the law of mass action. As an illustration they presented the neutralization of ammonia by boric acid, which leads to an equilibrium in which there is always present free ammonia even when boric acid is added to saturation. The hemolytic action of ammonia in such a neutralization mixture was found to parallel closely the neutralization of hemolytic toxins by the specific antitoxins.

Bordet looks upon the toxin-antitoxin neutralization as more like the neutralization of two colloids of different charges, and advanced an "adsorption theory." He does not accept the simple chemical explanation that toxin and antitoxin unite in fixed definite proportions, but believes that the process is more similar to the dyeing of a colloidal fabric by a colloidal dye. There seems to be a growing body of evidence in support of this theory as best explaining many of the observed phenomena of toxin neutralization, as well as other immunological reactions. For example, such typically colloidal phenomena as the "zone phenomenon" of inhibition or reversal of reaction by an excess of one agent, and the "Danysz effect" of unlike results produced by the same quantity of agent when added in single or in fractional doses, are exhibited in toxin-antitoxin neutralization.

However, the simple colloidal adsorption theory, as so far developed, fails to account for the specificity of the immunological reactions. Consequently, there is a tendency at present to look upon adsorption as only one part of the toxin-antitoxin neutralization, perhaps the first step that brings together the antigen and antibody, secondary chemical processes occurring in the colloidal complex accounting for the specificity of the reaction and the neutralization of the toxin.

Concluding Remarks

Time does not permit me to consider the many other known manifestations of the chemical principles that underlie the defense reactions. I have not even been able to speak of the interesting physical chemical processes that come into play in the important method of defense known as inflammation. But it is hoped that enough has been said to establish the main theme, that ***immunology is primarily a branch of biological chemistry***, and that there are problems of great interest and of the highest human importance awaiting solution by the chemist, either alone or in cooperation with the immunologist.

Up to the present time we have been enabled, through our study of immunity, to accomplish some tremendously important things, such as the curative treatment of diphtheria and scarlet fever and the prevention of these and a few other diseases, such as typhoid, small pox, tetanus, cholera and some others. But this is only a beginning. Even with these diseases our results are as yet far from 100 per cent of perfection, and with the rest of the great category of infectious diseases we have made little or no progress. Pneumonia still ranks high in the causes of death. Peritonitis and other acute infections still kill myriads. An influenza epidemic coming next month would find us no better prepared to prevent or to cure than did the last one, and so on. Our efforts to immunize accomplish at the best but partial success, and often they achieve nothing. Our few curative serums, of which only diphtheria has as yet a high proportion of success, are at best but feeble, dilute solutions of the protective agents, gleaned with difficulty from other animals and used in an unnatural environment. Our prospects of any great advance by means of improvement of the methods we now know and use seem but slight. We must have a new method and that is to be furnished by the joint action of the chemist and the immunologist. When we learn just what these mysterious protective agents are, how they act, and how to produce them artificially at will, we shall not have to depend on the dilute solutions of these agents that we can extract from a horse or can cause to develop in a man's own blood, but we can give him as much as he needs of the active agent that has been prepared by synthetic chemistry. That seems to be the goal we must work towards, and the organic chemist who is studying the linkage of pyrroles, piperazines and amino acids is probably quite as much a builder of the future immunology as the man who is investigating the serum reactions of the meningococcus. Emil Fischer and Kossel, through their study of the composition of the proteins, are perhaps quite as directly the founders of the future immunology as Metchnikoff, Ehrlich and Bordet, who worked more directly with the living processes of infection and resistance.

THE RARE GASES OF THE ATMOSPHERE

BY

RICHARD B. MOORE

PROFESSOR OF CHEMISTRY AND DEAN OF SCIENCE
PURDUE UNIVERSITY

Lecture delivered at Columbia University in the special course on Contemporary Developments in Chemistry given in the Summer Session of 1926 on the occasion of the opening of the Chandler Chemical Laboratories

COLUMBIA UNIVERSITY PRESS
NEW YORK
1927

Printed from type. Published March, 1927

Printed in the United States of America

THE RARE GASES OF THE ATMOSPHERE

by

RICHARD B. MOORE

Introduction

During the thirty or forty years immediately preceding the discovery of argon little work was done by chemists in connection with our atmosphere. No suspicion was entertained that the air was a veritable gold mine for chemical research, even after the evidence seemed to indicate that we had not yet found all there was to know about our atmosphere there was a feeling of doubt and uncertainty that made the work move slowly. The first clew to the undiscovered elements was partly accidental. In 1882 and several succeeding years Lord Rayleigh worked on the relative densities of hydrogen and oxygen. In 1893 he published his results on the densities of some other gases, notably those of oxygen, nitrogen and air.[1] He prepared oxygen and nitrogen by different methods and while the density of oxygen in all cases proved to be the same, that of nitrogen prepared by chemical methods differed from that of nitrogen obtained from the air. The latter was about 0.5 per cent heavier than the former. A difference of this magnitude could not be accounted for on the basis of experimental error and the riddle remained unanswered for some time. Finally, Professor (afterwards Sir) William Ramsay asked permission from Lord Rayleigh to take up the matter from the chemical standpoint. He had already in some previous work found that red-hot magnesium was a good absorbent for nitrogen. He concluded that if there was any difference between prepared nitrogen and atmospheric nitrogen separation might be made by passing the gas over a heated metal. This was tried, the oxygen first being removed by metallic copper. A slight increase in the density of the product was observed. Arrangements were now made to circulate the gas backwards and forwards over the heated magnesium. After several days of this treatment, the density was found to be 16.1. At this stage Ramsay did not know that he had to deal with a new element but thought the results were due to a triatomic form of nitrogen corresponding to ozone. The final absorption by magnesium of the gas gave a density of about 20. It was only when the spectrum of this product was examined that Ramsay became convinced that he had found a new element.

Meanwhile Lord Rayleigh was dealing with the problem in a different manner. He passed sparks through a mixture of nitrogen and oxygen in the presence of a solution of sodium hydroxide. The hydroxide absorbed the oxides of nitrogen formed, and after removal of the excess of oxygen, the resulting gas showed an increased density. From this point, Ramsay and Rayleigh continued the work together and announced the discovery of the new element "argon" at the British Association Meeting in 1894.

After the discovery of argon, Ramsay looked for other sources of the element. He heard through Sir Henry Miers of a paper by W. F. Hillebrand of the U. S. Geological Survey, in which it was stated that the mineral cleveite, on heating, gave off a considerable quantity of inactive gas which was supposed to be nitrogen. Ramsay suspected that this might be argon and obtained a sample of the mineral. From this he obtained a quantity of gas which on purification gave a residue with a spectrum entirely different from any that he had previously observed. This spectrum was mainly characterized by a bright yellow line which on measurement corresponded with the line observed by Janssen in the chromosphere of the sun in 1878. In this manner terrestrial helium was discovered.

It now seemed probable that other undiscovered gases existed beside argon and helium. Krypton and Xenon were found in the residue obtained from the slow evaporation of liquid air. Neon was discovered by the fractionation of 15 liters of argon, the new gas coming over with the first fraction on distillation.

In this manner, through the courage and foresight of Ramsay and his assistants, an entire new group of elements was added to those previously known. The main property common to all of them is that are absolutely inert to chemical action. So far no attempt to get them to combine with other elements has been successful and with our knowledge of these gases it seems reasonable to suppose that we may never be able to get them to react.

HELIUM

Occurrence

Kayser[2] was the first to detect the presence of helium in the atmosphere by spectroscopic methods, and his results were confirmed by Friedlander,[3] Ramsay's original determination of the quantity in the atmosphere being one volume of helium in 250,000 volumes of air.[4] Watson,[5] with better facilities at his command, calculated that the quantity is probably 1 part by volume of helium in 185,000 volumes of air. Helium is occluded in a large number of minerals; indeed it is contained to a greater or lesser extent in all minerals and rocks. The gas is found in uranium and thorium minerals as the alpha particle given off in radioactive changes is a charged helium atom. The helium found in the earth's crust, therefore, is partly, if not entirely due to the disintegration of radioactive types of matter in the earth's crust with the elimination of alpha particles. Generally speaking, the amount of helium in minerals is proportionate to the amount of uranium or thorium present. The Hon. R. J. Strutt has, therefore, sought to determine by this means the age of certain minerals and rock deposits with fair success. In some cases, however, the rate of leakage is greater than the rate of formation from radioactive changes.

Helium is found in the gases evolved from many mineral springs, notably the Bath springs in England, and the springs at Mazieres, at Monte Irone (Albano), at Railliere in the Pyrennes, at Wildbad in the Black Forest, and in a number of springs in Tuscany. In some cases the gases evolved carry as much as several per cent helium. Kings Well at Bath, England, is estimated to yield 1000 liters annually. (Table I.)

TABLE I

Locality	Spring	Nitrogen	Argon	Helium
Aix-les-Bains	Alun	......	1.153	.037
Badgastein, Austria	Gratenbüker	97.25	1.181	.169
Bagneres-de-Luchon	Bordeu No. 2	96.45	1.25	.15
Bagneres-de-Luchon	Ferras	96.085	1.23	.285
Bagneres-de-Luchon	Saul No. 2	94.826	1.271	.323
Bourbon-Lancy	Lymbe	91.96	1.20	1.84
Bourbon-Lancy	Reine	96.1	1.15	1.75
Cauterets	César	98.55	1.275	.165
Cauterets	Railliere	98.508	1.377	.115
Grisy	Ys	95.5	1.18	2.18
Mazieres	Romaine	91.46	.92	5.92
Nancy	Parc Sainte-Marie	95.36	1.29	1.60
Plombieres	No. 3	94.505	1.375	.120
Plombieres	No. 5	95.32	1.547	.104
St. Honore	Crevasse	95.76	1.147	.893
Santenay (Cote d'Or)	Carnot	......	.....	9.76
Santenay (Cote d'Or)	Fontaine Slaee	......	.....	8.40
Uriaae	....	93.98	.938	.932

Helium is found in certain volcanic and fumarole gases. The boracic acid fumaroles of Ladarello give off largely steam, associated with carbon dioxide, hydrogen, methane, hydrogen sulphide, nitrogen, etc. After the steam is condensed the remaining gas contains also about .0015 of 1 per cent helium. Helium is usually found in minerals containing uranium and thorium. One mineral, beryl, has been found to contain a considerable amount of helium and practically no radioactive matter from which helium is supposed to be derived. Boltwood explains this on the basis that in the crystallization of the mineral only one of the shorter lived products of disintegration such as radium or ionium, has remained with the mineral. This has decayed with the formation of helium, and as there was no uranium or thorium present, the mineral became inactive.

The volume of helium contained in any mineral depends upon several factors such as its geological age, the percentage of thorium or uranium present and the density of the mineral which will largely control the leakage of the helium from the crystals. Some minerals carry as high as eight or nine cubic centimeters of helium per gram.

Strutt[6] has examined the saline deposits of Stassfurt and found not only that they contain helium in appreciable quantities but also that the percentage of helium in the potash minerals was very many times larger than could be justified by the uranium content of the mineral. On the other hand the helium in the rock salt was much of the order that could be expected from its geological age if the helium originated from the radioactive material present.

Helium can be obtained from minerals either by strongly heating, or by solution in acids. A still better method is to fuse the mineral with acid sodium sulphate. Whichever method is used, the tube containing the mineral is connected with a Töpler pump, and the evolved helium may be taken through the pump and collected over mercury. A high temperature is required to get rid of all of the gas by heating alone. In the case of monazite, it has been shown that very little gas comes off at 500°C., and at 900°C., it took 24 hours of heating to liberate all of the gas.

In 1907 Cady and McFarland[7] published a report on the presence of helium in a number of natural gases, mainly from Kansas. Some of the samples tested ran as high as 1½ per cent helium by volume. Some analyses of gas from the western United States are given in Table II.

TABLE II

Number, Locality	Oxygen, O_2	Carbon Dioxide, CO_2	Olefines, C_2H_4	Carbon Monoxide, CO	Methane, CH_4	Ethane, C_2H_6	Hydrogen, H_2	Helium, He	Nitrogen, N_2
1 Dexter, Kansas..	0.20	.....	.....	.....	14.85	0.41	trace	1.84	82.70
6 Dexter (Greenwell well), Kansas............	0.10	.....	.....	.....	14.33	1.06	trace	1.64	82.87
41 Eureka new well, Kansas........	0.10	0.20	.....	.....	51.80	.....	.....	1.50	46.40
9 Fredonia, Kansas	trace	0.61	0.12	.....	82.25	.....	.....	0.616	16.40
4 Eudora, Kansas.	0.31	0.62	.....	.....	88.60	.....	.....	0.27	10.20
37 Augusta, Kansas	.....	.....	0.77	.....	79.10	7.44	.....	0.25	12.44
20 Moran, Kansas..	0.20	0.30	0.35	0.20	92.00	.....	0.39	0.214	6.35
24 Butler, Ohio.....	0.05	.....	0.40	.....	70.00	16.75	0.27	0.15	12.38

Moureu and Lepape [8] and Voller and Walter [9] have pointed out that very small quantities of helium are contained in certain gases of Europe, but the occurrences are unimportant.

The discovery of helium in the natural gases of Kansas opened the way for the large commercial developments by the United States Government which have taken place during the last eight years. After we entered the war, the Bureau of Mines, in coöperation with the Geological Survey, made a quick survey of our helium resources. After the armistice it was decided to continue this survey and the work was carried out under the writer's direction and under the immediate charge of Mr. C. W. Seibel. Every gas field in the United States was examined for its helium content and in a number of cases where the percentage of the rare gas was high and the volume of gas also high the field was examined by competent geologists to determine the actual volume of gas available. Most of these results are still held confidential by the Helium Board as necessary to the National Defense and therefore cannot be given in this paper. However, it is permissible to state that beginning with Texas in the southwest, the general trend of the helium in natural gas is to the northeast, going through the Osage in Oklahoma, crossing the southeastern portion of Kansas and going through southeastern Illinois, Indiana, Ohio, Pennsylvania, and New York, and coming back through West Virginia and Kentucky. There is no commercial helium in many of our largest gas fields, such as those of Louisiana, California, Wyoming, etc. There are one or two patches in the West which are hard to account for, but the amount of helium-bearing gas there is relatively not great. The following table gives some idea of the large number of gas wells in the United States which contain helium. (Table III.)

TABLE III

HELIUM IN NATURAL GAS
(Per Cent)

Number of Counties Sampled	Highest Per Cent	Lowest Per Cent	Average Per Cent
(1)	(2)	(3)	(4)
3	0.26	trace	0.13
7	0.11	none	none
5	1.04	none	0.20
2	0.12	0.04	0.08
9	0.61	0.11	0.34
17	2.13	trace	0.76
15	0.43	none	0.15
9	0.19	none	0.04
2	0.04	none	0.02
4	0.27	0.01	0.08
1	trace	trace	trace
13	0.31	none	0.11
1	0.17	0.17	0.17
15	0.55	none	0.30
24	1.04	trace	0.23
10	0.17	none	0.07
2	0.04	trace	0.03
1	0.36	0.35	0.35
26	1.73	none	0.43
2	trace	none	
3	0.12	none	0.05
2	0.06	trace	0.03
10	0.10	trace	0.01

Column 1 gives the number of counties in twenty-three states in which natural gas samples were examined for their helium content. As in some of these states, samples were taken from more than fifteen counties, and in many of the individual counties a large number of samples were examined, it can readily be seen how widely distributed helium is in our natural gas and what a tremendous volume of the rare gas could be obtained from this source.

Cady and McFarland pointed out that in all cases where helium was present nitrogen was also present and that there appeared to be a very rough connection between the helium and nitrogen contents. The above survey showed the same relation between these two gases.

S. C. Lind [10] has suggested that the radioactive origin of helium in natural gas is the most probable. He points out that the helium accumulation of the richest pools requires vertical migration from a depth of only 1 or 2 miles, over a period of 1-2 hundred million years, leaving out of account lateral migration and deep vents. The rarity of these rich helium pools is probably dependent on a combination of circumstances favorable to migration and local storage. He suggested that removal from the radioactive mineral and migration are both aided by the gases generated by certain chemical decompositions produced in the earth's crust by the original high-velocity alpha rays, such as the decomposition of water, nitrogenous material, or of carbonates to give CO_2.

There is no question that the amount of radium in the earth's crust is sufficient to give rise to all of the helium found in the richest pools of helium-bearing natural gas, provided we can visualize the diffusion of the helium through the earth's crust, and the necessary conditions for concentration. These are the difficulties which are not explained by Dr. Lind. We know that minerals hold helium and other gases, and it is extremely difficult to get such gases from them. It is necessary to heat a sample of powdered monoazite or thorianite to a temperature of 1000° and hold it at that temperature for from 24 to 30 hours to obtain all of the helium in the mineral. It is difficult to visualize, therefore, the conditions under which the minute quantities of helium in the earth's crust would diffuse 1 or 2 miles, and especially would diffuse into an area of concentration such as we have in natural gas. The radioactive origin is of course both logical and possible but it is by no means convincing and the difficulties of accepting it should stimulate us to look for other explanations which may lead us to great discoveries.

In only one case has a freak gas been found in the United States. This is the Woodside Well in Utah. It contains 27 per cent carbon dioxide over 1 per cent helium, a small amount of nitrogen, a small amount of oxygen, and only 2 or 3 per cent hydro-carbons. Whereas nitrogen is almost always a constituent of helium-bearing natural gas the presence of large quantities of nitrogen do not necessarily indicate the presence of helium. A natural gas containing 20 per cent or 30 per cent nitrogen is almost certain to contain from ½ to 1 per cent helium; if a natural gas contains as much as 80 per cent nitrogen, however, it rarely contains any helium at all.

Properties

Helium is a colorless gas, is odorless and tasteless.[11] Its density is 1.995, according to Watson.[12] The compressibility coefficient at zero is −0.0006 between pressures of 400 and 800 mm. of mercury.[13] The molecular weight and therefore, the atomic weight calculated from Heuse's value for the density is 4.0. Helium does not obey Graham's law of diffusion of gases, but passes through a porous diaphragm more slowly than is expressed by the law. The solubility of helium in water was determined by Antropoff,[14] who found for the absorption coefficient the value 0.0134 at zero, and 0.0226 at 50°, with minimum at 10°. The thermal conductivity of helium is 0.0003386.[15] The refractivity was first determined by Rayleigh and afterwards more accurately by Ramsay and Travers, who found it was 0.1238 times that of air, which gave the value for helium of 1.000361.[16] The viscosity was first determined by Rayleigh by measurement of the rate of flow of the gas through a capillary tube. He found the value 0.96; air being 1.0. Schultze redetermined the viscosity and found it to be 1.086 times that of air at 15° C.[17] The spectrum of helium has been studied by Runge and Paschen, by Watson and others. The lines fall naturally into groups, in each of which we have a principal series, and two subordinate series. One of these groups consists of a doublet series and the line (D3) is the principal line of this group. It is the lines of this group which characterize the solar spectrum of helium, and are seen when an ordinary vacuum tube containing helium is examined with a spectroscope. Generally speaking, all of the lines are seen in a vacuum tube, but the relative intensity of the lines varies with the pressure and, therefore, the color of the tube itself may vary materially with different pressures. The main helium line, usually called the "helium yellow," consists of a double line with the wave lengths 5876.21 and 5875.88.

Liquefaction of Helium

After several attempts by others, helium was successfully liquefied by Onnes. This was accomplished by the use of the Joule-Thompson effect after the gas had been pre-cooled by liquid hydrogen. The gas was circulated through an apparatus in which it was cooled first to the temperature of liquid air and then to − 258°C. by means of liquid hydrogen boiling under reduced pressure. It was then liquefied in an apparatus using the general principles of the Hampson type of machine. Liquid helium is colorless and very mobile; its density is 0.154, and its maximum density is at about 2° absolute.[18] The boiling point is 4.3° absolute; the critical temperature 5.25° absolute [19] and the critical pressure is 2.26 atmospheres. The critical volume (calculated) is 0.00271.

Helium has also been liquefied by Prof. J. C. McLennan and

his assistants at the University of Toronto. McLennan has obtained helium on several occasions in quantities of a liter or more. Recent reports from Onnes' laboratory indicate that helium has actually been solidified by subjecting helium to a pressure of 150 atmospheres at a temperature of 4½° absolute. Details are lacking at the time of writing.

Before the war, most of the helium which existed in the world was obtained by heating radioactive minerals or dissolving them in acid. The total amount in existence was probably not more than 10 or 12 cubic feet. Therefore the sudden determination to produce the rare gas in quantities sufficient for airships meant the starting of an entirely new and difficult technical project.

Following the operation of the three small experimental plants during the war it was decided to build a large production plant at Fort Worth using the Linde system of refrigeration. In order to get its cooling effect the Linde Company uses the nozzle expansion of the gas, getting thereby the Joule-Thompson effect. This method is not so efficient as adiabatic expansion in an expansion engine, except near the critical temperature of the gas, when the nozzle becomes nearly as efficient as the expansion engine.

In the production plant as first designed, the natural gas is treated for the removal of carbon dioxide, which is present in quantities varying from 0.2 to 0.3 per cent. This is done by bringing the gas under low pressure in contact with limewater in specially designed scrubbers. The natural gas free from carbon dioxide passes under low pressure to two gas-holders, which act as reservoirs for the compressors. The heat of compression is removed in the usual way through inter-coolers between each stage of compression and after-coolers in which the gas is cooled by circulating water. In addition to obtaining refrigeration by nozzle expansion, a carbon dioxide cycle is also used as a gas fore-cooler. The compressed gas is led through this fore-cooler, in which most of the water vapor is removed by freezing. The cooled gas goes from the fore-cooler to the gas interchangers, where it meets the cold gas returning from the still. This returning gas pre-cools the incoming gas to a temperature sufficiently low to give the desired effect after expansion. Arrangements are made in connection with these interchangers so that there is a progressive cooling first carried out by the returning vapors of the heavy hydrocarbons and ending up with the returning nitrogen which is liquefied in the upper portion of the still. From the interchangers the gas is expanded through a nozzle into the still and enters the bottom of the still partly in the liquid and partly in the gaseous state. The still consists of three units, each unit being a rectification column with a condenser at the top and a receiver at the lower end. The gas enters the lower unit and passes up through the middle unit and thence to the top unit. Each of these units is so designed as to remove a definite portion of the constituents of the gas in the liquid state and deliver a definite portion of the gas in the gaseous state to the next higher unit. A definite temperature gradient is maintained in each unit and the constituents are progressively removed by liquefaction and re-evaporation and go back through the interchangers. In order to get a reasonably high-grade product containing as much as 70 per cent helium, it is necessary to liquefy the nitrogen in the gas as completely as possible. In order to do this and not have the vapor pressure of the nitrogen too high the upper portion of the still should be near the temperature of liquid nitrogen. In order to accomplish this a separate nitrogen cycle has been used in which nitrogen is liquefied and surrounds the condensers of the upper unit of the still. The make-up for this nitrogen cycle in the original experimental plant was designed to obtain the necessary nitrogen from the natural gas in the operation of the plant itself, the nitrogen being obtained by the final fractionation of nitrogen from the methane in the upper portion of the still. Actually, the purity of this nitrogen has not been so high as expected, but this fact has not seriously affected the operations of the plant.

The liquids collected in the receivers of each unit are used as required in the condensers of the middle and lower units. All outgoing gases from the still pass through the gas interchangers and gas fore-coolers, thus pre-cooling the incoming natural gas. These outgoing gases are mixed at the final outlet and are returned to the mains which supply the cities of Dallas and Fort Worth. The gas, therefore, is simply robbed of its helium and a small portion of nitrogen, the rest of the gas being used for commercial heating and lighting purposes.

The cost of helium per thousand cubic feet in the small experimental plant during the war was from $300 to $400. Recently this cost has been reduced to from $25 to $30 a thousand when the plant is processing at least 150,000,000 cubic feet of natural gas per month.

Some recent improvements in connection with the plant have made it possible to do away with the auxiliary nitrogen refrigeration cycle, as well as other details which have helped materially in the reduction in cost.

Any gas used in an airship diffuses from the bags, therefore there is a steady actual loss of gas and in addition the gas remaining in the bags gradually decreases in purity. Therefore the purity of the helium gradually drops and the lifting power also drops. At a certain point, usually when the gas is about 85 per cent pure, the helium is pumped out and repurified, which simply means taking the air out, raising the purity, and using the helium over again.

One of the first methods of repurification used was by means of charcoal at low temperatures. The adsorption break with charcoal is very sharp, the purity of the gas obtained remaining at approximately 100 per cent until the saturation is reached when the purity drops very rapidly. A small plant was built by the United States Government using these methods but the actual cost of operation was greater than was the case with other methods, and therefore the plant was never operated as a commercial undertaking.

The repurification plant at Lakehurst, New Jersey, was designed by the Bureau of Mines for the Navy and has been a complete success. When the low-grade helium at Fort Worth was "stepped up" to 91 or 92 per cent it was done at a pressure of 1000 pounds on the cycle. In order to get a higher purity it was decided to use 1800 pounds pressure on the helium cycle at Lakehurst. This theoretically gave at the temperature of liquid air a purity of about 99 per cent helium and the actual purity obtained has averaged 98 per cent. The principle used is the simple one of pumping helium under high pressure through a coil or purifier to which is attached a "drain pot" for the liquefied impurities (air), the whole dipping in a bath of liquid air which is obtained by means of a separate cycle. All condensed liquids are re-evaporated to pre-cool the incoming gases and the refrigeration contained in the purified helium is also used to pre-cool the incoming helium. The air cycle is the regular Claude system. After being compressed to 600 pounds the air goes through the drying and carbon dioxide removal apparatus and then through a Claude interchanger, after which it divides, about 25 per cent going through the liquefier and 75 per cent through the expansion engine in the usual way. The impure helium is compressed as already stated to 1800 pounds and after being dried, goes through its own interchanger into the purifier.

The principal use for helium is as a substitute for hydrogen in airships. On account of its non-inflammability it has very distinct advantages over hydrogen. The disaster to the Shenandoah would have been much greater had it not been for the fact that the airship contained helium. With the present operating cost of $25 per thousand cubic feet and with further economies probable helium has come to stay as an airship gas. Up to the present time the U. S. Government has produced nearly 30,000,000 cubic feet in connection with its helium project.

Another use for helium which has been suggested is due to the fact that its heat conductivity is high. As a substitute for oil in transformers and similar cases some experimental work has been carried out.

Owing to the fact that its solubility in blood is less than nitrogen it has been used to some extent in a synthetic mixture of helium and oxygen in place of air for working under pressure in driving tunnels, etc. The time required in changing the workmen from the pressure at which they work to atmospheric pressure has been reduced several hundred per cent.

ARGON

Occurrence

Argon is found widely distributed in nature. It exists in the atmosphere in the proportion of 0.941 per cent by volume, after water vapor, and carbon dioxide have been removed from the air, and 1.845 per cent by volume of atmospheric nitrogen.[20] The proportion of argon in the air is constant, as samples taken from widely different localities show practically no variation in the argon content. Over the open sea, however, slightly higher proportions of argon have been found. It is found to a small extent in sea and river water, owing to its being dissolved with the air. Since argon is slightly more soluble than air itself, the proportion by volume in natural waters is a little higher. It has been found in many mineral springs, among them being those of Bath and Buxton; the old sulphur well at Harrogate, England; Monte Irone, Albano; the Poretta baths, Bolognian Apennines; Maizieres, Cote d'Or, etc. It is also present in certain natural gas in Kentucky, and in the helium-bearing gas at Petrolia, Tex. Only a few minerals yield argon on heating; the chief of these are malacone [21] and a mineral from Brazil described by Antropoff.[22] Both minerals contain zirconium in considerable quantities, and both are radioactive. The amount of argon present is probably too large to be derived from atmospheric sources and, therefore, there is at least a possibility that argon may also be a product of the disintegration of radioactive elements, although there is no direct proof of this statement. Argon has also been found in eliastite [23] and in a mineral carrying cerium from the South Caucasus.[24]

Recovery from the Atmosphere

Argon may be extracted from the atmosphere by passing air over heated metallic copper and then a strongly heated mixture of metallic magnesium and lime. The absorption of nitrogen by magnesium alone is a very slow process, but when mixed with lime the absorption is rapid and complete. The air is first passed over soda lime and phosphorus pentoxide, and then through a copper tube in which the oxygen is absorbed. From this it goes through the magnesium-lime mixture, over heated copper oxide, and finally through other soda lime and pentoxide tubes. In order to have the absorption complete, some care is required. First the apparatus must be completely exhausted and must remain air-tight. Connections may be made by antimony pressure tubing of small bore. Black and gray rubber do not stick tightly to the joints, and the apparatus is sure to leak. It is better to make glass joints wherever possible, and reserve the rubber for connections with the hard-glass tubes. The lime must be as free from carbonate as possible and, of course, perfectly dry. Before the air is passed through it, the magnesium-lime tube is strongly heated and soon turns black, owing to the glass being attacked by the magnesium. During this process carbon monoxide and hydrogen are formed in considerable quantities; the first by the reduction of the carbon dioxide obtained from the traces of carbon in the lime, and the second from the action of water vapor on the magnesium. These gases should be pumped off as fast as formed, and when the evolution is practically finished, air is slowly admitted to the apparatus. Traces of carbon monoxide and hydrogen formed during the actual experiment are oxidized by the copper oxide, and the products absorbed by the soda lime and pentoxide tubes.

When the nitrogen reaches the magnesium-lime tube, absorption is indicated by a sudden rise in temperature and the formation of a glowing ring at the rear end of the tube. Unless this ring is formed, absorption is not taking place. The glow gradually goes down the tube as the magnesium is used up, making it easy to know when to stop the experiment. If the apparatus leaks, small traces of air may be removed from the argon obtained by sparking with oxygen in presence of a strong solution of caustic soda. The excess of oxygen is then removed by means of phosphorus or calcium. For larger quantities of argon it is more convenient to use the method of Fisher and Ringe, according to which air is passed over a mixture of 90 per cent calcium carbide and 10 per cent calcium chloride heated to 800° C. In this case, the temperature is too high for glass, and iron tubes must be used. With a properly prepared apparatus, considerable quantities of argon may readily be made. The nitrogen and carbide give lime and carbon, while carbon dioxide and monoxide formed at the same time react with the carbide to give carbon. Argon can also be extracted from the atmosphere by the original sparking method of Lord Rayleigh, in which an inverted flask is used containing a strong solution of caustic soda. Through a two-hole rubber stopper and passing up through the solution are two glass tubes containing electrodes. The oxides of nitrogen formed by the sparking are absorbed by the caustic-soda solution.

Argon is now made commercially from the atmosphere by using the ordinary refrigeration method for liquefying air. The fractionation column—either the Claude process or the Linde process is used—is adopted so that a liquid oxygen containing 95 per cent oxygen and more than 4 per cent argon is obtained. This liquid oxygen is run into an auxiliary column and fractionated. In this manner, a product containing as high as 80 per cent of argon can be obtained. It is purified by chemical methods already described, and finally used in incandescent lamps instead of nitrogen. Liquid argon is colorless and transparent, and boils at -186.1° C. Its density at this temperature is 1.4046. If allowed to cool by means of its own evaporation it can be solidified to a white mass which melts at -188.9° C.

Physical Properties

Argon is an odorless, colorless and tasteless gas. Its density is 19.95 ($O = 16$), and the weight of a normal liter of gas is 1.782 g. Its atomic and molecular weights are taken as 39.91. The critical constants of argon are as follows: Critical density, 0.509 g./c.c.;[25] critical pressure, 47.996 atmospheres[26] and the critical temperature, -122.44 C.[27] The vapor pressure of solid argon is 411 mm. at -190.9°. The triple point is -189.3°. The molecular heat of vaporization of the liquid is 12.92 cal. at -125.49°; 24.01 cal. at -140.8°, and 35 cal. at -183.06°. The coefficient of expansion of the gas is 0.003668.[28] Its viscosity is 1.2 times that of air.[29] Its solubility in water is four volumes per 100 volumes of water at ordinary temperatures.[30] The refractivity ($\mu - 1$) of argon is as shown in Table IV.

TABLE IV

Wave Length γ	$(\mu - 1) \times 10$
6439	2796
5461	2823
4359	2851

The gas has a low dielectric cohesion, namely, 38, which is about one-fifth that of air. The sparking distance is about 40 per cent greater in argon than in air under similar conditions. The gas is diamagnetic. The thermal conductivity is $K = 0.00003894$ at 0° C. Argon gives two sets of spectrum lines, depending upon the pressure the gas is subjected to. For the sake of convenience, the two sets are called the red and blue spectrum. With a spark-gap and Leyden jar the blue spectrum is given.

Argon has two isotopes with mass numbers 40 and 36.

Neon

Neon is found terrestrially principally in the atmosphere. Watson's estimate is probably the most accurate [31] and is figured about one volume of neon in 55,000 volumes of air. Neon has been recognized in the gases from a number of thermal springs. Recently it has been detected in certain natural gases from Kansas and Texas which also carry helium.

Neon has usually been prepared by extracting the helium and neon together from the atmosphere. A convenient method of getting a volume of nitrogen enriched with neon and helium is to take the gas escaping from an ordinary Hampson liquefying machine and return it to the intake of the compression pump. In this manner, the gas is partially liquefied and the more volatile constituents, namely, neon and helium, become concentrated in the gas given off. A somewhat similar result can be obtained by reliquefying all of the gas escaping from the air liquefier, and if a current of air is now brought through this liquid, the first portion that evaporates will be extremely rich in helium and neon.

The enriched gas must then be freed from nitrogen and traces of oxygen in the usual manner of passing over heated metallic copper to remove the oxygen, and over a heated mixture of magnesium and lime or over calcium carbide in order to remove the nitrogen. The mixture of neon and helium can be separated by means of liquid hydrogen, or more cheaply by fractionation from charcoal at the temperature of liquid air. In this latter process, the neon is more readily adsorbed by the charcoal than the helium, and it is possible to get a reasonably pure sample of neon in this manner. By means of a modification of his apparatus for the fractionation, condensation and distillation of liquid air, Claude has been able to obtain the neon, helium and hydrogen which exist in the air in considerable quantities. The gas thus obtained contains about 50 per cent nitrogen.[32] The two gases may also be separated rather easily by diffusion.

Neon is now extracted from the air as a by-product in the obtaining of oxygen by liquefaction methods. During the process there is a partial separation of the helium and neon so that the two gases are in the mixture in approximately equal proportions. Whenever a demand for oxygen for blast furnaces becomes an accomplished fact, the tonnage of oxygen produced will be so large that the helium and neon in the atmosphere will become by-products which will give a considerable volume of these gases available from this source. At the present time no use for neon has been developed which would take care of such a supply. It is used to some extent for lighting purposes, and owing to its low dialectric cohesion is used in spark plug testers and for similar purposes.

The absorption coefficient of neon in water is 0.0114 at 0° C., and increases up to 0.0317 at 50°.[33] The density of the gas, according to Watson, is 10.08 (O = 16) the weight of a normal liter being 0.9002. The atomic weight is taken as 20.2. The compressibility coefficient at 0° between 0 and 1 atmosphere is − 0.00105, and neon, therefore, like hydrogen is an ultra-perfect gas.[34] The refractivity of neon is 1.00006716 at 0° and 760 mm. for the green mercury line. The dispersion at N. P. T. is about twice that of helium. The viscosity of the gas at O° C. is 2.981×10^{-4} in absolute units.[35] The dielectric cohesion of neon at 17° is 5.6; hydrogen being 205. The dielectric cohesion, therefore, is less than helium, whereas one would expect it to be between that of helium and argon. Very small traces of impurities raise the dielectric cohesion of neon considerably.

Neon can be liquefied and even solidified by means of liquid hydrogen. Its boiling-point is − 245.92° C. Its triple-point pressure is 35 cm. of mercury, and its critical pressure is 29 atmospheres. According to Onnes and Crommelin[36] the critical temperature of neon is approximately 42° to 45° absolute. The surface tension, density and refractive index of liquid neon have not yet been measured, but it has been calculated that the refractive index should be about 1.24. Neon has two isotopes with mass numbers 20 and 22.

Neon has a very characteristic spectrum, the most prominent visual lines being in the red and yellow. The wave lengths of the more prominent lines are as follows (Table V):

TABLE V

Wave-length	Intensity	Wave-length	Intensity	Wave-length	Intensity
6402.40	10	6183.79	10	5974.73	6
6383.15	8	6143.28	10	5944.91	10
6328.38	6	6128.63	8	5882.04	8
6304.99	8	6096.37	10	5852.65	20
6266.66	10	6074.52	10	5764.54	8
6217.50	8	6030.20	10	4259.53	6
6182.37	10	5975.78	8		

Krypton

Krypton occurs in the atmosphere in the proportion of one part in 20,000,000.[37] It has been found spectroscopically to exist in certain mineral springs, and also in certain natural gases in Kansas and Texas.

The original amount of krypton which Ramsay and Travers prepared had a volume of 11 cc. and was not by any means pure as it contained appreciable quantities of Xenon. As far as the writer is aware pure krypton has only been prepared once and this was by R. B. Moore.[38] The oxygen residues obtained in a Claude air liquefying plant were used after about 100 tons of air had passed through the plant. Fifty liters of oxygen left were allowed to evaporate slowly from Dewar flasks; the gas passing through a glass spiral dipping in liquid air. Under such treatment the oxygen and traces of nitrogen were not liquefied and passed on, but the krypton and xenon contained in the oxygen were solidified in the spiral. The latter was then removed from the liquid air and the gases thus obtained collected over mercury and fractionated from each other. This is possible since the vapor pressure of krypton at the temperature of liquid air (17 mm.) is nearly 10 times that of xenon. If fractionation takes place at the temperature of liquid air since both elements are solid, and the krypton must be slowly pumped from the xenon by means of a Töpler pump. The whole proceeding is tedious and involves a very considerable amount of work. 300 cc. of krypton was obtained and an almost equal volume of xenon. A good portion of the krypton was practically pure and the same statement applies to a part of the xenon. These samples were used for density determinations on which the present atomic weights are based.

Krypton and xenon can also be obtained by using charcoal cooled to − 120 C. as an absorbent medium. When enriched oxygen as used by Moore, is passed over charcoal at this temperature, the krypton and xenon are adsorbed but most of the other gases pass on. Careful refractionation is necessary.

Krypton is a colorless, odorless and tasteless gas. The density has been determined by Moore, using the best fractions obtained in the work described above. The value obtained was 41.506; oxygen being 16. The weight of a normal liter of krypton is, therefore, 3.708 g.[39] The compressibility of krypton was determined by Ramsay and Travers, and they showed that the value of pv decreases markedly with increase of pressure. Between 0 and 1 atmospheres, the compressibility is +0.00215

The solubility of krypton in water has been determined, the absorption coefficient is 0.1249 at 0°, 0.0788 at 20°, and 0.0823 at 50°. The viscosity at 10.6° C. is 1.361 times that of air; at 0° in absolute units, it is 2.344. Krypton is easily liquefied; it boils at – 151.7° C. The density of the liquid at its boiling-point is 2.155. The critical temperature of the liquid is – 62.5° C.; and its critical pressure is 54.3 atmospheres.[40] Krypton can easily be solidified, and the solid melts at – 169° C. The specific heat of krypton has not been determined. The ratio of the specific heats was determined to be 1.689. Krypton undoubtedly has a monatomic molecule, and the atomic weight is, consequently, identical with the molecular weight and equals 82.92. This figure is based on Watson's theoretical correction of Moore's density determination. Krypton probably has six isotopes whose mass numbers are given as 78, 80, 82, 83, 84 and 86.[41]

The spectrum of krypton depends materially upon the nature of the discharge. With the direct discharge in a Geissler tube, the number of lines is not large and is characterized mainly by a bright-green and a yellow line. When a spark-gap is used with a Leyden jar, a much larger number of lines show up in the blue. The aurora borealis gives a spectrum with a persistent green line; it was at one time thought that this was due to the presence of krypton in the upper regions of the atmosphere and the green line in question was the krypton green. It is now known however, that this line is due to other causes. The data shown in Table VI represent the principal lines in the two spectrums:

TABLE VI

FIRST KRYPTON SPECTRUM (DIRECT DISCHARGE)

Wave-length	Intensity	Wave-length	Intensity
5871.12	10	4454.12	10
5570.50	10	4400.11	6
5562.45	6	4376.33	10
4671.40	10	4362.83	9
4624.48	10	4319.76	10
4502.56	9	4318.74	8
4501.13	7	4274.15	10
4463.88	10		

SECOND KRYPTON SPECTRUM (CONDENSED DISCHARGE)

Wave-length	Intensity	Wave-length	Intensity
5633.17	6	4109.38	6
4765.90	6	4098.89	7
4739.16	7	4088.48	8
4619.31	6	4065.22	8
4577.40	6	4057.17	8
4475.18	7	3994.98	6
4355.67	10	3920.29	8
4293.10	6	3917.76	6
4145.28	6	3906.37	8

Xenon

Xenon exists in the atmosphere to the extent of one volume in 190,000,000. It is also found in very limited quantities in the gases evolved from certain mineral springs[42] and from certain natural gases from Kansas and Texas. The original volume of xenon obtained by Ramsay and Travers was only 3 cc. With this very small quantity of moderately impure gas they determined the first physical constants of the gas. R. B. Moore obtained xenon for the first time in reasonable quantity as described under krypton.[43] In fractionating xenon from krypton the latter can be gradually pumped off from the mixed solids since its vapor pressure is 10 times that of xenon. Finally a point is reached when the material seems to be pumped dry and no more krypton will come off. At this stage it is necessary to warm up the gases and thus vaporize them and then solidify them again by means of liquid air. A fresh surface is thus exposed and a small amount of krypton can then be pumped off. In this manner pure xenon can be obtained. F. W. Aston obtained considerable quantities of xenon by absorption with charcoal from oxygen residues rich in xenon.

It is colorless, tasteless and odorless gas. Its density as determined by Moore,[43] is 65.35 (O = 16). Therefore, a normal liter of xenon weighs 5.815 g. Xenon is reasonably soluble in water; its absorption coefficient being 0.2189 at 0° C. and 0.0878 at 50° C. A minimum solubility occurs at about 40° C.[45] The viscosity of xenon at 0° C. is

$$N_0 = 2.107 \times 10^{-4},$$

while its refractivity is $\mu - 1 = 705.49 \times 10^{-6}$ for the green mercury line ($\lambda = 5461$). Xenon can be easily liquefied and solidified.

The liquid is colorless and boils at —109.1° C. under atmospheric pressure. Its density at the boiling-point is 3.06 and, therefore, the molecular volume is 42.7. The critical temperature of xenon is +16.6° C., and the gas may be liquefied at this temperature by a pressure of 58.2 atmospheres. The melting-point of solid xenon is —140° C. Its atomic weight is the same as its molecular weight, namely, 130.2 (O = 16).[46] This figure is based on Watson's theoretical determination from Moore's density determination.

A Geissler tube containing xenon emits a blue light when a current is passed through it, and an examination of the spectrum shows that the principal lines are in the blue of the spectrum. If a spark-gap and Leyden jar be used, the color of the tube changes to green. The following (Table VII) are the principal lines in the two spectra:

TABLE VII

FIRST XENON SPECTRUM (UNCONDENSED DISCHARGE)

Wave-length	Intensity	Wave-length	Intensity	Wave-length	Intensity
4923.28	6	4624.46	15	4116.25	7
4916.63	6	4582.89	5	4109.84	5
4807.19	6	4524.83	6	4078.94	10
4734.30	8	4501.13	10	3967.74	10
4697.17	7	4193.70	8	3951.16	10
4671.42	10				

SECOND XENON SPECTRUM (CONDENSED DISCHARGE)

Wave-length	Intensity	Wave-length	Intensity	Wave-length	Intensity
5439.19	8	4603.21	10	4330.63	15
5419.40	10	4585.65	10	4245.54	10
5372.62	8	4545.34	8	4238.37	10
5339.56	9	4541.03	8	4193.25	8
5314.15	8	4462.38	20	4180.20	10
5292.40	10	4448.28	10	3950.70	8
4862.69	8	4395.91	10	3922.67	10
4844.50	10	4393.34	10		

Niton

This is the name given by Ramsay to radium emanation. In very many respects its properties are similar to those of the rare gases and therefore it really belongs to this group. The density was determined experimentally by Ramsay and Gray as 111.5 (H=1) which gives 223 as the atomic weight of niton. The gas is soluble in water and its critical temperature is 104.5°; its critical pressure is 62.5 atmospheres. It boils at - 62°, and freezes at - 71° C.

REFERENCES

1 Proc. Roy. Soc., 1893, Vol. 53, p. 146; 1894, Vol. 55, p. 340.
2 Chem. News, 1895, Vol. 72, p. 89.
3 Zeit. physik. Chem., 1896, Vol. 19, p. 657.
4 Proc. Roy. Soc., 1905, Vol. A, 76, p. 111; 1908, Vol. A, 80, p. 599.
5 Trans. Chem. Soc., 1910, Vol. 97, p. 810.
6 Strutt, Roy. Soc. Proc., A, Vol. 80, 1908, p. 572. *Ibid.*, A, Vol. 81, 1908, p. 278.
7 Jour. Amer. Chem. Soc., 1907, 29, 1524.
8 Compt. rend., 155, pp. 197–200, 1912.
9 Hamburger Wiss. Inst., 28, 1910.
10 Proc. Nat. Acad. Sci., 11, 772–9 (1925).
11 Ramsay and Travers, Proc. Roy. Soc., 1898, Vol. 62, p. 316.
12 Trans. Chem. Soc., 1910, Vol. 97, p. 810. It is 1.999, according to Heuse (Ber. deutsch. physikal. Ges., 1913, Vol. 15, p. 518).
13 Jaquerod and Scheuer, Mem. Soc. Phys. Nat., 1908, p. 35.
14 Proc. Roy. Soc., 1910, Vol. A 83, p. 414.
15 Schware, Ann. Physik., 1903, Vol. 11, p. 303.
16 Proc. Roy. Soc., 1901, Vol. 67, p. 331.
17 Ann. Physikal, 1910, Vol. 6, p. 302.
18 Onnes, Comm. Phys. Lab. Leyden, No. 119.
19 Onnes, Proc. K. Akad. Wetensch, Amsterdam, 1911, Vol. 14, p. 678.
20 Schloesing, Comp. rend., 1895, Vol. 121, p. 605; Kellas, Proc. Roy. Soc., 1896, Vol. 59, p. 66.
21 Ramsay and Travers, Proc. Roy. Soc., 1897, Vol. 60, p. 442.
22 Zeit. Elektrochem., 1908, Vol. 14, p. 585.
23 Lockyer, Proc. Roy. Soc., 1896, Vol. 60, p. 133.
24 Tschernik, J. Russ. Phys. Chem. Soc., 1897, Vol. 29, p. 291.
25 Crommelin, Proc. K. Akad. Wetensch, Amsterdam, 1910, Vol. 13, p. 607.
26 Crommelin, *ibid.*, p. 54.
27 Crommelin, *ibid.*, p. 54.
28 Kuenen and Randall, Proc. Roy. Soc., 1895, Vol. 59, p. 60.
29 Rayleigh, Proc. Roy. Soc., 1896, Vol. 59, p. 198.
30 Rayleigh and Ramsay, Phil. Trans. 1895, Vol. 186, p. 187.
31 Moureu and Biquard, Compt. rend., 1906, Vol. 143, p. 180; 1908, Vol. 146, p. 435.
32 Claude, Compt. rend., 1905, Vol. 141, p. 823; 1908, Vol. 147, p. 624.
33 Antropoff, Proc. Roy. Soc., 1910, Vol. A 83, p. 474.
34 Burt, Trans. Faraday Soc., 1910, Vol. 6, p. 19.
35 Rankine, Proc. Roy. Soc., 1910, Vol. A 83, p. 516; Vol. A 84, p. 181.
36 Proc. K. Akaa. Wetesnch, Amsterdam, 1915, Vol. 18, p. 15; 1911, Vol. 14, p. 163.
37 Ramsay, Proc. Roy. Soc., 1903, Vol. 71, p. 424; 1908, Vol. A 80, p. 599.
38 Trans. Chem. Soc., 1908, Vol. 93, p. 2181; Proc. Roy. Soc., 1908, Vol. A 81, 195.
39 Trans. Chem. Soc., 1908, Vol. 93, p. 2181.
40 Ramsay and Travers, Proc. Roy. Soc., 1901, Vol. 67, p. 329; Phil. Trans., 1901, Vol. 197, p. 47.
41 Aston, Phil. Mag. 39, 623 (1920).
42 Moureu and Lepape, Compt. rend., 1909, Vol. 149, p. 1174.
43 Proc. Roy. Soc., 1908, Vol. A 81, p. 195.
44 Trans. Chem. Soc., 1908, Vol. 93, p. 2181. Watson, *ibid.*, 1910, Vol. 97, p. 833.
45 Antrapoff, Proc. Roy. Soc., 1910, Vol. A 83, p. 474.
46 Moore, Trans. Chem. Soc., 1908, Vol. 93, p. 2181.

Synthetic Organic Chemistry

BY

E. EMMET REID
PROFESSOR OF ORGANIC CHEMISTRY
JOHNS HOPKINS UNIVERSITY

Lecture delivered at Columbia University in the special course on Contemporary Developments in Chemistry given in the Summer Session of 1926 on the occasion of the opening of the Chandler Chemical Laboratories

COLUMBIA UNIVERSITY PRESS
NEW YORK
1927

Printed from type. Published March, 1927

Printed in the United States of America

SYNTHETIC ORGANIC CHEMISTRY

By E. EMMET REID

This is the sesquicentennial of American independence and the semi-centennial of the American Chemical Society. Synthetic organic chemistry comes just about half way between these two, as it began about the time of Wöhler's famous synthesis of urea in 1832. Of all compounds benzene has been the most important as a starting point in syntheses. It was discovered by Faraday in 1825. This is a good time to pause a moment and take stock of the achievements of a century. Some of these have been spectacular triumphs as the synthesis of indigo by Baeyer, of alizarine by Graebe and of the sugars by Fischer. Sometimes a chemist has aimed at one thing and has hit something much bigger and has been quick witted enough to take full advantage of an unforeseen event, as Perkin's attempt to make quinine by the oxidation of aniline. Remsen discovered the freak saccharin in the course of a systematic investigation of sulphonic acids. Gomberg fell upon triphenyl-methyl in trying the Fittig synthesis on triphenylmethyl bromide.

By far the greater part of what we know as organic chemistry has been laboriously worked out by systematic, long continued investigation. An army of chemists has taken part in this conquest. The decennial index to Chemical Abstracts lists about seventy thousand names. True, all of these are not organic chemists. Many of the articles listed by our argus-eyed editors are not strictly on research and many of the authors belong to other divisions of our science, but even so the tillers of the field of organic chemistry are numbered by the thousands and these are the successors of other thousands that have gone before.

What has been accomplished by this army? Have the objectives gained been worthy of the efforts put forth? We may point to Richter's Lexikon in which are listed 144,150 compounds as having been made up to Jan. 1, 1910. In the 15 years since then, at least a hundred thousand have been added, so that we may now claim 250,000 compounds that have been studied. Some of those who are now students will probably live to see the number pass the million mark. Mere numbers tell little of the triumphs of synthetic organic chemistry since, in Richter, an ester which a student may put together in an afternoon counts just as much as indigo, the conquest of which required the efforts of a number of the most brilliant chemists for many long years. The variety of the organic compounds that have been made is more significant. We have acids comparable to nitric, bases almost as strong as caustic potash, compounds of carbon, hydrogen, oxygen and nitrogen far more toxic than mercury or arsenic, substances hundreds of times as sweet as sugar, odors by the score as delicate as that of the heliotrope, pharmaceuticals of all degrees of potency and dyes of every hue of the rainbow.

Of even more importance is the elaboration of the system of organic chemistry, so comprehensive that each of the 250,000 compounds finds its proper place in it, so extensive that niches are provided for millions of compounds that will be prepared by chemists yet unborn, so simple that any one of the hundreds of thousands of compounds can be quickly and surely located, and so orderly and knit together that the relations of any individual compound to all the others are directly shown. Not only are places provided for new compounds but methods of making them are apparent and the properties that they will have are indicated. In making these 250,000 compounds, methods have been perfected which can be applied, with slight modifications, to the preparation of many millions more.

As an achievement of the human mind organic chemistry is scarcely equalled by any other. Nowhere else can we find such an extensive collection of ascertained facts so well classified and coordinated. Nowhere else can we find so much knowledge so well organized and hypotheses so thoroughly verified down to minute details. Organic chemistry is a curious dual structure, both parts created by man, one elaborated in his mind and the other constructed by his hands. On the one side we have the most intricate structure of theories, equations, and molecular formulae, a picture gallery of molecules, X-ray pictures of their skeletons, pictures of things the eye has never seen and never can see. On the other side corresponding to these phantoms are realities, hundreds of thousands of organic compounds, each with its color, its odor, its solubilities, its reactivities and other properties. Each detail of the one structure has its counterpart in the other; each reproduces, explains and verifies the other. The two have been built up nearly simultaneously, at one time theory has progressed faster than practice and formulae of compounds have been written out before they were made; at another time, as in the case of the tartaric acids or of fumaric and maleic acids, more compounds have been discovered than could be accounted for by the then existing theory. Sometimes compounds have been made to fill out the predictions of the theory and sometimes the theory has had to be extended to account for compounds already in hand.

It is not necessary to recount in this place the contributions of synthetic organic chemistry to the material wealth of mankind. Drugs, dyes, rayon, bakelite, perfumes, nitro-glycerine, dynamite, smokeless powder, picric acid, T.N.T., lacquers and celluloid are only a few of its creations which have become necessary to our civilization.

Synthetic organic chemistry has taken on a new meaning within the last few years. The introductions to our text-books on organic chemistry contain sections on "sources of compounds." The distillation of coal, wood and bones are mentioned along with fermentation and other natural processes as furnishing us with a large number and variety of compounds from which, by various transformations, many others can be prepared. It is always pointed out that the name organic has lost its original significance and that we are no longer dependent on organisms for our compounds. It is shown that the compounds of carbon can be made from dead matter. Equations are given to show how methane can be formed by passing hydrogen sulphide and carbon disulphide over hot copper:

$$CS_2 + 2\ H_2S + 4\ Cu \rightarrow 4\ CuS + CH_4$$

It can also be made by the action of water on aluminum carbide but no one ever thought of preparing a thousand cubic feet of methane by these reactions. They were given to show that methane can be made from the elements as iron sulphide and copper sulphate can be.

Methane can now be made in quantities by the catalytic hydrogenation of carbon monoxide. The text-books have told us that traces of liquid hydrocarbons are set free when cast iron is dissolved in dilute acid. Liquid hydrocarbons can now be obtained by the hydrogenation of carbon monoxide with a suitable catalyst. In Germany the Bergius process for hydrogenating coal to give a mixture of liquid hydrocarbons similar to petroleum seems to be coming into large scale production.

Calcium carbide, from the electric furnace, yields acetylene which is combined with water, by the aid of a mercury catalyst, to form acetaldehyde. This is either reduced to alcohol, oxidized to acetic acid, or transformed to ethyl acetate. These are no longer text-book reactions given to show that it is possible to effect these syntheses, but actual manufacturing processes by which tons of the products are being turned out.

We can now make sodium cyanide direct from coke, nitrogen and soda ash. Calcium carbide and nitrogen give cyanamide which is a starting material for the preparation of a great variety of organic nitrogen compounds among them urea and cyanides.

The hydrocarbons of petroleum and of the gases from cracking stills are now being turned into alcohols, acids and other useful oxygenated compounds.

The latest triumph has been Patart's methanol process by which this alcohol is manufactured from water gas. Along with the methanol, higher alcohols up to hexyl are obtained. By modifications of this process these alcohols are produced in large amounts. Thus steady progress is being made toward a truly synthetic organic chemistry. Synthetic will mean what it says. It is not only possible to make a few grams of these substances in the laboratory but it is now practicable to manufacture tons. We may look forward to the time when many organic compounds will be made in plants rather than obtained from plants.

The great synthesis to which we all look forward is producing formaldehyde, or its equivalent, from carbon dioxide and water. The need of our present civilization is energy, energy to run our factories, our trains, our ships, our automobiles and more energy to keep our houses warm. A man requires 2000 Calories per day to support his internal activities and keep his body warm and about 2000 more when he does hard work. In ancient times this was about all the energy that he consumed but now energy for the things mentioned above and for many others has to be added to the modest 4000 Calories. If we plot the per capita energy requirements we will get an almost straight line for many centuries but about 100 years ago this starts to curve upward. The curvature increases from year to year. Where will this curve go in the future? We get the energy for our body processes from food. The energy for our machinery and for heating our houses comes almost entirely from the burning of carbon compounds, wood, coal, or petroleum. Wind mills and hydro-electric works furnish a very small percentage of the total energy required. Smelting processes are nearly all dependent on coal for heat and for taking oxygen away from the ores. It is not usual to think of the steel industry as dependent on organic reactions but so it is.

It is calculated that the fuel supply in sight will last for 1000 years at the *present* rate of consumption. A thousand years looks long to the individual but is short compared with the history of the race. But suppose the rate of energy consumption continues to increase exponentially as it has done in the last century. What then? Every stick of wood, every lump of coal, possibly every drop of petroleum is here as the result of what goes on in the green leaf under the influence of sunlight—of the photosynthesis of carbohydrates. Formaldehyde, CH_2O, is the *simplest* carbohydrate—many think it is the *first* carbohydrate to be formed from carbon dioxide and water, an endothermic reaction the energy for which comes from sunlight. So far as energy is concerned the hope of the future appears to depend on organic chemists finding a way to cause this reaction to take place in chemical plants rather than in Nature's plants.

We need better understanding of reactions and better means for carrying them out. What has been accomplished in organic synthesis is marvelous but when we are questioned about the reactions of organic compounds, even the best known reactions, the ones that are employed every day in our laboratories, we have to hang our heads and confess our profound ignorance. Hundreds of articles, comprising thousands of pages, have been written on esterification, yet we can not fully explain this, one of the common organic reactions. The future must bring intensive and extensive study of organic reactions. When we look at the syntheses that we use, the Friedel and Crafts, the Wurtz-Fittig, the Perkin, etc., we must admit that they are essentially crude. We are really synthesizing hydrochloric acid, sodium chloride and water and hoping that a fair percentage of the by-products will be what we want. How far inferior are our methods to some of Nature's! We eat a steak: in a few minutes the complex proteins which it contains are taken apart, the amino acids sorted out and then recombined to form the entirely different proteins that are suited to the building up of our bodies. We have yet much to learn about synthesis.

Constitution and properties. The relation of the properties of a compound to the structure of its molecule is the central problem of organic chemistry. Every time we take a melting point, a boiling point or a density or determine the solubility of a compound in any medium we are accumulating data for the solution of this problem. All the toxicity determinations that have been made in the laboratories of the Chemical Warfare Service, all the observations that are made on the pharmacological effects of drugs, all measurements of reaction velocities, strengths of organic acids, etc., are likewise contributions to the final solution of this problem. Already a vast amount of information has been accumulated. In limited fields many regularities have been observed. In the methyl alcohol series the toxicity increases at a regular rate while the solubility in water diminishes. If we write down a structural formula we can calculate the molecular weight and percentage composition, if it belongs to a well known series or group we can approximate its boiling point, and guess at its melting point. Usually we can say whether it will dissolve in water or not. We can describe its chemical reactions with considerable confidence. Frequently we can predict color, pharmacological effects, toxicity, odor and taste. Some day in the future we may be able to calculate the properties of a compound as now we calculate the molecular weight.

The highest development of synthetic organic chemistry comes in designing and constructing molecules for specified purposes. We have two important professions, architecture and engineering, which have to do with the design and construction of buildings of wood, stone, bricks, cement and steel. Engineering has been reduced to an exact science. Formulae have been worked out for all sorts of structures. To calculate the safe load for a bar we must know its length, its cross section and make certain determinations on samples of the material of which it is to be made. We simply substitute these in the proper formula and run the slide rule back and forth a few times and have the answer. If a bridge is to be constructed of a certain length to carry a certain load, the engineer draws his picture, figures out the strain and stress on every member and then calculates the necessary dimensions of each piece. It is not a case of making a free hand drawing and guessing at the size of the pieces to be put in. Each piece is fabricated according to the blue print. The pieces are assembled and bolted together and the bridge is opened for traffic. No one doubts that it will carry the load for which it is designed. Man has been building houses and bridges a long time and has accumulated the data required by the engineer.

We have been designing molecules for a short time only and our methods are correspondingly crude. We are greatly handicapped by the fact that so few chemical properties are capable of numerical expression. Odors and tastes can be described only in words. Suppose the problem is to produce a new compound having an odor of a certain type. The usual method is to study the constitution of the natural substance that has the desired odor and to build up something containing as nearly as possible the same groups. Ionone, the artificial odor of violets, is the best example of this method. It is isomeric with irone the natural odor of violets. Following this method we have all the labor of a tedious synthesis and have no idea till the purified substance is tested whether it is of any value. What is more discouraging is that we have no means of knowing by how much our compound falls short. Coming back to the bridge illustration, suppose the fabricator delivers a girder and is simply told that it is not of the right size but is given no information as to how much too long or too short it is. Suppose he were asked to send another one a little different and keep this up until one fitted. With our present incomplete knowledge we must synthesize one compound after another and test each one until we find one with a pleasing odor or one with a desirable pharmacological action.

In one field of organic chemistry, namely that of dyes, we have made so much progress that safe predictions can be made. We have amassed so much information as to the relation of the color and other properties of dyes to their constitution that the synthesis of new dyes is no longer guess work. This fortunate situation in which the dye chemist finds himself is due to two things. More study has been put on the relation of constitution to color than to any other property. Color chemistry has been more intensively studied than any other branch of chemistry. The other reason is that colors can be objectively studied, and their wave lengths and intensities can be measured and recorded in figures. Numerical values take the place of adjectives. If a phenyl group is introduced into a certain dye we can say that the absorption band has been moved from wave length so and so to such another value. Changes in constitution can be followed quantitatively.

Of course the field of dye chemistry has not been completely covered. In fact only limited areas have been at all well worked out. If we take any well known dye, such as malachite green, we can say pretty definitely what effect the introduction of methyl, ethyl or phenyl groups in certain positions will have. An inspection of the formula of a dye will tell whether it is probably fast or not, whether it will stick to cotton without a mordant and so on.

To evaluate the effects of all possible groups is a large task and requires the accumulation of numerous data. This kind of investigation is suited to university laboratories. For several years at the Johns Hopkins the effect of groups $—SCH_3$ and $—SO_2CH_3$ has been studied. In some cases ethyl, propyl, phenyl and other radicals have been used instead of methyl in these combinations. The results have been published:

Waldron and Reid, J. Am. Chem. Soc. **45**, 2399 (1923)
Foster and Reid, *Ibid.* **46**, 1936 (1924)
Palmer and Reid, *Ibid.* **48**, 528 (1926)
Holt and Reid, *Ibid.* **46**, 2329 and 2333 (1924)

Reference is made to these articles for the details of the preparations and for facts about the colors stated in words. The correct impression can be obtained only by an inspection of the colors which are shown by the skeins which are exhibited.

The first three references have to do with the effect of these groups on azo dyes, the groups being in the para, ortho and meta positions to the azo grouping. The work of Holt had to do with the same groups in malachite green and in phenolphthalein. The groups mentioned above have been contrasted with methyl, —CH_3, and methoxyl, —OCH_3, in the same positions. The bases $NH_2C_6H_4CH_3$, $NH_2C_6H_4OCH_3$, $NH_2C_6H_4SCH_3$ and $NH_2C_6H_4SO_2CH_3$ have been diazotized and coupled with a number of standard intermediates, salicylic acid, Schaeffer's salt, Neville and Winther's L-acid, R-salt and chromotrope. In addition to these single amines the double amines, benzidine, $NH_2C_6H_4{\cdot}C_6H_4NH_2$, thio-aniline, $NH_2C_6H_4SC_6H_4NH_2$, the corresponding sulphone, $NH_2C_6H_4{\cdot}SO_2{\cdot}C_6H_4NH_2$ and one which was prepared from "mustard gas,"

$$NH_2C_6H_4SCH_2CH_2SCH_2CH_2SC_6H_4NH_2,$$

have been diazotized and coupled to the same intermediates. The results are seen by comparing the colors of the dyed skeins. The effects of the groups are more pronounced and also more consistent in the para series. Here —SCH_3 is more batho-chromic than —OCH_3 which is stronger than —CH_3 while —SO_2CH_3 is decidedly hyposochromic. The groups are in the same order in the meta series but there effects are less intense. In the ortho series —OCH_3 appears to be stronger than —SCH_3.

In the —SCH_3 group the sulphur dominates. Changing the methyl to ethyl, propyl, butyl or even phenyl has little if any effect.

PERMEABILITY AND ELECTRIC PHENOMENA IN MEMBRANES

BY

LEONOR MICHAELIS

PROFESSOR OF BIOLOGICAL CHEMISTRY, UNIVERSITY OF BERLIN,
AND PROFESSOR OF RESEARCH MEDICINE, JOHNS HOPKINS UNIVERSITY

Lecture delivered at Columbia University in the special course on Contemporary Developments in Chemistry given in the Summer Session of 1926 on the occasion of the opening of the Chandler Chemical Laboratories

COLUMBIA UNIVERSITY PRESS

NEW YORK

1927

Printed from type. Published March, 1927

Printed in the United States of America

PERMEABILITY AND ELECTRIC PHENOMENA IN MEMBRANES

by

LEONOR MICHAELIS

1. Sieve and Homogenous Membranes

For the purpose of this lecture, let us define a membrane as a thin layer of a substance of a certain solidity separating two liquids from each other, and of such quality as to prevent the two liquids from mixing when shaken or stirred, or in other words, such as to prevent *convection,* while allowing diffusion of some or even all kinds of the molecules contained in the solutions. When only the one side of such a membrane is in contact with a liquid and the other with the air a membrane may function also as a filter. However, this case is not to be the proper subject of this lecture though closely related to it. The simplest way of realizing a membrane is the idea of a sieve the pores of which are narrow enough to prevent the convection. According to the size of the pores, larger or even smaller particles representing components of the liquids, are prevented from convection across the membrane. When the pores are narrow enough even some kinds of the dissolved molecules will not penetrate, and in this case it may happen that the membrane functions as an osmometer allowing the solvent to go through the membrane to such an extent that the difference of the levels compensates the osmotic pressure of the impermeable molecules. In fact, most of the membranes may be represented by sieves. However, there are membranes also of a homogeneous structure. Much has been said about the so-called lipoid membranes which are supposed to be a homogeneous membrane of a substance such as oil, wax, lecithine, etc., separating two aqueous solutions and being impermeable for any substance insoluble in the lipoid membrane substance. Though it must not be doubted that a membrane may work by this mechanism of selective solubility and though many instances have been shown by Overton and H. H. Meyer in membranes of living cells, yet the frequency of the occurrence of a membrane of this type without any sieve effect seems to me to be overestimated. Recently Kahlenberg described a selective dialysis through a membrane of stretched rubber dam. When different substances are dissolved in pyridine and separated from pure pyridine by a rubber membrane, only such substances are prevented from dialysing which are difficultly soluble in lipoids but easily soluble in water. This seems to be an example of a homogeneous membrane. However, it should be borne in mind that even in a rubber membrane the selective permeability of this kind occurs only in pyridine solutions. This membrane is not able to separate the easily water-soluble and the easily lipoid-soluble substances when these are dissolved in water. Later on I shall describe a kind of rubber membrane which quite certainly behaves like a sieve. Even in the pores of a sieve the so-called lipoid substances may sometimes behave differently from the water-soluble substances, because the lipoid substances in general are capillary active at the same time and have different properties with respect to the wetting of the walls of the pores. So, the sieve nature of certain membranes is quite obvious, while the homogeneous nature of a membrane very often cannot be definitely proven.

2. The Size of the Pores in Sieve Membranes

A pure sieve membrane exerts its effect by the fact that the substance of the membrane, itself, is practically impermeable to all substances concerned. It is only its holes or pores or channels which allow the penetration. Some examples of membranes in which the size of the pores differs are given here. A clay diaphragm prevents any convection but allows the diffusion of practically any dissolved substance, even of many colloids. It is used in electrochemical engineering as a means of preventing the mechanical mixing of the products of electrolysis without preventing the ions of the solution from following the forces of the electric field. Much narrower are the pores of parchment paper and of certain dried animal membranes such as gut, and pig's bladder, which are impermeable for colloids, or better, the impermeability through which seemed to Graham to be the characteristic of what he called colloids. Non-colloidal molecules permeate through these membranes and the presence of the membrane does not appreciably alter their rate of diffusion although, later on, we will discuss some smaller effects even in this regard. A third range in the size of pores is represented by such membranes as copper ferro-cyanide, discovered long ago by W. Traube and used by Pfeffer as an osmometer, which led van't Hoff to evolve the classic theory of osmotic pressure. This membrane is approximately impermeable to any dissolved substances and permeable for water alone. However, the discoverer himself and recently Collander in a new series of experiments found this membrane to be permeable for smaller molecules such as potassium chloride and urea. In the series of the aliphatic alcohols the permeability decreases with increase in the molecular weight. Larger molecules such as sugar are not all able to penetrate this membrane. In a certain sense the following kind of membrane which I found recently may be placed in the same range. This is a membrane of collodion which in contrast to the well-known collodion membrane usually used for dialysis and ultrafiltration, has been completely dried so as to remove the last traces of ether or alcohol before it is brought in contact with water. Such membranes were considered as completely impermeable and useless, hitherto. Yet they turned out to be of quite a peculiar interest. In regard to the permeability to the different kinds of dissolved molecules this membrane behaves very similarly to the copper ferrocyanide membrane. Generally all diffusion through these membranes proceeds very slowly but still there are great differences in the permeability according to the size of the molecules. Table I shows the approximate relative diffusion coefficients of some substances, the first series represents data obtained with a very easily permeable collodion membrane of the old, well-known type. These coefficients may be supposed to be approximately equal to those of free diffusion without any membrane. The second series gives the relative diffusion coefficients in the dried collodion membrane. The order of the substances arranged according to their diffusibility in a very dilute aqueous solution is the same in both cases, but the differences between the substances are enormously increased in the case of the dried

TABLE I

	Methyl-alcohol	Ethyl-alcohol	Propyl-alcohol	Butyl-alcohol	Urea	α-Mono-Chlor-hydrine	Glucose
Relative diffusion coefficient for the ordinary, easily permeable collodion membrane..............	1.22	1.15	1.0	1.0	1.0	0.7	0.5
Relative diffusion coefficient within a *dried* collodion membrane..	9.2	4.1	3.0	0.82	1.0	0.07	0.000

collodion membrane. Note the great differences between methyl, ethyl, and prophyl alcohols. This membrane is completely impermeable to molecules of the size of any sugar. The only most evident contrast to the copper ferrocyanide membrane consists in that the dried collodion membrane is extremely difficultly permeable even to water, which fact certainly has something to do with the fact that dried collodion is not easily wetted by water.

3. Electric Potential Differences of the Two Sides of a Membrane

A remarkable phenomenon in membranes which is not only interesting by itself but also leads to important conclusions, is the difference of the electric potential on the two sides of the membrane when in contact with two different solutions. In general, when two solutions of different composition are in contact even without any membrane, a potential difference is established, which fact plays an important part in the determination of the electromotive force of galvanic chains in which liquid junction potentials sometimes give rise to an inconvenient complication. When the two solutions are separated by a membrane, the potential difference is, as a rule, different from the one established in free contact. Let us restrict ourselves to the most simple cases, as such simple cases wholly suffice for recognizing the essential mechanism of the membrane in this respect. There are two such simple cases: either the two solutions contain only one and the same electrolyte but in different concentrations, or the two solutions each contain a different electrolyte but in equal concentration. When there are two solutions of the same electrolyte in different concentrations the potential difference is produced by the difference of the mobility of the cation and the anion. While diffusion is taking place the more mobile ion moves ahead, charging the more dilute solution with electricity of its own sign. The potential difference thus established amounts to

$$E = \frac{u - v}{u + v} \, 0.057 \log \frac{c_1}{c_2} \text{ volts,}$$

where c_1 and c_2 are the concentrations of the two solutions, and u and v are the mobility of the cation and of the anion. There are two limiting cases. When the mobilities u and v are equal, then the potential difference is zero. On the other hand, when the mobility of one of the two ions is relatively so small as to be neglible with respect to the other, the potential difference reaches its maximum value, amounting to ± 57 millivolts if the ratio of the two concentrations is 1 : 10 (the sign depends on whether the mobility of the cation or of the anion is negligible). Generally the effect of a membrane separating such solutions is to change the potential difference from that obtained when the two solutions are in contact without a membrane. That may be interpreted by admitting the ratio of the mobilities of the two ions to be different within the pores of the membrane and in a free solution. The simplest case is a chain with two solutions of KCl, 0.1 N and 0.01 N. As the mobilities of K^+ and Cl^- are equal, there is no potential difference in free contact. Almost every membrane interposed between these solutions produces a potential difference, mostly in the sense that the more dilute solution becomes positive. Hence we are led to the interpretation that the mobility of the Cl^- ions becomes relatively smaller than that of the K^+ ions. The absolute value of the mobility being of no importance for the potential difference, the conclusion only concerns the change of the mobilities of K^+ and Cl^- relatively to each other. There is sufficient reason for assuming that the *absolute* value of the mobility is decreased by the membrane even with the K^+ ion, but that the mobility of the Cl^- ion must be decreased still more. An ordinary clay diaphragm shows only traces of such an effect. The usual large-pored collodion membrane may produce a small potential difference of 5–10 millivolts, a parchment paper membrane one of 10–15 millivolts. When one prepares a membrane of solid paraffine by impregnating filter paper with molten paraffine, the paraffine when solidifying seems to shrink a little and to form very fine ultramicroscopical pores, and a real sieve membrane arises which is permeable even to sugar; from this we obviously recognize that the permeability of a substance does not have any connection with its solubility in paraffine. Such a membrane gives a potential difference of about 13 millivolts under the conditions mentioned before. A membrane of wax prepared in the same manner gives 21 millivolts; filter paper impregnated with mastic gives 25 millivolts, with India rubber about 30 millivolts, and finally a membrane of dried collodion gives 50–55 millivolts, thus almost reaching the theoretical maximum value of 57 millivolts which is reached when the membrane is entirely impermeable for the anions. Chains with other salts give results which always lead to the same conclusion that the anions become relatively less mobile than the cations when permeating the membrane.

The impermeability for any anion, of the dried collodion membrane appears to be the realization of an old idea of Ostwald, quoted many times by biologists in order to explain certain properties of the cell membrane. This impermeability ought to be proved by direct diffusion experiments. Such an experiment is more difficult than may be anticipated, because the speed of diffusion of even the most mobile ions is so small that with membranes of sufficient thickness to secure a certain mechanical solidity some weeks are necessary to get definitely determinable amounts across the membrane, and within this period a change of permeability of the membrane often takes place which can be recognized by the fact that the maximum effect of the potential difference in the chain described before is no longer obtained. If we were able to make membranes as thin as a cell membrane and yet sufficiently mechanically resistant, the diffusion experiment would be easy. But that unfortunately is not the case. However, such an experiment which is quite reproducible in a series of selected, good membranes can be effected. When a .1 N solution of hydrochloric acid is separated from pure water by such a membrane, no acid diffuses through the membrane even in 2–3 weeks. The anions do not diffuse and do not allow the H-ions to diffuse either on account of the electrostatic attraction. However, when a solution of hydrochloric acid is separated by the membrane from a solution of potassium chloride, the potassium chloride solution becomes acid, this acidification progresses slowly but markedly from day to day. Here the H^+ and the K^+ ions are exchanged through the membrane. There is no diffusion of the hydrochloric acid in the usual meaning of the word. I hope to be able to improve the technique of such diffusion experiments in general in order to support the assumption of the immobility of the anions.

The second kind of simple chains is obtained when the two solutions each contain a different salt but in equal concentration. For the sake of simplicity let us take two salts with a common ion; e.g., HCl and KCl, or KCl and K_2SO_4. In such a case the potential difference should be

$$E = \pm\ 0.057 \log \frac{u_1 + u_c}{u_2 + u_c},$$

where u_c means the mobility of the common ion of the two salts, and u_1 and u_2 the mobilities of the two different ions. According to whether the common ion is the cation or the anion, the positive or the negative sign in this formula is valid. Such chains, simple as they may appear, yield results less easily interpreted in general, as there are three different mobilities to be dealt with. The conditions may be easily understood only in the case of the dried collodion membrane, as the mobilities of the anions may be set = o. Now, when a *cation* is the common ion of the mobility u_c, then $u_1 = u_2 = 0$, and hence

$$E = 0.057 \log \frac{0 + u_c}{0 + u_c} = 0.$$

This calculation is practically confirmed by experiment for any combination of salts with any kind of ions provided the cation is the common ion. In opposition to the case when an *anion* is the common ion and the cations are different and possess the velocities u_1 and u_2, the potential difference becomes

$$E = 0.057 \log \frac{u_1}{u_2} \text{ volts.}$$

So we may approximately compute the relative mobilities of the two cations within the membrane, from the measurements of the potential difference, of course only when using a membrane in which the mobility of the anion has actually proven to be negligible by the measurement of a chain of the first type. Table II shows the relative values of the mobility of the univalent alkali

TABLE II

	Li^+	Na^+	K^+	Rb^+	H^+	Ratio $Li^+ : H^+$
Relative mobility in a simple aqueous solution..........	0.52	0.65	1.00	1.04	4.9	1 : 9.4
Relative mobility within the dried collodion membrane...	0.048	0.14	1.00	2.8	42.5	1 : 890

metal ions. The first line gives the values for the free diffusion as known for a long time by the determinations of transfer numbers. The second line shows the relative values of the mobility within the membrane. Obviously the order of increasing mobility is the same, but the differences in the mobilities of each two members of the series is largely increased in the membrane. The greatest difference is the one between Li^+ and H^+. The ratio of their mobilities in free diffusion is 1 : 10, but in the membrane almost 1 : 1000. Even the ratio of Na^+ and K^+, which in free diffusion is 1 : 1.5 reaches in the membrane about 1 : 7. Some dried collodion membranes have also been found in which the effects are somewhat smaller.

All of the kinds of membrane give the common effect of retarding the anions. When we consider all these membranes we become aware of the fact that all of them consist of a material which under different experimental conditions has proved to be electronegative against most of the aqueous solutions. But there are also electropositive membranes. It is very remarkable that we do not know of any membrane which is electropositive against an aqueous solution under all or at least most conditions but only of amphoteric membranes such as membranes of proteins, casein or coagulated albumin, alumina, animal membranes consisting of gelatine—or protein-like substances. Such membranes take a positive charge in acid solutions, when the hydrogen ion concentration is greater than the one of the isoelectric point of the ampholytic substance. It is a remarkable coincidence that the influence of these membranes on the diffusion potential when in contact with two different solutions is also shifted when the hydrogen ion concentration passes from the one side of the isoelectric point to the other. This has been pointed out by Höber and Mond for diaphragms of casein or globulin, by Fujita in my laboratory for gelatine and coagulated egg albumin. The effect of such membranes investigated by the method of measuring the diffusion potential is such as to lead to the conclusion that these membranes retard the anions when in contact with a solution on the alkaline side of the isoelectric point and retard the cations when on the acid side of it. Another example of the reversion of the properties in an amphoteric membrane was shown by Girard a long time ago. When barium chloride is separated by an animal membrane from pure water and the barium chloride solution is slightly acidified, the Cl^- diffuses much faster through the membrane than Ba^{++} does. But when the solution is slightly alkaline, the Ba^{++} migrates faster than Cl^- does.

These effects on the diffusion potential differences may be explained as follows. It has been known for a long time, since electroendosmosis was experimentally investigated by Quincke and theoretically explained by Helmholtz, that these membrane-forming substances take an electric charge when in contact with an aqueous solution, this charge generally being negative but in some cases also positive. Nowadays we represent the electric double layer at the boundary of two phases as two sets of ions. Take for instance collodion which is negatively charged against an aqueous solution. We presume that the collodion absorbs a layer of negative ions from the solution, hydroxyl ions or even other anions, and thus by electrostatic attraction, causes the positive ions of the solution to form another layer adjacent to the absorbed layer of anions. The first absorption layer of the anions is best thought of as a monomolecular layer of ions. Harkins and Langmuir's idea of a monomolecular layer of molecules at absorbing interfaces may justly be expanded likewise to absorbed *ions*. This layer may be considered to be strongly attached to the absorbing wall so as to be practically immovable against the wall. However the adjacent layer of positive ions is located in the fluid and movable part of the liquid. It is not a real plane but a layer of a certain thickness. It consists only in that the concentration of the cations in the liquid is higher on the side facing the boundary of the liquid than in the main part of the liquid. The channels or pores of the membranes possess the same distribution of ions in the form of a double layer. The anions of the solution inside the channel are partially fixed by the wall of the channel losing their mobility while the cations remain freely movable. Let us assume as a first approximation, that the mobilities of the movable ions within the capillary channel are the same as in a free solution, still those of the anions which are fixed by the wall will not participate in any diffusion. Thus the average mobility of the anions is decreased. This average mobility is the only magnitude which can actually be measured in terms of the potential differences described above. Therefore u and v in the above equations mean average mobilities. According to a general law of absorption the percentage rate of absorption of the ions will be the greater the more dilute the solution. When the channels are narrow enough and the electrolyte concentration of the liquid is small enough, it may be that practically all of the anions are absorbed and immovable. In this case we attain the maximum value of the diffusion potential, such as described before for the extremely fine pored dried collodion membrane, this effect being approximately 57 millivolts for a concentration range of 1 : 10. It is in best agreement with this theory that this maximum effect is most easily reached in dilute solution, when for instance we have a 0.01 N and a 0.001 N solution, while in the combination 1 N : 0.1 N the maximum effect even in very small pored membranes can never be reached.

We assumed in the beginning that the mobilities of those ions which are freely movable in central portions of the channels are the same as in a free solution. This assumption is, of course, only very rough and we have to correct it according to the results of experiments with a collodion membrane between two solutions of different electrolytes such as hydrochloric acid and potassium chloride in equal concentration. These experiments were interpreted by us on the assumption that the difference of the mobilities in the pores is greatly exaggerated as compared with the differences in free water. That may be conceived by applying our knowledge on the causes of the differences of mobilities in general. Schematizing a little we may imagine every ion dragging along a shell of water when moving through the solution, the diameter of the water shell dragged along varying with the chemical individuality of the ion. Among the alkali ions the water shell becomes greater according to the order Li^+, Na^+, K^+, Rb^+, Cs^+. The velocity of an ion is determined by the extent of the surface of the dragged water shell, this water shell being compelled to overcome the frictional resistance of the surrounding water. I should like to emphasize that this idea is a little schematized since there is no definite water shell sharply confined to the surrounding water molecules. However our manner of representation though not perfectly adequate is sufficient for the present purpose. Now, when the channels are wide enough to allow the cations themselves but not the surrounding water shell to enter, the frictional resistance will increase with the magnitude of the water shell which the ion drags along, or as the attractive force for water by the ion becomes greater. There is a layer of water fixed at the wall in any case, such as in any capillary which is wetted by water, and the attraction of this unmovable water layer for the moving ion represents an additional frictional force against the motion of the ion.

4. The Glass Membrane

A membrane very remarkable in its electromotive effect is a very thin membrane of common glass which can be easily made by blowing a glass tube in the usual manner of the glass blower. The conductivity of such a membrane is very poor. Potential differences can be measured in general only by electrostatic methods such as by the quadrant electrometer. Especially between two solutions of different hydrogen ion concentrations, high potential differences can be observed at the two sides of the glass. The first one to describe this effect was Cremer and particularly through the investigations of Haber and Clemensievitsch it has been pointed out that the potential difference is approximately the same as in a hydrogen ion concentration chain with gas electrodes. Haber's theory is as follows. Traces of water but not of the dissolved electrolytes penetrate the glass wall which may be considered as a colloid with a certain, though small faculty of swelling in water. So within the membranes hydrogen and hydroxyl ions are present as a consequence of the dissociation of the water, but these ions are in a quite constant concentration not depending on the concentration of the hydrogen ions in the outside solution. This condition being given a simple thermodynamical consideration shows that solutions of different hydrogen ion concentrations on the two sides of the glass should behave just as they do in a hydrogen gas chain. The perfect realization of the potential difference expected from this theory meets some difficulties and seems to depend on the kind of glass. However sometimes it may be realized in fact. Some authors even recommended the glass for *p*H measurements in physiological liquids. It has to be supposed, for this purpose, first that the theoretical maximum value of the potential difference is really obtained, secondly that no other ions except the hydrogen ions control the potential difference, particularly not those of sodium and calcium. On the other hand several authors described the effect as not quite reaching the expected value and pointed out that the potential difference depends on other cations also. The anions were without any influence at all. Thus it seems that sometimes the glass membrane may give effects rather similar to the dried collodion membrane. I should suggest, therefore, to apply the theory developed before for the collodion membrane, likewise for the glass membrane, as a preliminary trial which ought to be proven by further investigations. Even in the collodion membrane the hydrogen ions turned out to be by far the most permeable ions. If we assume the glass membrane to behave as a sieve membrane with still smaller pores than collodion it may happen, in the limiting case, that only hydrogen ions can penetrate the membrane. If that occurs the maximum effect such as obtained by Haber can also be explained by the theory developed for collodion, and these theories are not contradictory at all but only different representations of the same idea. However the theory which I suggested is more general, including the other as a special or limiting case.

5. Phenomena of Electric Polarization in Membranes

When a membrane is in contact with solutions of the same composition on either side no potential difference is set up and nothing happens, of course. When we close such a system into an electric circuit the current flowing across the membrane will in general produce some alterations in the composition of the solutions. In order to understand this phenomenon we may simply consider the membrane as a third phase interposed between the two solutions and assume the relative mobilities of the different ions to be different from the one in the aqueous phases. It is of no importance for this consideration whether the membrane is really a new homogeneous phase, or a channel system, and whether the considered mobility is but an average mobility.

Many years ago Nernst and Riesenfeld showed that a phase interposed between two solutions of the same composition must produce a difference in the electrolyte concentration on the two sides of the middle phase when an electric current flows, whenever the transfer numbers of the ions in question are different within the membrane and in the solutions. A special case of that is a phenomenon first described by Hittorf and carefully studied by Bethe and Toropoff, a change of the hydrogen ion concentrations. The one solution becomes acid, the other alkaline, in the neighborhood of the membrane. This effect plays an important rôle in the practice of electroendosmosis and of the electrodialysis recently recommended by W. Pauli for purifying colloids. When the membrane is electronegative the liquid of the cathode side of the membrane becomes acid, the one of the anode side becomes alkaline, opposite to the changes taking place at the two metal electrodes. The effect is greater the more dilute the solutions are and it can no longer be observed at all in stronger solutions, such as, let us say, 0.1 normal salt solutions. In membranes which are electropositive against the solution, the effect is reversed. When the solutions are such as to correspond to the isoelectric point of the membrane—supposing it is an amphoteric membrane—this effect vanishes completely. This alteration of the hydrogen ion concentration is a part of the general phenomenon of the change in concentration of any ion through the effect of a membrane. However it must necessarily be supposed that the concentration of the dissolved electrolytes is so small that the participation of the ions of the water itself in conducting the current is appreciable. For that reason this effect vanishes in electrolyte solutions of higher concentrations.

6. Anomalous Endosmosis

I may assume that the phenomenon of electroendosmosis through membranes is so well known, that a description of it is unnecessary here. It may suffice to state that endosmosis consists in the transport of water through a porous membrane when an electric current flows through the membrane interposed between any aqueous solutions. But another apparently related phenomenon may be mentioned a little more in detail. When a membrane separates a solution from pure water, a stream of water flows from the pure water into the solution. This arrangement can be used for measuring the osmotic pressure, if the dissolved substance cannot diffuse at all through the membrane. But also when the dissolved substance is permeable, such a stream of water can flow in the beginning of the diffusion process provided the permeability for water is greater than the one for the dissolved molecules, *e.g.*, when a common, well permeable collodion membrane separates pure water from a sugar solution, sugar diffuses into the pure water. But in the beginning of the process the diffusion of water into the sugar solution not only takes place at the same time but it persists as well and is the most obvious phenomenon. The cause of this phenomenon needs no further explanation. However sometimes it happens that this water flow has the opposite direction, *e.g.*, from the solution into the water, in opposition to what may be expected according to the osmotic pressure. This so-called anomalous osmosis occurs only in electrolyte solutions, never in such solutions as sugar. Several examples of anomalous osmosis have been on record for a long time, but this problem reached a stage of a better experimental treatment only a few years ago when Jacques Loeb described it for a usual, relatively large-pored collodion membrane which has previously been in contact with a protein solution such as gelatin. The pores of such a membrane are covered with an absorbed layer of the protein which cannot be removed by washing. Now, there are some solutions which show anomalous endosmosis in such a membrane, for instance a very dilute solution of hydrochloric acid or of a salt of a trivalent cation, that means a solution of such a kind as to charge the gelatinized membrane positively. This phenomenon can only be observed when the solutions are very dilute. But also, in such electrolyte solutions which do not cause a water stream in the direction opposite to the osmotic pressure, the osmosis is sometimes anomalous in such a membrane insofar as the speed of the flow of water is greater or smaller than the speed of the water current which is produced by the osmotic pressure of a non-electrolyte such as sugar of the same osmotic pressure. This effect, also, disappears when the concentrations are higher, then the usual osmotic effect takes place again. Loeb recognized that besides the usual osmotic effect, in electrolyte solutions there is another effect which is of electrical nature and interferes with the first one and according to the conditions may increase or decrease the velocity which would be produced by the osmotic pressure alone. A very careful study has been published on this subject by Bartell, who investigated pure, collodion membranes which were not coated with protein. The effect is smaller in this case, as a rule, not reversing the direction of the expected water stream, but only diminishing or increasing it. The whole phenomenon is very interesting and may ultimately explain many processes of resorption and secretion in living organisms, which processes often take place against the osmotic pressure. The explanation given by Loeb and by Bartell consists in assuming an electroendosmosis interfering with the common osmosis. Now, electroendosmosis requires the existence of a steady electric current through the membrane. There is a potential difference across the membrane, as we have shown before. But which conducting material closes this potential difference to a real electric current? Freundlich and Nathanson assume that the substance of the membrane may function as the conductor. This explanation does not hold at all for the collodion membrane. For collodion owes its conductivity only to the water content of the pores. Dry collodion happens to be an ideal insulator, I dare say a better one than even paraffine. Now, an electric current in a liquid medium means the movement of cations in one direction without an equivalent stream of another kind of cations moving in the opposite direction, or without the moving cations dragging along the equivalent amount of anions in the same direction. How can such a process take place in a membrane simply separating two liquids? Speaking candidly, we do not know yet. There is an explanation given by Bartell. He assumes, the outer layer of the double layer at the wall of the pores which is freely movable in the liquid, to be the conductor. The electric current is supposed to be a circuit moving along the axis of the cylindric channel in one direction and along the wall of the cylinder in the opposite direction. We may be certain that there is something right in this hypothesis, but we cannot understand it completely. Perhaps this lecture may be an incentive for some of you to ponder a little on that problem and to attempt to solve it yourselves.

RADICALS AS CHEMICAL INDIVIDUALS

BY

CHARLES A. KRAUS
PROFESSOR OF CHEMISTRY
BROWN UNIVERSITY

Lecture delivered at Columbia University in the special course on Contemporary Developments in Chemistry given in the Summer Session of 1926 on the occasion of the opening of the Chandler Chemical Laboratories

COLUMBIA UNIVERSITY PRESS
NEW YORK
1927

Printed from type. Published March, 1927

Printed in the United States of America

RADICALS AS CHEMICAL INDIVIDUALS

By Charles A. Kraus

I. Introduction

Historical

The terms radical and element were introduced into chemistry at the very beginning of the development of chemistry as a science. Originally, the concepts underlying these terms were not sharply differentiated. Any kind of matter that retained its properties in a series of transformations and that imparted characteristic properties to compounds in which it was present, was termed a radical. As the concept of an element became more sharply defined, however, the term radical was used to designate combinations of elementary substances which retained their identity in various reactions and which imparted more or less specific properties to compounds in which they were present.

The earliest radicals that attracted the attention of chemists were groups of elements which, according to our present day notions, possess a great tendency to function as ions. Thus the nitrate and sulphate groups were radicals simulating electronegative elements while the ammonium group was a radical simulating the electropositive elements. In inorganic chemistry, the importance of the electroaffinities of the radicals was recognized at a very early date, but as organic chemistry developed and numerous organic radicals were introduced, the electroaffinity of radicals was in a large measure lost sight of. We shall have occasion to return to this point later on.

It was found that various organic groups are capable of a great many interactions in which apparently they retain their identity and such groups were spoken of as radicals. Indeed, many of these groups were given special names and it was even proposed to abbreviate these names in writing chemical formulas, and to some extent this practice is followed even today. The radical theory proved very fruitful in organic chemistry at a time when the whole subject was still in a chaotic state, but it obviously was not adapted to representing the extensive and complex phenomena of organic chemistry later on. With the introduction of valence concepts, and the hypothesis of the tetravalence of carbon, the radical theory fell into disuse, particularly in organic chemistry. It was only after Gomberg's capital discovery of triphenyl methyl, a compound which did not fit in with the conventionally accepted tetravalence of carbon, that the radical theory again attracted the serious attention of chemists.

If a radical is, indeed, what might be termed a compound element, we should expect that it would simulate an element not only in its compounds but also in the free state. However, all efforts to obtain radicals in the free condition met with failure and this proved a stumbling block in the path of the general acceptance of the radical theory in the latter half of the nineteenth century. It is true that a few radicals such as cyanogen, for example, were isolated, but the significance of the result was not understood. Other typical radicals such as ammonium, or the nitrate radical, could not be obtained in the free state.

Chemists have not always had a clear conception as to what the properties of a given radical should be in the free state, nor have they always realized that the properties of radicals, like those of elements, are profoundly influenced by conditions imposed upon them by surrounding elements or groups of elements. Finally, it was to be expected that, generally, radicals would not prove to be stable in the free state, for the constituent atoms of which they are composed are always more or less reactive and, since a free radical is necessarily in a state of unsaturation, internal reaction may be expected to occur. This holds particularly in the case of strongly electropositive radicals, in which there are mobile electrons present, with which the constituent atoms have a tendency to react.

Classifications of the Elements

The properties of radicals or groups of elements, like those of elements, are due to the tendency of such radicals to lose or gain electrons, or to share electrons with other elements or groups. We should expect to find radicals simulating elements in all their more characteristic chemical and, perhaps, physical properties. Therefore, before proceeding to a discussion of the properties of radicals, it is necessary to consider broadly the properties of the elements themselves.

The elements, with the exception of the inert gases, belong to one of three types:

1. Electropositive elements,
2. Electronegative elements,
3. Amphoteric elements, elements which are neither strongly electropositive nor strongly electronegative.

PERIODIC SYSTEM OF THE ELEMENTS.

				H_4X	H_3X	H_2X	HX	O	MH	MH_2	MH_3	MH_4								
	X_2O	XO	X_2O_3	XO_2	X_2O_5	XO_3	X_2O_7	O	M_2O	MO	M_2O_3	MO_2	M_2O_5	MO_3	M_2O_7	MO_4				
	1	2	3	4	5	6	7	8–0	1	2	3	4	5	6	7	8				
I							1 H	2 He	3 Li	4 Be	5 B	6 C								II
II				6 C	7 N	8 O	9 F	10 Ne	11 Na	12 Mg	13 Al	14 Si								III
III				14 Si	15 P	16 S	17 Cl	18 A	19 K	20 Ca	21 Sc	22 Ti	23 V	24 Cr	25 Mn	26 Fe	27 Co	28 Ni	29 Cu	IV
IV	29 Cu	30 Zn	31 Ga	32 Ge	33 As	34 Se	35 Br	36 Kr	37 Rb	38 Sr	39 Y	40 Zr	41 Nb	42 Mo	43 —	44 Ru	45 Rh	46 Pd	47 Ag	V
V	47 Ag	48 Cd	49 In	50 Sn	51 Sb	52 Te	53 I	54 Xe	55 Cs	56 Ba	57–71	72 Hf	73 Ta	74 W	75 —	76 Os	77 Ir	78 Pt	79 Au	VI
VI	79 Au	80 Hg	81 Tl	82 Pb	83 Bi	84 Po	85 —	86 Nt	87 —	88 Ra	89 Ac	90 Th	91 Pa	92 U						VI
	Rare Earth Group:				57 La	58 Ce	59 Pr	60 Nd	61 —	62 Sm	63 Eu	64 Gd	65 Tb	66 Ds	67 Ho	68 Er	69 Tm	70 Yb	71 Lu	

1. If the affinity of an element for one or more electrons in the neutral atom is weak, it is necessarily electropositive; in the pure condition it exhibits metallic properties. Its vapor is monatomic; it forms stable compounds with electronegative elements, and it does not, as a rule, form compounds with other electropositive elements. In general, the reactions of an element of this type are such as to enable it to pass into the condition of a positive ion and the compounds of such an element invariably exhibit the properties characteristic of salts. These elements belong to the first type, of which the alkali metals are examples.

2. If the affinity of the atoms of an element for the electron is great, it is necessarily electronegative. It is a non-metal; its vapors are polyatomic; it forms stable compounds (salts) with strongly electropositive elements and slightly stable compounds (non-salts) with other strongly electronegative elements. In general, it tends to react so as to form a negative ion. These elements belong to the second type mentioned above, of which the halogens are examples.

3. Finally, we have elements whose affinity for the electron is not very strong. In the presence of other elements having a very low affinity for electrons, they combine with these electrons to form anions while, on the other hand, in the presence of elements having a greater affinity for electrons, they lose their electrons to the latter elements and go into a condition that is relatively electropositive, although they do not usually form positive ions. Some of these elements are metals in the free state, while others are non-metals. Certain elements, such as carbon, arsenic, etc., are metals in one condition and non-metals in another. Their vapors are polyatomic and they usually combine with other elements only little more electropositive or electronegative than themselves. They form rather stable compounds with the alkali metals as well as with the halogens. Such elements are said to be amphoteric, this term being intended to express their tendency to act electronegatively toward electropositive elements and electropositively toward electronegative elements. Carbon, tin, germanium, arsenic, etc., are typical examples of this class of elements.

It will be noted that the valence of the amphoteric elements is greater than unity. There is only a single exception to this rule, namely hydrogen, which is a typical amphoteric element and whose valence is unity. Since the properties of the elements are more or less influenced by their condition of valence, a comparison among different elements is complicated owing to their difference of valence. As we shall see presently, most of the common radicals are univalent and comparison of their properties, therefore, is uninfluenced by valence change. The resemblance which many univalent radicals bear to hydrogen is very striking and is, in part, due to the fact that they have the same valence as that element.

If we arrange the elements according to their atomic number, writing the elements of the helium group underneath one another, and writing elements of higher atomic number to the right and elements of lower atomic number to the left, we obtain a representation of the periodic system, as shown in the figure, which is particularly adapted to bring out the relationship of the elements to one another.

The elements lying to the right of the helium group have an excess of one, two, or more, electrons over that corresponding to the configuration of an inert gas. The affinity of these elements for the electron is low, so much so, that, even in the pure state, the electrons are not permanently combined with the atomic nuclei. These elements, when in combination with other elements, whether strongly or weakly electronegative, lose their electrons and form positive ions. The compounds of these elements are typical salts.

The elements immediately to the left of the helium group have a deficiency of one, two, or more, electrons over that corresponding to the configuration of an inert gas, and, as such, have a tendency to add electrons thus going into the negative ionic condition. When these elements combine with strongly electropositive elements, they invariably form anions and the compounds are true salts, but when they interact with other elements having a considerable affinity for electrons they do not form ions; the electrons are in some way shared between the atoms of the elements in question.

As we proceed to elements further to the left of the helium group, their affinity for excess electrons decreases, while at the same time, their affinity for the electrons that they normally possess as neutral atoms likewise decreases. The atoms of such elements, therefore, either add or lose electrons with more or less equal facility, depending upon the nature of other atoms present. In the presence of other strongly electronegative elements, they tend to act electropositively, while, in the presence of strongly electropositive elements, they tend to act electronegatively. They thus constitute what we may properly term the amphoteric elements.

According to the periodic system as here represented, hydrogen occupies a unique position. This element is at times classed with the alkali metals and at others, with the halogens. Properly, it should be classed with neither of these groups. Its chief resemblance to either the alkali metals or the halogens, lies in the fact that it is univalent. So far as its chemical properties are concerned, it much more nearly resembles the amphoteric elements, being neither strongly electropositive nor strongly electronegative, forming true salts with strongly electropositive elements and forming compounds which are not true salts with strongly electronegative elements. It closely resembles the univalent amphoteric groups which we are about to discuss.

II. Radicals

Classification. Like the elements, radicals may be divided into three main groups, namely:

1. Strongly electropositive radicals,
2. Strongly electronegative radicals,
3. Amphoteric radicals.

Like the elements, the radicals may be arranged in a series in the order of their electroaffinities. If we arrange the radicals in such an order, with decreasing electropositive affinity to the right, or decreasing electronegative affinity to the left, then we find that the amphoteric radicals constitute an intermediate group, to the right of which lie the strongly electronegative radicals and to the left the strongly electropositive radicals. The dividing line between the amphoteric radicals and the strongly electronegative radicals on one hand, and the strongly electropositive radicals on the other, is not a sharp one. This intermediate group of radicals thus forms a transition group between the two extreme divisions.

The series of electroaffinities is a two-fold one. In the case of the positive radicals, the electronegative affinity has become so weak that it is not apparent, while conversely, in that of the strongly electronegative radicals, the electropositive affinity is so weak that it is not in evidence. In all the intermediate radicals, however, their dual tendencies are marked.

In addition to the radicals which fall within the above named three divisions, there are certain other groups of elements which properly should not be classed as radicals. They enter into certain well defined reactions to form characteristic compounds from which, in general, the original groups may readily be obtained, and in certain respects they simulate elements of the amphoteric type. However, they do this only in a limited sense. These groups will be discussed under the heading, 4 Exceptional Groups.

1. Strongly Electropositive Radicals

This group includes a great many important radicals both inorganic and organic. The most familiar examples of this type of radicals are the ammonium group and the various substituted ammoniums. In general, the amphoteric elements of the 4th, 5th, 6th and 7th groups of the periodic system form compounds of the type $R_{n+1}M^{n}X$ where R is an organic group or hydrogen, M is an amphoteric element of valence *n*, and X is a strongly electronegative element or group. All these radicals of the type $R_{n+1}M^{n}$ are strongly electropositive, many of them approaching the alkali metals in this respect. We should expect that in the free condition these groups would exhibit metallic properties, that is, that they would contain electrons not firmly bound to the positive nuclear group. Thus far, however, it has not been possible to obtain any of these radicals in the free state. Nevertheless, the evidence is quite conclusive that if the proper conditions for stability could be obtained, these radicals would actually exhibit metallic properties.

The ammonium group is the oldest and best known example of a radical of this type. However, all attempts to isolate the ammonium radicals have failed. Nevertheless, ammonium is readily obtained in the form of a somewhat unstable amalgam, whose properties correspond to those of amalgams of the alkali metals. Certain of the substituted ammoniums have been obtained in the form of stable amalgams. These amalgams are metallic and their physical properties closely resemble those of corresponding amalgams of the alkali metals. Tetramethyl, tetraethyl and tetrapropyl ammonium have been obtained as solutions in liquid ammonia. These solutions, both in their physical and their chemical properties, correspond precisely with solutions of the alkali metals in the same solvent. Since it has been shown that such solutions of the alkali metals are in fact true metallic solutions, it follows that the corresponding solutions of the substituted ammoniums are likewise metallic.

Since the free radicals of the ammonium type contain negative electrons in a mobile condition and since such groups contain hydrogen or organic groups which have a considerable affinity for electrons, it is not surprising that such radicals should prove to be unstable in the free state. While corresponding groups of other elements of the 5th group of the periodic system, such as phosphorus, arsenic and antimony, and of the 6th group, such as sulphur, and the 7th group, such as iodine, are well known as positive ions, their stability in the free state is lower than that of the corresponding derivatives of nitrogen and we know less about them. It may safely be inferred, however, that all these groups are strongly electropositive and that if the conditions for stability could be fulfilled, they would exhibit metallic properties in the free state. As indicating the electropositiveness of these various groups, it may be pointed out that the hydroxy-derivatives of these groups are strong bases, comparable with potassium hydroxide in this respect.

Certain of the elements of the 2d, 3d, and 4th groups of the periodic system form compounds of the type $R_{n-1}M^{n}X$. Typical examples are methyl mercury chloride, CH_3HgCl, trimethyl tin chloride, $(CH_3)_3SnCl$, and triphenyl germanium chloride, $(C_6H_5)_3GeCl$. The groups of the type $R_{n-1}M^{n}$, are, for the most part, amphoteric in their properties but, if the element M is sufficiently electropositive, the whole group may be sufficiently electropositive to be classified with the electropositive rather than with the amphoteric radicals. This is particularly true of the alkyl-mercury group, RHg. Methyl mercury chloride exhibits rather marked electrolytic properties in various solvents and the corresponding hydroxide CH_3HgOH is a decided base in aqueous solution. Groups of the type RHg have been obtained in the free condition by electrolysis of their halides and these groups are, in fact, metals. They thus constitute the only known examples of electropositive radicals which actually appear in the metallic condition.

Most of the groups of the type $R_{n-1}M^{n}$ are truly amphoteric radicals and as such do not exhibit metallic properties in the free state. This is the case for example with trimethyl tin, $(CH_3)_3Sn$, as it is also with triphenyl methyl $(C_6H_5)_3C$. Many chemists appear to hold the view that groups of this type are compound metals. This, however, is an erroneous notion. These groups in the free state exhibit no metallic properties and in their chemical properties resemble, not the characteristically metallic elements, but, rather, the amphoteric elements which lie on the border line between metallic and non-metallic elements, many of which are non-metals. These groups will be discussed below.

The strongly electropositive radicals, like the corresponding elements, exist in their compounds only as positive ions and these compounds, therefore, without exception, are true salts. With strongly electronegative elements or groups of elements, the electropositive radicals form stable salts, that is, salts stable toward oxidizing and hydrolizing agents. With weakly electronegative elements, or radicals, the strongly electropositive radicals form true salts, that is, compounds which are ionized in solution as well as in the pure liquid state. Such compounds, however, are not stable with respect to strong oxidizing or hydrolyzing agents, for the reason that the anions with which they are combined are reactive toward such reagents. The strongly electropositive radicals do not form compounds with one another nor with strongly electropositive elements. Combined with certain of the heavy metallic elements, they form compounds which exhibit metallic properties, that is, they form alloys.

2. Strongly Electronegative Radicals

This group includes a large variety of radicals both inorganic and organic, all of which tend to act as anions. Familiar examples of radicals of this type are: nitrate, NO_3, sulphocyanogen, SCN, cyanogen, CN, sulphate, SO_4 and, in fact, the negative constituent of all the inorganic and organic acids. These groups invariably contain electronegative elements such as oxygen, nitrogen, chlorine, etc., at times combined with other elements. In general, the more electronegative the elements in a group, the more strongly electronegative is the radical in question.

Only a small number of radicals of this type have been obtained in the free condition. Yet a sufficient number of them have been studied to show clearly that they simulate the halogens in almost every respect. The earliest known example of a group of this type is cyanogen, CN, which is a rather weakly electronegative radical. Sulphocyanogen, SCN, was isolated by Söderbäck, who studied its properties extensively and found them to correspond very closely to those of iodine or bromine, the group being somewhat more electronegative than iodine. Gomberg has obtained the perchlorate radical, ClO_4, in solution and has found that in many respects it resembles the halogens.

The strongly electronegative groups are probably dimolecular in the free state or in solution, or otherwise, they form even more complex molecules. In other words, the free radicals of this type resemble the halogens which, under ordinary conditions exist as di-atomic molecules. The strongly electronegative radicals combine with other strongly electronegative elements or groups of elements. The more strongly electronegative radicals replace less strongly negative elements or radicals in their compounds. These radicals combine directly with electropositive elements to form true salts. With water they hydrolyze according to an equation of the type:

$$X_2 + HOH = XH + XOH.$$

With weakly electropositive elements or groups, they form compounds that exhibit some, but not all, of the properties of salts. In such compounds the electrons are doubtless shared between the electropositive and the electronegative constituents. Such compounds are not electrolytes in the pure condition, but they often exhibit electrolytic properties when dissolved in suitable solvents. This point will be discussed below.

Altogether, these radicals, in the free state, as well as in their compounds, simulate certain of the electronegative elements so closely that they may truly be considered as compound electronegative elements, so far as their ordinary chemical properties are concerned.

3. Amphoteric Radicals

The amphoteric elements of the 4th group of the periodic system yield radicals corresponding to the formula R_3M, where R is an organic group or hydrogen. These radicals are neither strongly electropositive nor strongly electronegative and they combine readily with more strongly electronegative elements or groups. They are readily obtained in the free condition by simple processes of reduction or, less often, oxidation.

These radicals in the free state are non-metallic. They are readily reactive both toward oxidizing and reducing agents. In the free state they generally exist in the dimolecular condition, the unsaturated valence between a pair of radicals being mutually satisfied. The strength of the bond M—M between two radicals of this type is dependent on the nature of the central element M, as well as on that of the substituents R. The more electronegative the central element, the more stable, in general, is the bond M-M. Thus trimethyl tin exists in the dimolecular condition in benzene at its freezing point but is appreciably dissociated into uncombined radicals in boiling benzene. The compound is oxidized somewhat slowly by atmospheric oxygen and is oxidized with ease by the halogens. Triphenyl germanium is not oxidized by atmospheric oxygen and is halogenated only with great difficulty. The Si—Si bond is very stable although we lack definite information which would make a comparison possible. The C—C bond is in general a very stable one, both toward oxidizing and reducing agents. Only in the case that aryl groups act as substituents do we find that the C—C bond weakens. Thus triphenyl methyl is slightly dissociated in various solvents and is readily reactive toward oxygen as well as toward halogens or toward strong reducing agents.

According to this view, ethane is a normal compound of two methyl groups. In principle, there is no difference between ethane and triphenyl methyl save as regards the stability of the C—C bond. These two radicals may well be considered as extreme examples of the same class of substances. Naturally, in complex groups of atoms, such as we have in the case of carbon compounds, division may occur at some other point than at that of symmetry, thus leading to the formation of new radicals. Thus, ethane may lose hydrogen to form an ethyl group. Such groups as ethyl and methyl, while their compounds with hydrogen are inert toward various reagents, nevertheless, form compounds whose properties correspond very closely to those of other amphoteric radicals. As examples we may mention methyl chloride and sodium methyl.

Let us consider a few of the typical reactions of these amphoteric radicals. If we treat such a radical with a strong reducing agent, such as, for example, metallic sodium, we have the following reaction:

$$(R_3M)_2 + 2Na = 2NaMR_3.$$

The reaction consists merely in a combination of the negative electron, due to sodium, with the group R_3M and the formation of an anion R_3M^-. Considering the reaction as an ionic one—which it is—it may be written:

$$(R_3M)_2 + 2Na^+ + 2e^- = 2R_3Me^- + 2Na^+,$$

where e^- represents the electron. Sodium in such a compound functions as a positive ion and the compound has all the properties characteristic of a salt. For example, sodium triphenyl germanide, $NaGe(C_6H_5)_3$, is readily soluble in liquid ammonia as well as in many organic solvents. In liquid ammonia, it is, if anything, more highly ionized than most common salts, while its benzene solution is a fairly good electrical conductor.

Because of the fact that the anion R_3M^- is weakly electronegative, such a salt is quite reactive toward more electronegative elements or groups of elements. So, for example, when sodium triphenyl germanide is treated with oxygen we have the reaction:

$$(C_6H_5)_3GeNa + \tfrac{1}{2}O_2 = (C_6H_5)_3GeONa.$$

The oxygen atom combines with the electron on the triphenyl germanyl anion and forms a germanolate anion. Similar reactions occur with other electronegative elements.

When sodium triphenyl germanide is treated with even a weakly acidic substance, hydrolysis occurs because sodium triphenyl germanide is the salt of the exceedingly weak acid triphenyl germane, $(C_6H_5)_3GeH$. We thus have the reaction:

$$(C_6H_5)_3GeNa + HOH = NaOH + (C_6H_5)_3GeH.$$

This reaction is typical of compounds of this type.

On treating amphoteric radicals with a strongly electronegative element, combination occurs according to the equation:

$$(R_3M)_2 + X_2 = 2R_3MX.$$

For example, if we treat trimethyl tin with bromine we have the reaction:

$$[(CH_3)_3Sn]_2 + Br_2 = 2(CH_3)_3SnBr.$$

This reaction is strictly analogous to that between hydrogen and chlorine according to the equation:

$$H_2 + Cl_2 = 2HCl.$$

Many writers speak of compounds of the type R_3MX as salts. This is not in general correct. It is true that, when dissolved in suitable solvents, namely: hydroxy and amino- solvents, electrolytic solutions are formed, but it has been shown that this is due to interaction of the compound with the solvent to form a new type of compound. In fact, the reaction is closely analogous to that between the halogen hydrides and the same solvents. The halogen hydrides, such as hydrogen chloride or hydrogen bromide, are not salts nor are they electrolytes in the pure state. They are not even electrolytes when dissolved in many solvents of high dielectric constant. In general, it is only when the halogen hydrides are dissolved in hydroxy- or amino- solvents that they yield electrolytic solutions. If hydrogen chloride is dissolved in liquid ammonia, we have the reaction:

$$HCl + NH_3 = NH_4Cl.$$

If trimethyl tin chloride is similarly dissolved in liquid ammonia we have precisely the corresponding reaction:

$$(CH_3)_3SnCl + NH_3 = (CH_3)_3SnNH_3Cl.$$

The compound formed in this reaction exhibits electrolytic properties in solution. Like hydrogen chloride, trimethyl tin chloride is comparatively volatile while the corresponding compound with ammonia is rather difficultly volatile. When dissolved in water, or an alcohol, or ammonia, or an amine, trimethyl tin chloride, like hydrogen chloride, forms an electrolytic solution, but, when dissolved in nitrobenzene, it does not yield an electrolytic solution.

We see here a striking similarity between the properties of trimethyl tin and hydrogen. Under suitable conditions hydrogen combines with lithium according to the equation:

$$H_2 + 2Li = 2LiH.$$

Lithium hydride is a typical salt in which hydrogen acts as anion. It is an electrolyte in the solid as well as in the liquid condition. Similarly sodium trimethyl tin $NaSn(CH_3)_3$ is a solid at ordinary temperatures and is an excellent electrolyte in solution.

Hydrogen being weakly electropositive, its chloride, HCl, is a relatively weak electrolyte except when dissolved in water. Its solutions in the alcohols are much weaker electrolytes than corresponding solutions of typical salts. Similarly, trimethyl tin, being weakly electropositive, its chloride, trimethyl tin chloride, is a weak electrolyte when dissolved in such solvents as the alcohols. It is an interesting fact that, just as many acids in water and the alcohols conform approximately to the law of mass action, so trimethyl tin chloride, dissolved in ethyl alcohol, conforms to the same relation.

Amphoteric elements of the 5th group form radicals corresponding to the formula $(R_2M)_2$. Such groups are known in the case of nitrogen and arsenic. They have not as yet been investigated as extensively as the corresponding radicals of the elements of the 4th group but clearly they belong to this same class of compounds. Hydrazine, $N_2H—NH_2$ is a simple example of a group of this type. However, in the case of nitrogen, as in that of carbon, the bond between the central atoms is rather stable except in the case that the substituents R are aryl groups. Thus, tetraphenyl hydrazine exhibits properties resembling those of free triaryl methyl radicals. Salts corresponding to $NaMR_2$ have been prepared, particularly in the case of nitrogen. Thus, we have the salts, $NaN(C_6H_5)_2$ and $NaNHC_6H_5$. The latter is sodium anilid.

Correspondingly, compounds of the type R_2MX are known. In the case of elements of high atomic number these compounds are comparatively stable. It is scarcely necessary to mention that the cacodyls belong to this type of radicals.

A few divalent groups of the type R_2M^4 are known and without doubt many more may be obtained. While they have not been studied extensively, they clearly are true radicals. Tin derivatives corresponding to the formula R_2Sn are known. These compounds exist in the poly-molecular condition in solution and in general show a great tendency to polymerize. They are very reactive toward electronegative elements. We thus have the reaction:

$$(CH_3)_2Sn + I_2 = (CH_3)_2SnI_2.$$

Dimethyl tin di-iodide is a compound which does not exhibit electrolytic properties in the free state but which does exhibit such properties when dissolved in hydroxy- or amino- solvents. The same group reacts with strong reducing agents, such as sodium, as follows:

$$(CH_3)_2Sn + 2Na = Na_2Sn(CH_3)_2.$$

Di-sodium dimethyl stannide is a salt with a highly colored anion which is readily soluble in liquid ammonia. It is extremely reactive toward oxidizing agents or toward compounds containing strongly electronegative elements. We may consider that di-sodium anilid, $Na_2NC_6H_5$ is a derivative of the divalent group C_6H_5N, although this group is not known in the free state.

4. Exceptional Groups

As previously stated, certain groups of elements simulate radicals in certain of their properties but not in others. Groups of this kind are met with chiefly in the case of elements of the 3d and 5th groups of the periodic system. For example, triphenyl arsine, $(C_2H_5)_3As$, exhibits certain of the properties of a radical or element. It combines with the halogens, chlorine, for example, to form a compound according to the equation:

$$(C_2H_5)_3As + Cl_2 = (C_2H_5)_3AsCl_2.$$

On treating triphenyl arsenic dichloride with suitable reducing agents, the original compound is obtained. The above reaction is analogous, in many respects, to that between zinc and chlorine according to the equation:

$$Zn + Cl_2 = ZnCl_2.$$

So far as this type of reaction is concerned, triphenyl arsine behaves like a strongly electropositive divalent element, but it is, in fact, not a strongly electropositive divalent radical. The chloride is not a salt and we do not even know that the two halogen atoms are combined with arsenic in the same manner. Certainly, this group is not amphoteric for, on treating it with sodium, the electrons of this element do not combine with the group in question. The tendency of such compounds as triphenyl arsene to add halogens (or other negative elements) is apparently due to the fact that the group contains two electrons to which it tends to add other electrons with the formation of electron pairs but such groups are not capable of adding electrons that are not shared.

When we come to elements of the 3d group of the periodic system, we again find a dissymmetry with regard to the possible number of electrons that may be lost or added. Boron is the only element with regard to which we have any substantial information. We should expect that boron might share 3 electrons, since it is only weakly electronegative, or that, in the limit, it might add 5 electrons. Boron forms compounds of the type BR_3, where R is an organic group. Boron triphenyl is a typical compound of this type. Boron triphenyl does not combine with electronegative elements. If reaction takes place with such an element, substitution occurs. Thus, in the case of boron, we have compounds of the type R_2BX.

On the other hand, boron, when treated with strong reducing agents, such as metallic sodium, tends to add one electron. Thus, according to Krause, on treating triphenyl boron with metallic sodium in ether, there is produced a salt-like substance soluble in ether, corresponding to the formula $NaB(C_6H_5)_3$. This compound exhibits salt-like properties, its solutions are highly colored and it is very reactive toward oxygen. Apparently, in sodium triphenyl boride, an electron, due to sodium, is rather loosely held by the boron triphenyl molecule, forming a very weak anion. This is indicated by the fact that sodium triphenyl boride reacts with triphenyl methyl according to the equation:

$$(C_6H_5)_3BNa + (C_6H_5)_3C = (C_6H_5)_3B + (C_6H_5)_3CNa.$$

Electronically, the reaction may be written:

$$(C_6H_5)_3Be^- + Na^+ + (C_6H_5)_3C = (C_6H_5)_3B + Na^+ + (C_6H_5)_3Ce^-,$$

where e^- represents the electron. In other words, triphenyl methyl is more electronegative, or has a greater affinity for the electron, than triphenyl boron and when sodium triphenyl boride is treated with triphenyl methyl, the triphenyl methyl group acquires the electron from triphenyl boron and goes into the condition of the triphenyl methyl anion.

In connection with triphenyl boron, it is interesting to note that this group forms a remarkably stable compound with ammonia corresponding to the formula $(C_6H_5)_3B{\cdot}NH_3$. This compound is readily soluble in liquid ammonia and when treated with metallic sodium in that solvent, reaction takes place according to the equation:

$$(C_6H_5)_3B \cdot NH_3 + Na = (C_6H_5)_3B \cdot NH_3Na.$$

Apparently, triphenyl boron amine has a considerable affinity for an additional electron and accordingly, in the presence of sodium, takes up one electron to form the anion $(C_6H_5)_3B{\cdot}NH_3^-$. It is not known with which atom in this group the electron is associated but presumably it is boron. Sodium triphenyl boron amine is a salt which is readily soluble in liquid ammonia as well as in benzene and other organic solvents. It is remarkably stable, not being acted upon by oxygen. On electrolysis of this compound in liquid ammonia solution, triphenyl boron amine, $(C_6H_5)_3B{\cdot}NH_3$, is deposited at the anode.

While triphenyl boron amine thus has the power to add an electron to form an anion, it does not, at the same time, have the power to combine with electronegative elements or groups. Certainly it cannot be classed as an amphoteric group. On the basis of such information as we have, we are not justified in classing this group as a radical, unless we wish to broaden the concept embodied within this term.

III. Conclusion

To conclude: radicals are not merely convenient fictions that we employ in the formal representation of the facts of chemistry. They are, rather, physical entities in the same sense as the elements themselves. The study of radicals is of great importance in the development of our knowledge of chemical phenomena, for, while the number of elements having given properties is limited, and while the properties of these elements do not admit of alteration on the part of the experimenter, the number of radicals is practically unlimited and their properties may be varied, almost at will, through suitable variation of the substituent atoms. As yet, the surface of this interesting field of discovery has scarcely been broken.

It may safely be predicted that, in the further development of the subject, the amphoteric radicals will play a more important rôle than either the strongly electropositive or the strongly electronegative radicals. The latter two types of radicals are stable only as positive or negative ions, respectively, and as such, offer little that is new. On the other hand, the amphoteric radicals make it possible to study the properties of compounds as a function of the electroaffinities of their substituents, while corresponding studies of compounds of the elements are necessarily so limited in scope, that it is difficult to arrive at general conclusions. All the amphoteric elements have a high valence, many of their compounds are metallic in the pure state and usually insoluble, so that the number of compounds of this type available for study is very limited indeed. On the other hand, the amphoteric groups are univalent, extremely numerous, their compounds are nonmetallic and are readily soluble in a large variety of solvents. It may be anticipated that a systematic study of amphoteric radicals, particularly from the physical chemical point of view, will lead to many new and interesting results.

The Influence of Pressure upon Chemical Transformations

BY

ERNST COHÉN

PROFESSOR OF PHYSICAL CHEMISTRY
UNIVERSITY OF UTRECHT

Lecture delivered at Columbia University in the special course on Contemporary Developments in Chemistry given in the Summer Session of 1926 on the occasion of the opening of the Chandler Chemical Laboratories

COLUMBIA UNIVERSITY PRESS
NEW YORK
1927

Printed from type. Published March, 1927

Printed in the United States of America

THE INFLUENCE OF PRESSURE UPON CHEMICAL TRANSFORMATIONS

By Ernst Cohen

Although the study of the influence of temperature upon physico-chemical reactions has for many decades enjoyed a progressive development, a fact that has found expression in the construction of special laboratories dealing only with high or low temperature work, exact quantitative investigations concerning the influence of high pressures, a study which we have termed Piezochemistry (a word derived from the Greek πιέζειν meaning to press), have been taken up only quite recently. It was not until engineering science was able to produce an apparatus by means of which very high pressures could be attained and also maintained constant for an indefinite period of time that it became possible to study various problems whose solution is of importance not only for pure chemistry, but also for other sciences.[1]

Before discussing the question of the influence of pressure upon reaction velocity, it might be well, first of all, to describe briefly the apparatus employed in this kind of piezo-chemical investigations in particular, as they have been employed during the past years at my laboratory at Utrecht.

The apparatus [2] by means of which the influence of pressure may be studied is contained in a steel compression bomb *AA* (Fig. 1), filled with mineral oil or mercury, into which oil may be forced by means of a high pressure pump. When the apparatus inside the bomb is to be used for electrical measurements, the necessary wires may be connected to the outside instruments by means of the binding posts *CDEC′D′E′* (a fourth is not shown in the diagram) which pass through the cover plate *LKL*. As it is necessary in most cases that the pressure remain constant for indefinite periods of time, the gauge, Fig. 3, is so arranged that upon change in pressure an electric motor is shunted in automatically which puts the pump, Fig. 2, into operation. The latter then forces oil into the bomb until the original pressure has been restored. The variations amount to about one per thousand. Fig. 3 shows the manner in which the movement of the electric motor is transmitted to the pressure mechanism of the pump. Since the ordinary spring pressure gauges change in the course of time, the pressure is checked by means of a weight pressure gauge as depicted in Fig. 4. The principle upon which it operates is very simple. A steel block, Fig. 5, is bored out in the form of two coaxial cylinders of unequal diameters. The upper cylindrical part has an area of, for example, 3 sq. cm., the lower of

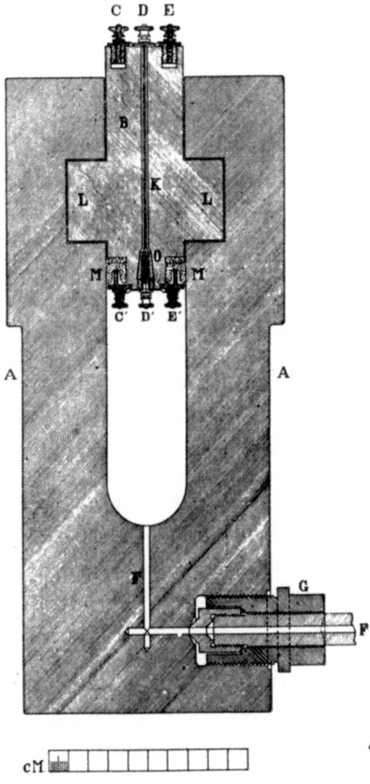

FIG. 1.—Bomb for studies under high pressure.

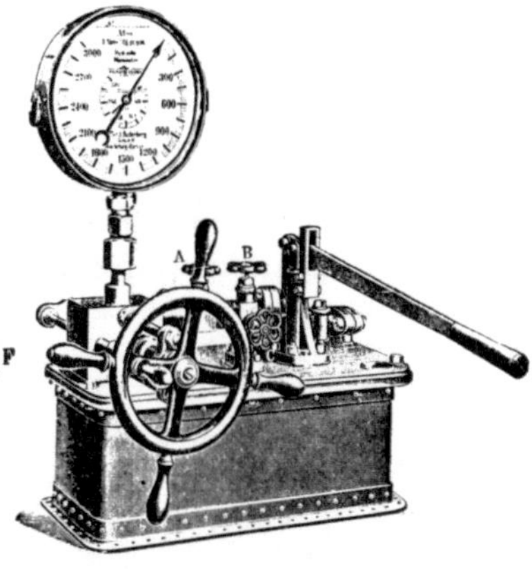

FIG. 2.—Pump and gauge.

2 sq. cm. A piston of the same form is ground into this opening in such a manner that a viscous liquid like oil may be forced between it and the walls of the steel block. If then there be forced through the side channel *B* under the piston some liquid such as oil, say at a pressure of 1000 atmospheres, this pressure will be exerted upon a surface of 1 sq. cm. and the piston will be raised up with a force of 1033 kg. In order that the balance be restored a weight of 1033 kg. must be added to the piston. This is done by placing weights upon the plate *E* located

FIG. 3.—Equipment for high pressure investigations.

at the bottom of the rod *A*. If for the present the friction of the piston in the cylinder be disregarded, it will become evident that by placing 1034 kg. on *E*, the balance will be destroyed. The piston will move downward until the liquid is compressed to a point where it also is under a pressure of 1034 kg. If on the other hand, the plate *E* be loaded with 1032 kg., the piston will rise.

Until now the frictional resistance of *C* has been left out of consideration. However, it is of utmost importance that errors

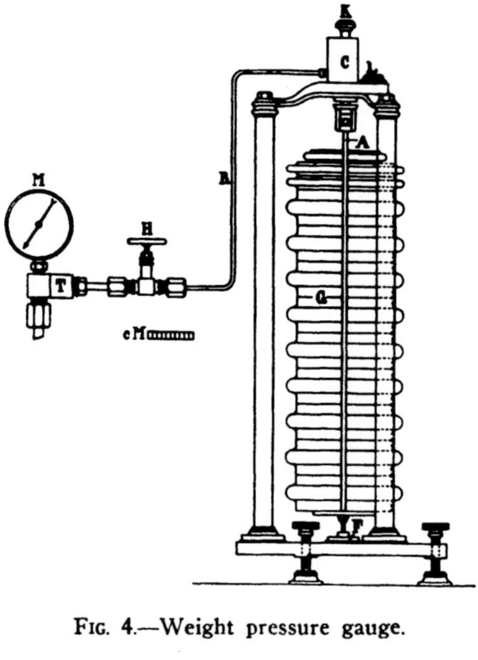

Fig. 4.—Weight pressure gauge.

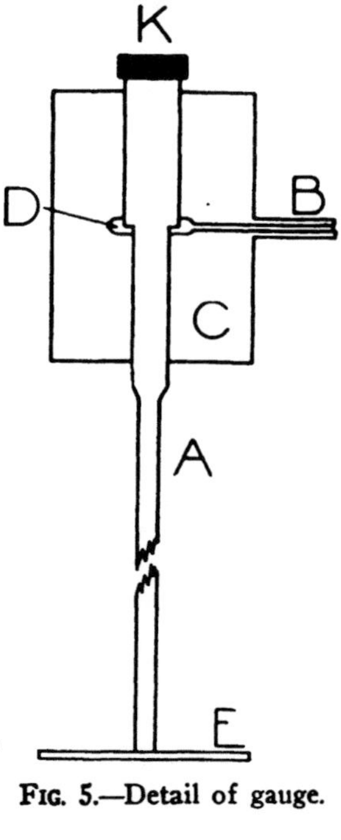

Fig. 5.—Detail of gauge.

from this source in the measurement of pressure be avoided. This may be accomplished by using as the compression liquid one which is at the same time not too viscous to be forced between the piston and the wall of the cylinder. This in itself will not do away with friction, but only when the piston as well as the attached weights are put into rotation on its own axis.

The readings of the spring gauge may be checked very conveniently before and after each experiment by means of the weight pressure gauge, to which it is connected by means of a T connection, Fig. 4. The stopcock *A*, Fig. 2, is closed; *H*, Fig. 4, is opened, and after *C*, Fig. 2, has been closed, by simply turning the wheel of the hand pump, Fig. 2, oil is forced into both gauges. The weight pressure gauge has been loaded beforehand with weights equivalent to the desired pressure. As soon as the rod *A*, Fig. 5, together with the attached weights begins to rise, the whole is brought into rotation by simply turning the knob *K* by hand. While *A* rotates, the pressure on the spring gauge is noted. The difference in pressure between the actual reading and that equivalent to the weights yields the error of the gauge at that particular pressure.

The pressure bomb is set in a thermostat whose temperature can be kept constant by means of an electric regulator within a few thousandths of a degree.

On the basis of the state of matter of the reactants which take part in a reaction we may consider our problem from three different points of view. This question may be submitted: What influence does external pressure exert upon reaction velocity in the case of homogeneous gaseous, solid, or liquid systems?

The whole procedure is theoretically quite simple where the process involves a mixture of dilute gases. If n different kinds of molecules take part in the reaction the velocities at various pressures are proportional to the n-1 power thereof.[3] Consequently the velocity of a monomolecular reaction is independent of the pressure, a fact which I was able to verify in the case of the decomposition of arsine.[4] To my knowledge no other quantitative data on this subject have ever been presented, and the same is true where high pressures are concerned. "The perfect gas laws and empirical equations of state that we are accustomed to using around atmospheric pressure are hopelessly inadequate at pressures between 100 and 1000 atmospheres."[5]

Quite as inadequate is the quantitative information concerning the influence of pressure upon the velocity of transformation of solid systems, and for that reason I can go no further into this subject.[6]

In the case of homogeneous liquid systems conditions are somewhat more satisfactory. Therefore, we shall deal with these in particular in this lecture. But even here only a beginning has been made and the solution of numerous problems must be relegated to the future. I am quite convinced that the courteous invitation on the part of Columbia University to speak to you concerning the influence of pressure upon chemical transformations was given to me principally, because I, perhaps, on the basis of the investigations which have been carried out in my laboratory at Utrecht, am in a better position than anyone else to prove to you how meager our knowledge is in this most interesting field. After the exact quantitative experiments carried out there had shown that pressure exerted a profound influence not only upon the saponification of ethyl acetate but also upon the inversion of cane sugar, and that this effect could be either positive or negative, the query arose as to whether this effect were primary or secondary. For elucidation of this question it may be said that a change in pressure changes several factors which have an important bearing upon reaction velocity, such as viscosity of the reacting medium, the degree of dissociation of the reacting substances, their association, etc. Since it formerly was not possible to introduce corrections for the above factors in the values obtained experimentally for the effect of pressure, the question as to whether pressure exerted a primary or secondary effect remained unsettled. To obtain answer to this question two ways are open to us. The influence of pressure upon a given reaction may first be studied, and next its effect upon the above-mentioned factors, and the results may then be introduced in our calculations. It must also be known how the reaction velocity varies with a change in these factors.

Or we may study the influence of pressure upon the speed of such reactions as closely resemble one another. In such cases the corrections will be approximately equal. Any difference in the velocity of the reactions must then be due to those conditions which were not the same in the specific cases, or to the nature of the reaction itself. For instance, the influence of pressure upon two ionic reactions may be investigated in very dilute aqueous solution. Under such conditions the influence of pressure upon the viscosity and the association of the solvent will be the same, as well as changes in ion mobility and degree of dissociation, provided that similar electrolytes be employed. Thus only the specific nature of the reaction remains as a factor. If pressure has no bearing upon the reactions as such, but is only secondary, then the effect should be quite similar in both cases.

For the solution of this problem A. L. Th. Moesveld [7] studied the influence of pressure upon the reaction

$$5\,HBr + HBrO_3 = 3\,Br_2 + 3\,H_2O$$

and compared the results which he had obtained with those found for the saponification of ethyl acetate

$$CH_3COOC_2H_5 + NaOH \rightarrow CH_3COONa + C_2H_5OH$$

in aqueous solution under practically identical conditions.

In both cases conditions are very much alike in that the same solvent is employed and an ionic reaction takes place in which very strongly dissociated electrolytes take part. Except for the reactions themselves the only difference between them lies in the fact that in the saponification of the ester

$$OH^- + CH_3COOC_2H_5 = CH_3COO \quad + C_2H_5OH$$

one ion is replaced by another, whereas the twelve ions present originally in the equation

$$5\,HBr + HBrO_3 = 3\,Br_2 + 3\,H_2O$$

disappear entirely leaving only non-electrolytes as the products. Because of the disappearance of the ions this reaction is accompanied by an appreciable positive volume change. The saponification reaction is likewise attended by a positive volume change,[8] but probably to a lesser degree. In the latter case this is due to the fact that sodium hydroxide is hydrated to a greater extent than the resulting sodium acetate. Thus, if the action of pressure upon reaction velocity is connected with a change in volume taking place during the reaction, then this effect should reveal a difference in magnitude, but not in sign.

In the study of the influence of pressure upon reaction velocity the experimental electrical methodology employed by Ernst Cohen and H. F. G. Kaiser,[9] first described by Bogojawslensky and Tammann [10] was followed. Because of its universal application in this field I shall outline it briefly, and at the same time take up with you the velocity of saponification of ethyl acetate by sodium hydroxide.

The velocity constant of this reaction may be represented by the equation

$$k = \frac{1}{t_2 - t_1} \cdot \frac{C_1 - C_2}{C_1 C_2},$$

where $t_2 - t_1$ gives the time during which the reaction has taken place, C_1 is the concentration of the reactants, or of the sodium hydroxide when equivalent quantities are used, at the time t_1, and C_2 the concentration at t_2. The actual procedure which was employed by us is actually nothing more than the modified Kohlrausch method for measuring the electrical resistance of solutions. From the latter values the concentrations of the solutions may readily be determined. This method possesses many obvious advantages. After the mixture to be investigated has been placed under pressure at constant temperature, the concentrations $C_1, C_2 \ldots$ at the times $t_1, t_2 \ldots$ may be determined by measuring the resistance from time to time using the ordinary Wheatstone bridge, alternating current and telephone. One complicating factor does enter despite the fact that very dilute solutions, 0.02 N, were employed. James Walker [11] once wrote as follows concerning this experimental procedure, "The chief conditions for the convenient application of the method are: first, that there should be a considerable difference in conductivity between the initial and final systems, and second, that the change in conductivity should be proportional to the progress of the reaction. It occurred to me that these conditions would be well fulfilled in the saponification of an ester by a caustic alkali. The conductivity of the alkali, say sodium hydroxide, is much greater than that of the sodium salt produced by the saponification owing to the high velocity of the hydroxyl ion as compared with the salt ion. Since, too, sodium hydroxide and sodium salts of monobasic acids are approximately equally ionized under the same conditions, the ionization in dilute solutions remains practically the same throughout the saponification, and thus the alteration in the conductivity is almost exactly proportional to the progress of the reaction." It has been proved that the latter is not applicable. Therefore it has been necessary at a given temperature to determine empirically for each of the various pressures and employing known solutions the relationship between the concentration of the reactants and the electrical conductivity. If the saponification is allowed to proceed at the same temperature and under a known pressure, and if the conductivities of the reacting mixture be taken at the times t_1, $t_2 \ldots$, the concentrations corresponding to the measured conductivities may readily be deduced from the conductivity-concentration curve for that pressure.

The apparatus which was used for these experiments is shown in Fig. 6. It is composed of the vessel *A* made of Jena glass, containing a dipping electrode *C*. By means of a capillary, *A* is connected with the container *D* which is attached to the lower end of the tube. The glass stopper *B* is ground into *A* and is fused to *C*. Thin platinum wires are connected to the platinum electrodes 1 and 2. The wires pass through the glass wall of *C*, as well as the hard rubber stopper *F* which is cemented into the glass stopper of the dipping electrode. The platinum wires lead to the small mercury capsules *HH*. *G* is a hole, and if the electrode is put under pressure, the surrounding oil passes through *G* into the stopper so that its various parts are all under the same pressure. The electrodes are covered with platinum black by the method of Lummer and Kurlbaum.

The apparatus is manipulated in the following manner. After the two solutions, in this case 0.02 N ethyl acetate and 0.02 N sodium hydroxide, have been mixed in specially constructed pipettes, the resulting liquid is poured into a 50 cc. beaker. *A*, which has a capacity of about 28 cc., is then dipped into the solution until *D* has been entirely filled and *A* but partly. It is then removed from the beaker, the lower end of the capillary is closed by placing a finger over it and *A* is filled full. The stopper *B* is put into place. The whole apparatus is then placed into the basket-like contrivance *KLL*, made of a copper-tin alloy, which contains a small glass dish of mercury. The mercury acts as a seal for the lower capillary of *E*. The basket *KLL* is attached to the cover of the pressure bomb, Fig. 1, in such a

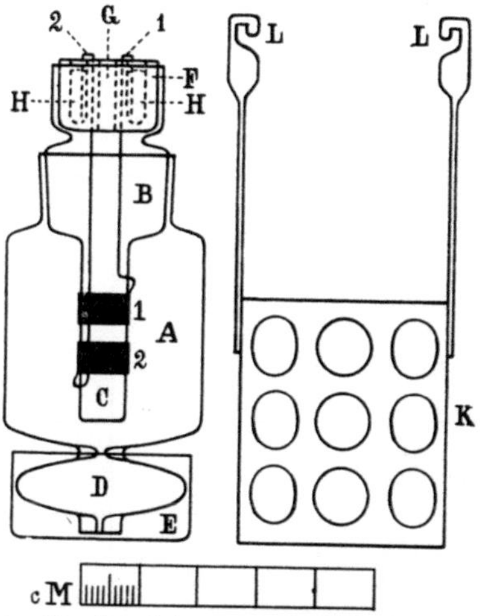

Fig. 6.—Conductivity vessel and accessories.

manner that the two thin copper wires which are soldered to the binding posts on the cover of the bomb dip directly into the mercury filled cups *HH*. The purpose of the bulb *D* is to prevent the entrance of mercury into that part of the apparatus containing the electrodes. Naturally the presence of mercury within the conductivity apparatus would materially affect the resistance. Since the compressibility coefficient of the reaction mixture at 1500 atmospheres, the highest pressure attained, amounted to 39.9×10^{-6}, the volume change of the liquid in *A* was equal to $1500 \times 39.9 \times 10^{-6} \times 28 = 1.7$ cc. As *D* had a capacity of 3 cc., the mercury could not enter *A*.

After the apparatus has been placed under pressure and allowed to assume, after dissipation of the heat of compression, the temperature of the experiment, which was 2.4° C. for the saponification of the ester, the resistances, W_1, W_2, W_3 .... of the reaction mixture are measured at the times t_1, t_2, t_3 .... by aid of which the corresponding concentrations could be had by reference to curves mentioned above. In the various experiments $t_2 - t_1 = 60$ minutes. The results which were obtained are given in the following table:

TABLE I

THE VELOCITY OF SAPONIFICATION OF 0.01 N ETHYL ACETATE BY 0.01 N SODIUM HYDROXIDE AT 2.40° C., AND VARIOUS PRESSURES, AS DETERMINED BY THE ELECTRICAL METHOD

Pressure in Atmospheres	Concentration, $C \times 10^6$	*k* Calculated from C_1 and C_2 by the Electrical Method	*k* Determined by Titration	Percentage Influence of Pressure
1	C_1 393 C_2 305 C_1 398 C_2 309	1.22 1.21	1.22	
500	C_1 354 C_2 274	1.37	1.35	11.9
1000	C_1 357 C_2 269	1.53	1.52	25.5
1500	C_1 339 C_2 253	1.67	1.67	37.4

The values for *k* given in the fourth column were determined by another method. The close numerical agreement leaves nothing to be desired. These measurements show that the rate of saponification increases directly with the pressure, so that at 1500 atmospheres the reaction proceeds 37 per cent faster than at one atmosphere.

The effect of pressure upon the reaction

$$5\,HBr + HBrO_3 = 3\,Br_2 + 3\,H_2O$$

was studied by A. L. Th. Moesveld in identically the same manner, but at four different temperatures, 1.04°, 15.04°, 25.04° and 39.04°C. The individual details of this investigation may be found in the original article. However, let me emphasize the fact that this is a quadri-molecular reaction which probably proceeds according to the following equations:

$$H^+ + Br^- + H^+ + BrO_3^- = HBrO + HBrO_2. \quad (1)$$

$$\left.\begin{aligned} HBrO + HBr &= Br_2 + H_2O \\ HBrO_2 + 3HBr &= 2Br_2 + 2H_2O \end{aligned}\right\} \quad (2)$$

Reaction 1 goes very slowly, whereas 2 takes place at high speed.

In this case the concentration may easily be deduced from the conductivity of the reaction mixture. It is unnecessary to plot a known graph depicting the relationship between conductivity and concentration as must be done where one ion is simply replaced by another. At the completion of this reaction ions are no longer present, only bromine and water. Since both acids are practically entirely dissociated at the dilution employed—hydrobromic acid 0.05 N, and bromic acid, 0.0102 N, and as neither causes a repression of the ionization of the other, and furthermore, since the percentage dissociation does not change appreciably in the course of the reaction during which the concentration falls by two-thirds, we may make the assumption that the fall in conductivity is a linear function of the concentration and that zero conductivity will be obtained when the concentration is zero. Special measurements showed that bromine, even in the presence of Br_3^- ions, exerted no influence upon the conductivity of hydrobromic acid solutions.

In regard to the calculations which must be made in order that the value for k may be derived from the measured resistance of the reaction mixture, we know that the equation $\frac{dx}{dt} = k(A - x)^4$ is applicable to quadri-molecular reactions, providing the reactants are present in equimolecular proportions. A represents the initial concentration and x the amount transformed at the time t. Thus,

$$k = \frac{1}{3}\frac{1}{t_n - t_0}\left(\frac{1}{(A - x)^3} - \frac{1}{A^3}\right).$$

The conductivities are directly proportional to the concentrations, whereas the reverse holds in respect to the resistances. Thus [12]

$$k^1 = \frac{1}{t_n - t_0}(W_t^3 - W_0^3).$$

Of course the volumes of the solutions change if measurements are made at different temperatures, and correction factors must be introduced to do away with any errors resulting from such variations. However, these will not be discussed here in any further detail.[13]

The general results of the measurements at 25.04°C. are given below:

TABLE II

Pressure in Atmospheres	k Corrected	k Average Value	Percentage Influence of Pressure
1	195.5 194.9	195.0	
500	184.3 184.6	184.4	− 5.4
1000	174.6 175.2	174.9	−10.3
1500	166.8 166.4	166.6	−14.6

The experiments at other temperatures demonstrated that the pressure coefficient of reaction velocity varies but slightly with temperature.

If the results in Table II be compared with those of Table I, it will at once become evident that the above case presents a marked contrast with that of the saponification, both in sign as well as in size. We may therefore conclude that pressure, as such, exerts a profound influence upon reaction velocity, irrespective of any secondary differences.

In closing let me direct your attention to the fact that these results offer a method whereby the more recent theories concerning the interrelationship of radiation and reaction velocity may be controlled.

REFERENCES

1 Piezochemie kondensierter Systeme by Ernst Cohen and W. Schut, Leipzig, 1919, gives a summary of all of the available information up to that time on the subject.

2 Ernst Cohen, Katsuji Inouye and C. Euwen, Zeitschr. f. physik. Chemie, 75, 257 (1910).

3 van't Hoff, Lectures on Theoretical and Physical Chemistry. Translated by R. A. Lehfeldt, London, 1898. Vol. 1, p. 236.

4 Zeit. f. physik. Chem., 25, 483 (1898).

5 See J. A. Almquist, J. Chem. Ed., 3, 385 (1926).

6 Ernst Cohen and W. Schut, Piezochemie Kondensierter Systeme, Leipzig, 1919, p. 411.

7 Zeitschr. f. physik. Chemie, 103, 486 (1923).

8 Benrath, Zeitschr. f. physik. Chemie, 67, 501 (1909).

9 Zeitschr. f. physik. Chemie, 89, 338 (1915).

10 Zeitschr. f. physik. Chemie, 23, 13 (1897).

11 Proc. Roy. Soc., London, 78, *A* 157 (1906).

12 For calculations of k see the work of A. L. Th. Moesveld, Zeitschr. physik. Chemie, 103, 481 (1923).

13 See A. L. Th. Moesveld, Zeitschr. f. physik. Chem., 103, 486 (1923) in particular page 18.

A DEVELOPMENT IN THE CHEMISTRY OF SANITATION

BY

JOHN ARTHUR WILSON

CHIEF CHEMIST, A. F. GALLUN AND SONS COMPANY
AND CONSULTING CHEMIST, MILWAUKEE SEWERAGE COMMISSION

Lecture delivered at Columbia University in the special course on Contemporary Developments in Chemistry given in the Summer Session of 1926 on the occasion of the opening of the Chandler Chemical Laboratories

COLUMBIA UNIVERSITY PRESS
NEW YORK
1927

Printed from type. Published March, 1927

Printed in the United States of America

A DEVELOPMENT IN THE CHEMISTRY OF SANITATION

Mechanism of the Activated Sludge Process of Sewage Disposal as Developed at Milwaukee

By JOHN ARTHUR WILSON

The city of Milwaukee has just completed the building of a sewage disposal plant that ranks as a great chemical as well as engineering achievement. The city draws its water supply from Lake Michigan, which is also the only practical outlet for its sewage. The problem confronting the city was the very difficult one of finding a method for separating all of its sewage into two parts: (1) dry solid matter in such form as to be salable as a fertilizer that must be shipped long distances from the city economically and (2) water pure enough to be run into the source of its drinking water. After ten years of intensive research, a plant capable of doing this was developed and built. Each day the plant receives about 85,000,000 gals. of sewage, which it separates into practically pure water and over 100 tons of dry fertilizer for which a demand has been created in many parts of the country since its high fertilizing value was clearly established. The sewage is actually freed from more than 99 per cent of all the pollution it contained before it enters the lake. The scope of this lecture is limited to a brief description of the process finally adopted and a discussion of the chemical principles involved.

The sewage of a great city is an extremely complex mixture of domestic and industrial wastes. In Milwaukee it includes refuse from tanneries, packinghouses, breweries, paint factories, steel mills, and many other kinds of industry. Much of the material contained in sewage is extremely resistant to separation from water. Any simple method, such as distillation, is prohibitive because of the enormous volumes to be treated. Settling alone is ineffective because some of the polluting materials are so finely divided that they will not settle out upon standing. Centrifuging merely separates the sewage into layers of varying composition, but all of which contain polluting materials. Ordinary evaporation on open ground would require more land than could be had, besides being very offensive to the community around. Filtration of the sewage is unsatisfactory because of the prohibitive cost of handling such large volumes and because the filtrates obtained under ordinary conditions are never pure enough to be run into the lake. In fact, the only practical method of dealing with the sewage so as to produce a reasonably pure effluent to be run into the lake was found to be the so-called activated sludge process. But the use of this method required the development of various other operations in order to make it workable, so that the process became a series of operations.

The method finally adopted as practical consists of the following major steps:

1. Coarse screening of sewage.
2. Settling of coarse materials of sewage.
3. Fine screening of sewage.
4. Incineration of coarse screenings and settled matters.
5. Aeration of sewage mixed with activated sludge.
6. Settling of activated sludge from aerated mixture.
7. Decantation of supernatant water into lake.
8. Drawing activated sludge from bottoms of settling tanks.
9. Return of part of activated sludge to screened sewage.
10. Chemical treatment of waste activated sludge.
11. Filtering treated sludge.
12. Return of filtrate to screened sewage.
13. Drying sludge cake.
14. Grinding, storing, and shipping sludge as fertilizer.

All of the operations are continuous and individual pieces of apparatus are stopped only for cleaning or repairs while others take their places.

Water-Sewage Cycle

The pumping of the city's water supply from Lake Michigan is really the beginning of the process. In serving the various sanitary and industrial purposes, the water becomes polluted with waste materials, which it conveys to the disposal plant, acting as a vehicle for carrying these wastes from many places to one designated point. It is now the function of the disposal plant to remove these waste materials from the water, sending it back to the lake essentially as pure as when drawn, thus completing the cycle.

Removal of Coarse Material

The sewage is carried to the plant on Jones Island by a system of intercepting sewers. It is first passed through bar screens having a space of about two inches between the bars. This removes only the very large materials that might choke the pumps. The velocity of flow of the sewage is then reduced to less than one foot per second by allowing it to flow into open concrete channels 8 feet wide and 8 feet deep. This gives the heavier materials time to settle out. Each of these channels, or grit chambers, is capable of isolation from the rest for the purpose of draining and cleaning, which is done by means of a clamshell bucket operated from a locomotive crane.

The sewage then passes through fine screens consisting of revolving drums covered with manganese bronze plates with openings 2.4 millimeters wide through which the sewage flows. Brushes operating over the surface of the drums at right angles to the direction of rotation sweep the screenings off continuously onto a moving conveyor.

The settled materials and screenings consist of about 10 tons of dry solid matter out of a total of 110 tons to be removed from the 85,000,000 gals. of sewage received per day. The present plan of handling the screenings is to press out the loosely held water and then to burn them in an incinerator.

Removal of Finely Divided Material

The removal of coarse material from the sewage involves only simple and obvious mechanical operations, but by far the larger part of the pollution is the finely divided material which is not so easily separated from the water. The great discovery underlying the activated sludge process is that this finely divided material can be rendered capable of settling out from water quickly by bubbling air or oxygen through it for a sufficiently long time.

The screened sewage passes first into a mixing channel and is kept well agitated by air blown through it. Here activated sludge returned from the system is mixed with the sewage and the mixed liquor then flows on into the aeration tanks through feed channels, the rate of flow being controlled by sluice gates.

The aeration tanks consist of 24 concrete tanks, each having two parallel compartments 236 feet long and 22 feet wide with 15 feet effective depth. In concrete containers resting on the bottom of each tank there are 2514 filtros plates, each 1 foot square. Air compressed to 10 pounds per square inch is supplied to each container to be diffused through the filtros plates and allowed to bubble up through the mixed sewage and sludge in tiny streams as the mixture passes through the aeration tanks. The flow is so regulated that it takes 6 hours for the mixed liquor to travel through the aeration tanks, during which time 1.5 cubic feet of air have been bubbled through it for each gallon of raw sewage entering the system.

From the aeration tanks the mixed liquor flows into large sedimentation tanks, of which there are eleven. Each is a concrete tank 98 feet in diameter with 15 feet effective depth. The mixture enters the tanks continuously at such rate as to allow an average detention period of 30 minutes. The sludge rapidly settles to the bottom carrying with it more than 99 per cent of all the pollution originally contained in the sewage. Practically pure water continually overflows into effluent troughs and from thence into the lake.

The sludge drawn off from the bottoms of the sedimentation tanks contains about 2 per cent of solid matter. Under the hydrostatic head it flows out into a sludge well. About 90 per cent of the sludge is pumped back into the incoming screened sewage and the remaining 10 per cent to the filter house. The sewage furnishes about 100 tons per day of dry activated sludge, but the equivalent of about 900 tons is kept circulating through the aeration tanks with the sewage. The operations involved in handling the waste sludge will be better appreciated after a discussion of some of the principles involved in the entire process.

Mechanism of Aeration

The reactions involved in aerating the mixed liquor are not yet understood, although certain facts are clear. When insoluble materials are suspended in a liquid, the rate at which they will settle depends, among other things, upon the size of the individual particles, the greater the size the more rapid the rate of settling. In the case of sewage, the settling does not follow the simple law of Stokes because many of the particles have a strong attraction for water and carry electrical charges. Nevertheless the particles present in sewage do settle rapidly when they unite to form relatively large particles. Much of the finely divided materials of sewage originally consisted of animal and vegetable tissues presenting a relatively coarse structure. But through putrefaction and other chemical changes, these tissues have been broken down into particles which will not settle out from water upon standing. When oxygen or air is bubbled through sewage, these tiny particles unite with each other, tending to form coarse particles analogous to the original tissues. If the bubbling is continued for several days, the particles become large enough to settle out fairly quickly upon standing.

If some of this settled sludge is now added to another portion of raw sewage and oxygen bubbled through it, the tiny particles tend to combine with the larger ones rather than with each other. But the large particles are already of sufficient size to settle rapidly when the bubbling has ceased. Where two tiny particles combine, the double-size particle is still too small to settle out, but where one large and one tiny particle combine, the resulting particle will settle at least as quickly as the original large one.

This explains why it is desirable to mix sludge already capable of settling rapidly with the incoming raw sewage before aerating it. The time required to bring the sewage to the point where purification can be effected by settling is thus reduced to a fraction of that otherwise required. The term activated sludge is applied to the coarse sludge because it is supposed to have been activated in some way to make it capable of combining with the impurities in sewage and removing them from the water. Decision as to whether the name is an appropriate one may have to await a better understanding of the mechanism of the reactions.

The effectiveness of aeration in building settleable sludge from sewage is a function of temperature and pH value. The temperature relations have not yet been thoroughly worked out, but it has been demonstrated that the action in the aeration tank is completed very much more quickly in warm sewage (25° C.) than in cold sewage (7° C.).

The pH value of the sewage and sludge mixture was found to be of tremendous importance in aeration; optimum effects are obtained at the pH values 7.4 and 3.4. The effect at 3.4 seems to be a coagulating one not involving the same type of reactions as obtained at pH = 7.4, which is the value always found in the aeration tanks. At pH values near 5.5, the sludge breaks down even in the presence of a plentiful supply of oxygen, but this may be due to the fact that 5.5 seems to be the optimum point for putrefaction of the sludge.

Oxygen is necessary to the reaction. Aerating with a gas like hydrogen is ineffective even at optimum pH value, whereas oxygen or air is very effective. It has been claimed that living microorganisms are essential, but this claim so far is purely speculative.

There is an apparent analogy between the reactions in the aeration tanks and the building of tissues in the animal body. Food is broken down in the animal's digestive tract into more finely divided materials, which pass into the blood streams. In the presence of oxygen, entering by way of the lungs and blood streams, these finely divided materials are rebuilt into tissues.

In this operation, the tiny particles tend to fix themselves to tissues already built rather than merely to combine with each other. If the reaction of the blood shifts much from its normal pH value of 7.4, tissue building ceases. Similarly, in the aeration tanks, the building of sludge proceeds at an optimum rate at this same pH value of 7.4 and sludge building ceases when the pH value is shifted much from 7.4. When the animal dies and oxygen is no longer available tissue building ceases, the tissues begin to atrophy and the pH value falls to values lower than 7.4. When the air supply is cut off from the aeration tanks, the sludge begins to break down, and the pH value falls to values lower than 7.4. The processes of the animal body tend to keep the pH value of the blood constant at 7.4 and the aeration of mixed liquor tends to keep the pH value constant at 7.4 in the face of fairly wide fluctuations in pH value of the raw sewage. Finally, definite tissues in the animal body derived from foods of widely different kinds have properties nearly constant and activated sludge obtained from a great variety of sewage has some properties which are remarkably constant; for example an apparent isoelectric point at pH = 3.4.

Considering the nature of tissue building in the animal body by means of cellular activity, the analogy may be a questionable one, but it seems worthy of some careful thought until some further light is thrown on the mechanism of aeration.

The Sludge Problem

The most difficult and serious problem encountered in the work at Milwaukee was to find a practical method of disposing of the 1,275,000 gals. of waste activated sludge produced daily. The success of the entire project depended upon devising some means of reducing this to a dry material that could be shipped and sold as fertilizer. But before it can be subjected to ordinary drying operations economically, it is necessary to reduce its water content from 98 to 80 per cent or less by some kind of filtering device, which means removing from it more than 1,000,000 gals. of water. Only then is it practical to dry it further by the application of heat.

Raw sewage is extremely difficult to filter. As it is aerated more and more, it becomes increasingly easy to filter it satisfactorily. But the reactions in the aeration tanks are functions of temperature and in the winter, when the sewage is cold (less than 10° C.), sludge building proceeds at a very slow rate. The sludge formed will settle out reasonably well in the sedimentation tanks, permitting purification of the sewage, but it is still so finely divided that it can be filtered only with great difficulty; the tiny particles of sludge work their way between the fibers of the filter cloth and adhere to them, soon clogging the cloth and rendering it practically impervious to water, even under pressure. The particles of sludge formed in summer, when the sewage is warm (above 20° C.), are relatively large and filter easily because they are too big to work their way into the interstices of the filter cloth to clog it.

The problem was to find a method of treating winter sludge in such manner as to make its filtration practical. In spite of the high state of development of filtration engineering, no filter could be found capable of economically removing the bulk of water from sludge produced in winter. It was necessary to treat the sludge chemically before it could be filtered.

Degree of Dispersion

From the standpoint of colloid chemistry, the problem of the activated sludge process appears to be practically one of regulating the degree of subdivision of the solid matter derived from the sewage and the degree of swelling, or water absorption, of the jelly-like constituents. In the aeration tanks, purification of the sewage is made possible by the highly dispersed material joining the much coarser activated sludge, which settles out quickly upon standing, permitting the clarified solution above to be decanted off into the lake.

The further removal of water from the settled sludge by filtration is also dependent upon the degree of subdivision of the solid part of the sludge. It is usually found with complex suspensions of this sort that certain changes in chemical reaction of the mixture result in material changes in degree of subdivision of the solid matter.

Part of the sludge consists of organized jellies, such as fibrous protein matter. The ease with which sludge can be freed from water by filtration depends not only upon the fineness of subdivision of a portion of the sludge but also upon the extent to which the jelly constituents are swollen with water. A mass of fibrous protein matter may hold water in two very different ways: (1) In the interstices between the fibers and from which it may be forced out by the application of relatively slight pressure and (2) within the substance of the fibers themselves, like the water absorbed by a block of gelatin, and from which the water cannot be removed mechanically, except by the application of enormous pressure. The sludge will be in its best condition for filtering when the individual particles are as large as possible and the water content of the jellies is at a minimum value.

In order to draw a working picture to be used as a guide in experimental work, both in this field and in leather chemistry, it was found desirable to try to develop theories accounting for the behavior of suspensions of particles and the swelling of jellies. In working with complex mixtures like sewage sludge, it is not always possible to make quantitative applications of theories, but the mere picture of a theory proved for simpler conditions may lead to the solution of an otherwise baffling problem. In trying to adjust the degree of subdivision of the solid matter of the sludge and the degree of swelling of its jelly constituents so that it might be filtered, two theories somewhat related served as a guide. They deal with the stability of suspensions and the swelling of jellies. A brief outline of each may serve to make the story clearer.

Stability of Suspensions

As a rule, colloidal dispersions become more stable with increasing value of the electrical charge on the surface of the particles. This may be due in part to the repulsive action of like charges, but it is probably due also, and perhaps in an even greater degree, to the forces which tend to cause wetting of the surface of the particles. With increasing difference of potential between the solid surface and the liquid phase, there is an increased attraction between solid and liquid which resists any decrease in surface such as would result from two or more particles coalescing.

It was pointed out by the writer[1] in 1916 that a peculiar type of equilibrium must exist at the surface of contact between an aqueous solution and any electrically charged solid. The nature of this equilibrium is pictured in the example of a particle of gold in a sol stabilized by a trace of potassium chloride. The work of Beans and Eastlack[2] indicates that the stability of a gold sol formed in very dilute potassium chloride solution is due to the formation of a complex between chloride ions and the atoms of gold at the surface of the particles. In combining with the gold, the chloride ions impart their negative electrical charges to the particles.

This view makes it very easy to picture the theory about to be described, but it should be recognized that the theory holds regardless of the mechanism responsible for the electrical charge on the particles.

For each chloride ion combining with the gold surface, there is left in solution one potassium ion, whose field of motion is now restricted to the thin film of solution wetting the gold surface because its outer electron is now in the gold surface. The volume of this film of aqueous solution enveloping any particle will be measured by the surface area of the particle and the average distance that the potassium ions are free to travel from the oppositely charged surface. This film will also contain ionized potassium chloride diffusing in from the main bulk of solution, but the concentration of potassium ions will be greater than that of chloride ions, since some of the potassium ions are balanced only by the electrical charges on the surface of the particle. In the solution far removed from the surface of the particle, potassium and chloride ions will be present in equal concentration.

The system described consists of three phases: (1) the electrically charged gold surface, (2) the film of solution wetting the surface and in which the concentration of potassium ions is greater than that of chloride ions, and (3) the main bulk of solution, in which the potassium and chloride ions are present in equivalent concentration. The conditions obtaining in such a system are strikingly analogous to those described by Donnan[3] in his theory of membrane equilibria and, in fact, the various relations derived by Donnan are applicable here.

The first problem is to determine how the ionized potassium chloride will be distributed between phases 2 and 3 at equilibrium. When equilibrium is established, if a small virtual change is made reversibly at constant temperature and volume, the free energy will remain unchanged; that is, no work will be done. The change here considered is the transfer of dn mols of K^+ and Cl' from phase 3 to phase 2. The work, which equals zero, is

$$dn \cdot RT \cdot \log_e \frac{[K^+]_3}{[K^+]_2} + dn \cdot RT \cdot \log_e \frac{[Cl']_3}{[Cl']_2} = 0$$

whence

$$[K^+]_3 \times [Cl']_3 = [K^+]_2 \times [Cl']_2$$

(The brackets indicate concentration in mols per liter.) Equilibrium will be established only when the product of the concentrations of K^+ and Cl' has the same value in both phases 2 and 3. Similarly, it can be shown when any other ionogen is added to the system, provided it does not alter the absolute value of the electrical charge on the gold surface, that the product of the concentrations of all ions required to make an undissociated molecule has the same value in both solutions. For example, if sodium sulfate is added to the system, the product

$$[Na^+] \times [Na^+] \times [SO_4''] \quad \text{or} \quad [Na^+]^2 \times [SO_4'']$$

will have the same value in phase 2 as in phase 3.

The following system of notation will simplify further discussion. In phase 3, the main bulk of solution, let

$$x = [K^+] = [Cl']$$

In phase 2, the film wetting the gold surface, let

$$y = [Cl']$$

and

$$z = [K^+]$$

balancing surface electrical charge whence total concentration of K^+ is equal to $y + z$.

The equation of products may now be written

$$x^2 = y(y + z)$$

But here we have the product of equals equated to the product of unequals, from which it is apparent, mathematically, that the sum of the unequals is greater than the sum of the equals, or that

$$2y + z > 2x$$

The reasoning thus indicates that the concentration of ions in phase 2 is greater than that of phase 3 at equilibrium by an amount that we may call e. Then

$$2y + z = 2x + e$$

and

$$x = y + \sqrt{ey}$$

which indicates also that x is greater than y, or that the concentration of ionized KCl is greater in phase 3 than in phase 2. The potassium chloride does not distribute itself uniformly throughout all of the water of the system, but is less concentrated in the surface film containing additional potassium ions.

This unequal distribution of ions between phases 2 and 3 will result in an electrical potential difference measured by

$$E = \frac{RT}{F} \log_e \frac{x}{y} = \frac{RT}{F} \log_e \frac{2x}{- z + \sqrt{4x^2 + z^2}}.$$

If now the concentration of KCl be increased without limit, while the absolute value of the electrical charge remains constant, since z is constant, the value of E must decrease without limit, for

$$\lim_{x=\infty} E = \frac{RT}{F} \log_e \frac{2x}{\sqrt{4x^2}} = 0$$

It is thus evident that the difference of potential between the enveloping film and the surrounding solution will be a maximum when there is no free KCl present and will decrease, approaching zero, as the concentration of salt is increased without limit. The writer believes that the forces opposing coalescence of the particles is determined by this potential difference rather than by the absolute value of the electrical charge on the particles, which may remain constant while the potential difference varies, because the surface film completely envelops the particles and tends to endow them with its own properties. With a lowering of the potential difference between phases 2 and 3, there is an increasing tendency for the particles to coalesce and precipitate out. It is easy enough to develop equations for the addition of other ions of any valency, so long as the effect on the absolute value of the electrical charge is known.

It is also quite apparent from the equations that any lowering in the value of z, or of the absolute value of the electrical charge, will result in a lowering of the value of E. With this picture of the factors effecting the stability of suspensions in mind, it is desirable to consider the mechanism of the swelling of jellies.

Swelling of Jellies

The only quantitative theory of the mechanism of the swelling of protein jellies in acid and alkaline solutions to be found in the literature is that of Procter and Wilson.[4] It can be described briefly by considering a hypothetical protein in contact with an aqueous solution of HCl. The protein G is a jelly insoluble in water, but completely permeable to water and all dissolved ions in solution, is elastic and follows Hooke's law, and combines chemically with hydrogen ion, but not with chloride ion, forming a completely ionized protein chloride according to the equation

$$[G] \times [H^+] = K[GH^+] \quad . \quad . \quad . \quad . \quad . \quad (1)$$

One millimol of G is immersed in a solution of HCl, which penetrates G, where some of the hydrogen ions combine with the protein and are thus removed from solution, leaving the solution absorbed by the jelly richer in Cl′ and poorer in H^+ than the external solution. When equilibrium has been established, the HCl will be distributed between the external solution and the jelly phase, or solution absorbed by the jelly, in a manner similar to that of KCl between phases 3 and 2 in the gold sol described above.

To make the similarity more striking, let

$$x = [H^+] = [Cl'] \text{ in the external solution}$$

and

$$y = [H^+]$$

and

$$z = [GH^+] \text{ in the jelly phase,}$$

whence

$$[Cl'] = y + z \text{ in the jelly phase.}$$

From the reasoning given above, it is apparent that

$$x^2 = y(y + z), \quad \ldots \quad (2)$$

and

$$e = 2y + z - 2x, \quad \ldots \quad (3)$$

where e is defined as the excess of concentration of diffusible ions of the jelly phase over that of the external solution. From (2) and (3) we get

$$x = y + \sqrt{ey}, \quad \ldots \quad (4)$$

which shows that the hydrogen ion concentration is greater in the external solution than in the jelly, and the chloride-ion concentration is greater in the jelly than in the external solution.

Since [Cl′] is greater in the jelly than in the external solution, the chloride ions of the protein chloride will tend to diffuse outward into the external solution, but this they cannot do without dragging their protein cations with them. These protein cations, however, are not in solution in the generally accepted sense of the term, but form an elastic structure which resists the pull of the anions, which are actually in solution. The actual measure of the pull is that of the difference between the total energy directed outward from the jelly and that directed from the external solution towards the jelly. This is obviously represented by the value e and, according to Hooke's law

$$e = CV \quad \ldots \quad (5)$$

where C is a constant corresponding to the bulk modulus of the protein and V is the increase in volume, in cubic centimeters, of 1 millimol of G.

Since we have taken 1 millimol of G,

$$[G] + [GH^+] = \frac{1}{(V + a)},$$

or

$$[G] = \frac{1}{(V + a)} - z, \quad \ldots \ldots \quad (6)$$

where a is the initial volume of 1 millimol of the protein. From (1) and (6)

$$z = \frac{y}{(V + a)(K + y)}, \quad \ldots \ldots \quad (7)$$

and from (2) and (3)

$$z = e + 2\sqrt{ey}. \quad \ldots \ldots \quad (8)$$

From (5) and (8)

$$z = CV + 2\sqrt{CVy}, \quad \ldots \ldots \quad (9)$$

and from (7) and (9)

$$(V + a)(K + y)(CV + 2\sqrt{CVy}) - y = 0, \quad \ldots \quad (10)$$

where the only variables are V and y.

If the molecules or atoms of the protein are not themselves permeable to the ions of the solution, the quantity a should not be taken as the whole of the initial volume of the jelly, but only as the free space within the original, dry jelly through which ions can pass. For the hypothetical protein G the limiting case is considered where $a = 0$. In most cases of protein swelling, this assumption introduces an error less than probable experimental error because of the relatively large values for V over the significant swelling range. Equation (10) thus reduces to

$$V(K + y)(CV + 2\sqrt{CVy}) - y = 0. \quad \ldots \quad (11)$$

Given the values of the constants K and C, one can plot the entire equilibrium as a function of any one variable; for, given y, we can calculate V from (11); then having V we can calculate e from (5); now having both e and y, we get z from (8); and with z, e and y, we get x from (3).

Procter and Wilson tested the theory for gelatin and HCl. They obtained the value $K = 0.00015$ by titrating gelatin solutions with HCl and measuring the increase in hydrogen-ion concentration with the hydrogen electrode. This was done independently of any swelling experiments. The value $C = 0.0003$ was obtained for the gelatin used in the experiments from the swelling data. This made it possible to calculate all of the variables of the equilibrium over the hydrogen-ion range covered as well as to measure them. Figure 1 shows that the agreement

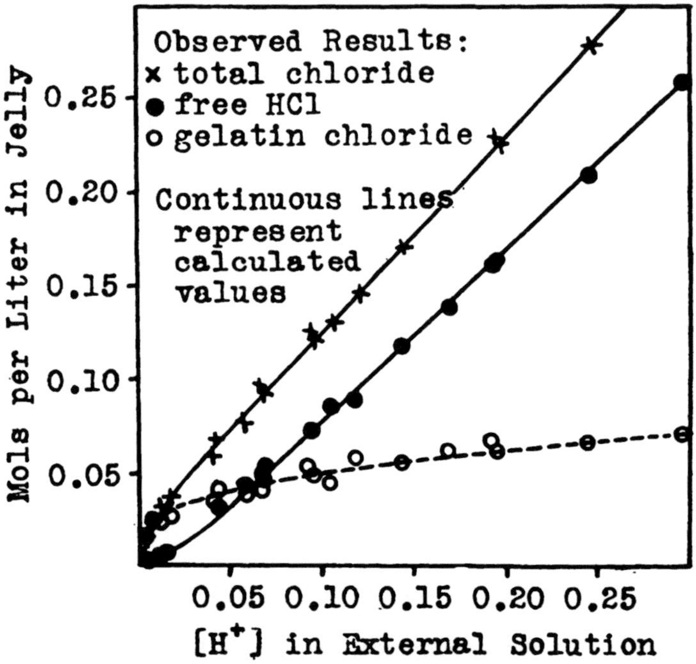

FIG. 1.—Observed and calculated values for the distribution of HCl in the system gelatin jelly—HCl—water.

found between calculated and observed results was as nearly perfect as the experimental conditions would permit. Figure 2 shows the variation in volume of 1 gram of gelatin and the calculated values assuming an equivalent weight of 768 for gelatin; here too the agreement is practically perfect and serves to place great weight on the value of the theory.

Extension of the theory to include the addition of other electrolytes to cover any special case is not difficult. It is also apparent that an electrical difference of potential will be set up between the external solution and jelly phase exactly similar to that described in the discussion of the gold sol. It is possible here to give only the briefest picture of the theories.

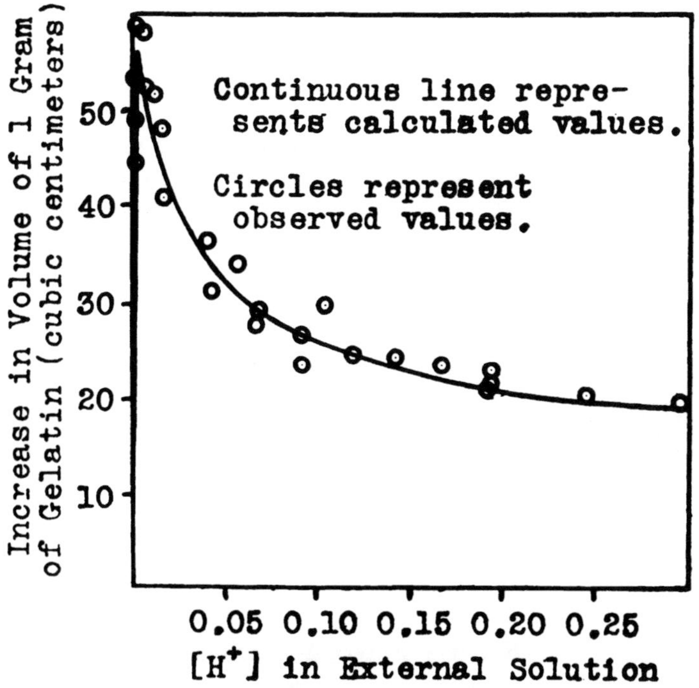

FIG. 2.—Observed and calculated values for the swelling of gelatin jelly in HCl solution.

Isoelectric Point

Proteins are amphoteric substances, reacting both as weak acids and as weak bases. In this respect, they retain the properties of the amino acids from which they are formed. Hydrated aminoacetic acid is capable of assuming either a positive or negative charge, or both, by ionizing as acid or base, thus

$$H^+ + {}'OOC\cdot CH_2\cdot NH_2\cdot HOH =$$

$$HOOC\cdot CH_2\cdot NH_2\cdot HOH = HOOC\cdot CH_2\cdot NH_2H^+ + OH'$$

The ionization constant of a protein as an acid may thus be represented as follows:

$$[H^+] \times [G'] = K_a[GH]$$

But

$$[H^+] \times [OH'] = K_w,$$

or

$$[H^+] = K_w/[OH'],$$

from which

$$[GH] \times [OH'] = k[G'],$$

where

$$k = K_w/K_a.$$

But this is essentially the same equation as (1) except for the fact that $[H^+]$ is replaced by $[OH']$. It is thus apparent that proteins will behave in solutions of increasing concentration of alkali much as they do in solutions of increasing concentrations of acid, so long as they undergo no chemical change other than that indicated in the hypothesis.

In passing from strongly alkaline to strongly acid solution, the electrical charge on a protein becomes decreasingly negative, zero, and then increasingly positive. The zero point is that at which the number of positive charges on the protein structure is exactly equal to the number of negative charges, and the quantity z of our equation either equals zero or represents the sum of equal concentrations of oppositely charged ions, which would have the same effect as though it were actually zero. This point is called the isoelectric point and is usually recorded in terms of pH value.

Relative Filtering Efficiency

Returning now to the problem of filtering activated sludge, the first step is to find some method of measuring its physical state. Activated sludge does not lend itself readily to measures of particle size or degree of swelling of the jelly-like constituents, but since the practical problem to be solved was the preparation of the sludge in condition suitable for filtering, it seemed logical to use as a measurement of this condition the rate of filtration of the sludge under standard conditions.

The apparatus used to make this measurement consists of a battery of 20 Buechner funnels set in filter flasks, all connected to one pump furnishing a high vacuum. As a standard, a sample of sludge was selected of such condition that its rate of filtering was considered on the border line between satisfactory and unsatisfactory. It was found that 500 cc. of such sludge diluted to contain 1 per cent of solid matter could be dewatered on a standard laboratory filter in just 20 minutes. This made it convenient to define the "relative filtering efficiency" of any sample of sludge as 2000 divided by the number of minutes required to filter 500 cc. of 1 per cent sludge through a standard filter.

This gives the sludge which was only just satisfactory a relative filtering efficiency of 100. This number is of course purely arbitrary, but it serves as well as any other since it is only the comparison of rates of filtration that counts. Changing the size of the funnels or the kind of filter paper used merely changes the constant in the equation of proportionality between the laboratory measurements and the large-scale, plant operations.

Several years of operation of large-scale, filter presses showed that the efficiency obtained for any given type of sludge is directly proportional to the value defined as relative filtering efficiency, provided the filter press is operated under fixed conditions, including the delivery of the same amount of sludge per unit area of filter cloth in every case.

Having adopted a standard method of measuring the condition of the sludge, the next step was to study the effect of each of the variable factors known to affect the stability of suspensions and the degree of swelling of protein jellies. Without introducing any special treatment, it was found that the relative filtering efficiency of the sludge varied from a minimum of 5 in February to a maximum of 100 in September. This represents the effect of temperature of the mixed liquor in the aeration tanks. Quantitatively, the problem resolved itself into one of so treating February sludge as to raise its relative filtering efficiency from 5 to 100, since heating 85,000,000 gals. of sewage per day to summer temperature was obviously prohibitive. A number of accounts of this work have already appeared in the literature.[5]

Effect of pH Value

Activated sludge may be a very complex mixture of many substances, but it behaves much like insoluble protein matters. If it were a single protein substance, the logical first step in trying to increase the rate of filtration of its aqueous dispersion to a maximum would be to bring the reaction of the water to the isoelectric point of the protein, for at this point the degree of dispersion and swelling would both tend to be at minimum values. as, for example, in the case of dispersion of gelatin. The fact that sludge may be a mixture of many things complicates matters, but the procedure to follow is much the same as it would be with a single protein substance. Although each constituent of the sludge may have its own isoelectric point, there must be some pH value at which the total sum of all positive charges on the insoluble phases must be equal to that of all the negative charges and such a point might be referred to as the iso-electric point of the sludge. At this point the positively charged particles would tend to coprecipitate the negatively charged ones, resulting in a neutral mass more easily filterable than at any other pH value.

Fig. 3 shows the effect of pH value upon the relative filtering efficiency of February sludge having an initial pH value of 7.4. Higher values were produced by adding sodium hydroxide and lower values by adding sulfuric acid. The curve shows a clearly defined point of maximum at pH = 3.4 and a point of inflection at pH = 7.4, which is an optimum point for aeration. Various precipitation tests made upon activated sludge indicate that pH = 3.4 actually does correspond to an isoelectric point.

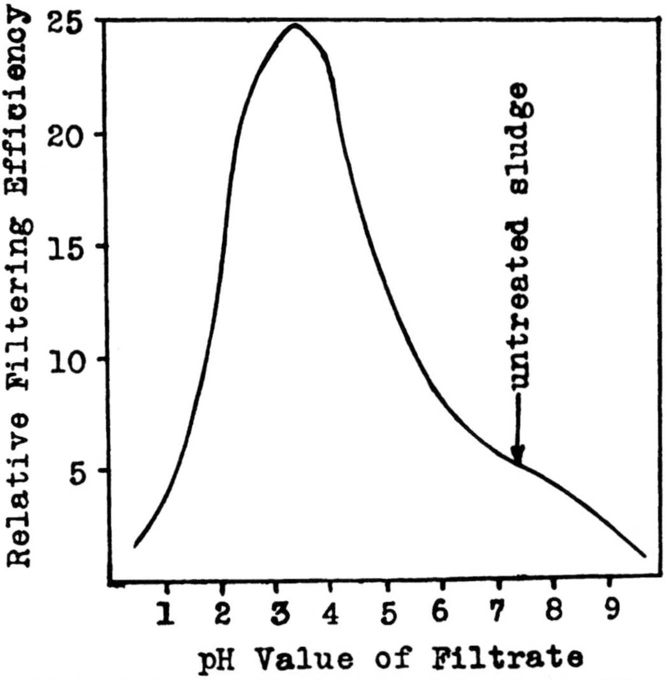

FIG. 3.—Effect of change of p^H value upon relative filtering efficiency of February sludge. pH value lowered from 7.4 by adding sulfuric acid and raised from 7.4 by adding sodium hydroxide.

Bringing the sludge to its apparent isoelectric point increases its relative filtering efficiency from 5 to 25, a very large gain, but still not enough to permit satisfactory filtering in winter. This measurement has been made daily for a number of years and two very curious facts have been noted. The first is that the relative filtering efficiency is multiplied by 5, regardless of its initial value, by bringing the pH value to 3.4 with sulfuric acid. For example, when the pH value of September sludge is dropped from 7.4 to 3.4, its relative filtering efficiency is increased from 100 to 500. The second curious fact is that the apparent isoelectric point always seems to be at pH = 3.4, regardless of known large variations in the composition of the raw sewage. This strengthens the view that the building of sludge in the aeration tanks is analogous to the building of tissues in the animal body.

Effect of Aluminum Sulfate

Where precipitation of sols is effected by mixing equivalent concentrations of oppositely charged particles, the nature of the

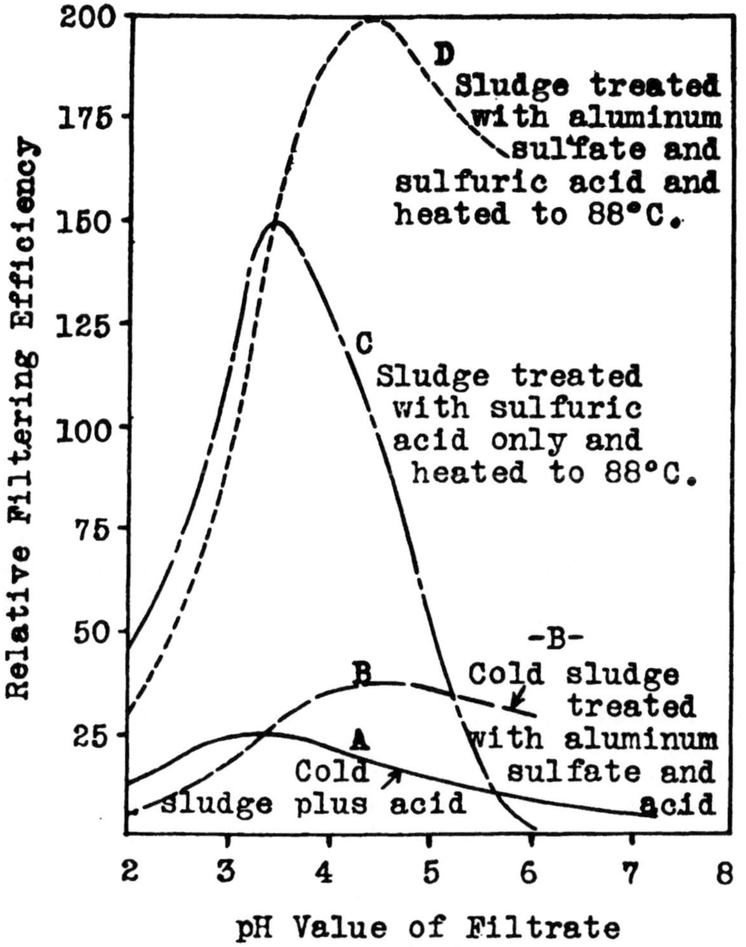

FIG. 4.—Effect of damage of pH value upon the relative filtering efficiency of February sludge treated in different ways.

precipitate depends to some extent upon the nature of the particles. If precipitation by bringing the pH value to the apparent isoelectric point does not produce a precipitate easily filtered, it is possible that coprecipitation with oppositely charge particles may. The most promising precipitant examined was aluminum sulfate, which when sufficiently diluted yields positively charged particles of colloidal alumina. When planning the tests with aluminum sulfate, it was obvious that the sludge-alumina complex would have an apparent isoelectric point, so that pH value as well as alumina content would have *to be* controlled. It was apparent that this isoelectric point must lie at some value greater than 3.4, since a mixture of neutral sludge at pH = 3.4 with positively charged alumina must give a positively charged mass. At pH values less than 3.4, adding aluminum sulfate should do more harm than good, since the alumina particles would tend to add their positive charges to those of the sludge.

The effect of adding aluminum sulfate to February sludge is shown in curve *B* of Fig. 4. For each gram of dry solid matter in the sludge, there was added 0.1 gram of aluminum sulfate in diluting to standard conditions. The pH value was varied by adding sulfuric acid. The point of maximum occurs at a pH value greater than 3.4, as expected, and actually lies at 4.4. At this point the relative filtering efficiency has been increased from 5 to about 40, a greater increase than with acid alone, but still not enough to permit satisfactory filtering of winter sludge. The lowered relative filtering efficiency at pH values less than 3.4 also confirms the theory.

The same curious facts were found with the addition of aluminum sulfate that were noted when sulfuric acid alone was used. The rate of filtration was always increased about 8 times by this treatment, regardless of the initial rate, and the optimum always occurred at pH = 4.4.

Effect of Common Salt

The reasoning given in the theory indicates that the addition of any salt not reacting chemically with the sludge will tend to decrease the degree of swelling of the jellies and the degree of dispersion of the other materials. But the coalescence of particles depends, not only upon lowering the electrical difference of potential at the surface, but also upon the attraction of the substance of the particles for water. With material having a small attraction for water, like gold particles, precipitation is brought about by reducing the potential difference at the interface to less than about 15 millivolts, whereas precipitation of a dispersion of gelatin by the addition of sodium chloride requires a lowering of the potential difference to an almost immeasurable value, because the attraction of gelatin for water tends to keep the gelatin dispersed. In this respect, activated sludge resembles gelatin; such great quantities of salt are required as to make the ordinary process of salting out a prohibitive one for conditioning activated sludge.

Effect of Temperature

After a suspension has been brought to its isolectric point, precipitation is not instantaneous; a certain amount of time is

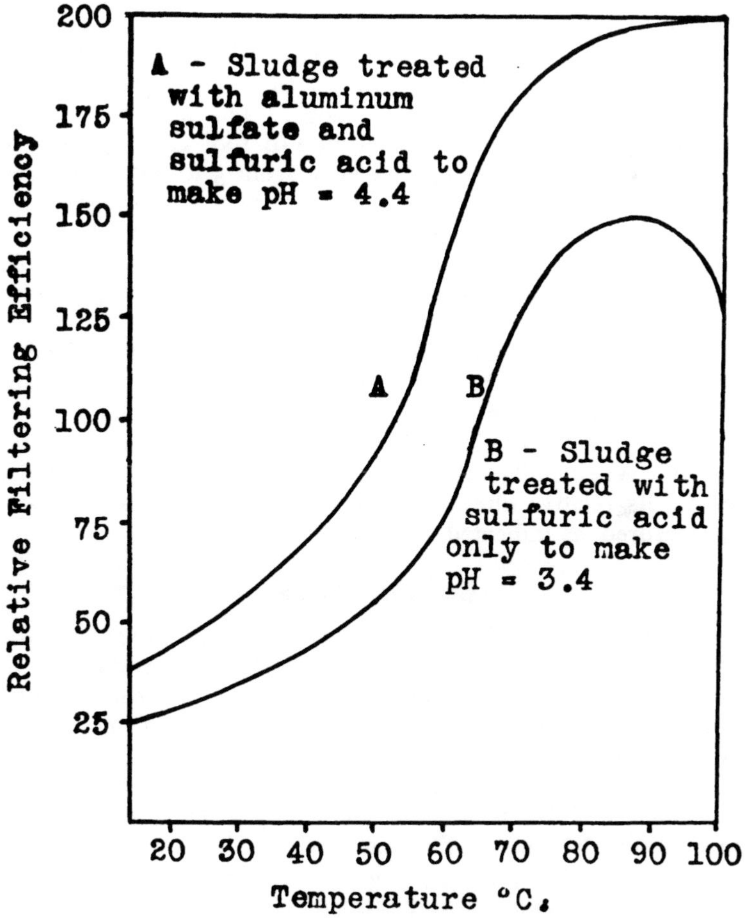

FIG. 5.—Effect of heating upon relative filtering efficiency of February sludge previously treated with aluminum sulfate and sulfuric acid or with acid alone.

required for the particles to come together. With activated sludge this time element is very important, because the sludge tends to become septic and break down upon standing without aeration. In fact, letting activated sludge stand very long after acidifying does not bring about a further increase in relative filtering efficiency, but rather a decrease. But rise of temperature tends to bring about an increase in rate of coalescence when colloidal particles are brought to their precipitating point.

Fig. 5 shows the effect of increase of temperature upon the relative filtering efficiency of February sludge with and without the addition of aluminum sulfate, as noted above, and with the pH value adjusted to the optimum point in each case. By acidifying to make pH = 3.4 and heating to 88° C., we get an increase in relative filtering efficiency from 5 to 150. Higher temperatures seem to break down the sludge. The use of aluminum sulfate at pH = 4.4 and a temperature of 100° C. bring about an increase in relative filtering efficiency from 5 to 200. This represents an increase of about 4000 per cent and brings winter sludge into a condition such that it filters twice as rapidly as untreated summer sludge. Thus a very simple and practical solution was found for what seemed to be a very perplexing problem of vast importance.

In order to get a large increase in relative filtering efficiency by the application of heat, it is necessary to adjust the pH value to that of the apparent isoelectric point. If sludge far removed from its isoelectric point is heated, it becomes more highly dispersed. When sludge at its normal pH value of 7.4 is heated to 88° C., its relative filtering efficiency drops to values too low to be measured, as indicated in curve *C* of Fig. 4.

Seasonal Changes

Fig. 6 illustrates the seasonal changes in the sludge treated in different ways. The variable factor is the temperature of the mixed liquor in the aeration tanks, as was demonstrated by a large scale test made during one winter, in which one aeration tank was allowed to treat sewage at normal winter temperatures and another to treat only sewage heated to summer temperature. It should be noted in the curves how any given treatment seems to multiply the relative filtering efficiency by a constant factor, regardless of the initial condition of the sludge. Fig. 6 formed the basis for planning the treatment of the sludge for filtering.

Dewatering the Sludge

The waste activated sludge flows first into a 42,000-gal. tank equipped with diffuser plates and baffles to facilitate mixing. Sulphuric acid is mixed with the sludge in this tank and the flow is so regulated as to keep the pH value of the filter effluent constant at pH = 3.4. A control laboratory is placed at one end of the tank, where pH values of the filter effluent are being made continuously and the acid valve adjusted accordingly. It is planned to replace the manual work here by a recording potentiometer equipped to move the valve adjustment to maintain

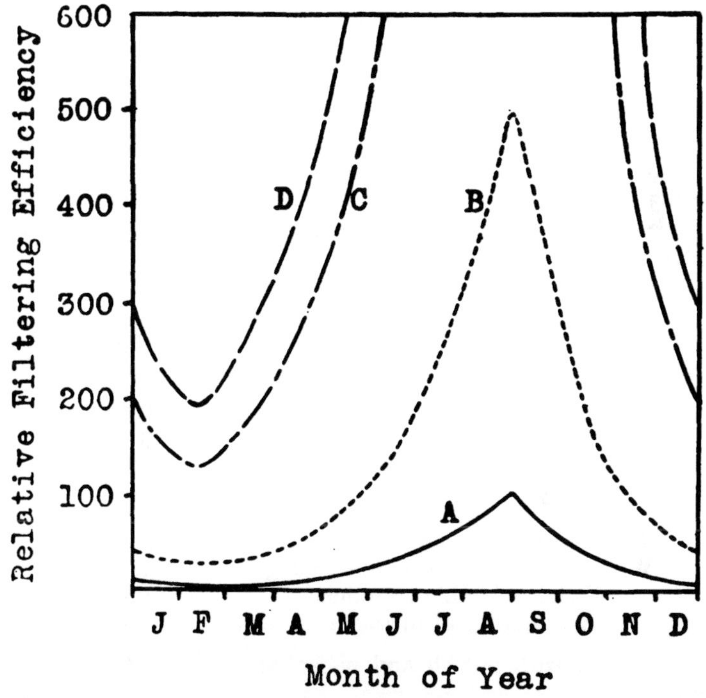

FIG. 6.—Seasonal changes in relative filtering efficiency of activated sludge treated in various ways. A — cold sludge not treated. B — cold sludge treated with sulfuric acid to make $p^H = 3.4$. C—sludge acidified to make $p^H = 3.4$ in filtrate and then heated 88°C. D—sludge treated with aluminum sulfate and sulfuric acid to make $p^H = 4.4$ in filtrate and then heated to 88°C.

constant voltage in a quinhydrone electrode vessel through which filter effluent is flowing continuously.

Where the use of sulfuric acid alone and heat are sufficient to make the sludge filter easily, it is to be preferred to the use of aluminum sulfate in addition, because of the higher cost of the alum mixture. The acidified sludge passes into a heating system where its temperature is raised to the desired point varying from no applied heat in September to about 80° C. in February. The heated sludge then passes on to the 24 Oliver continuous vacuum filters, each with about 500 square feet of effective filtering surface. The hot filtrate is returned to the heat exchangers, where it gives up its heat to the incoming sludge, and from where it flows back into the sewage channel.

The hot sludge cake, containing about 80 per cent of water, is conveyed to 6 Atlas direct-indirect-heat continuous rotary driers, each 7 feet in diameter by 60 feet long, enclosed in brick settings. As the drum revolves, the hot gases from the furnace, mixed with large volumes of outside air, circulate around and into the drum through numerous air valves attached to the drum shell. After passing through the driers, the sludge is conveyed to rotary screens, from which the finer particles are transported directly to the storage house by belt conveyors, while the coarser particles are first passed through a pulverator. The powdered sludge is shipped away as fertilizer when ordered.

Analysis of Sludge

The following is a typical analysis of the powdered sludge as a fertilizer:

	Per Cent
Water	5.75
Total phosphoric acid	2.34
Total nitrogen, as ammonia	7.32
Water-insoluble nitrogen	6.65
Water-soluble nitrogen	0.67
Active water-insoluble nitrogen	3.94
Total available ammonia	4.61
Availability of nitrogen	62.97

When mixed with potash and phosphate, this sludge becomes a complete fertilizer and experience has proved that such mixtures compare very favorably with the best commercial fertilizers on the market.

Thus by means of chemical as well as engineering experiments, Milwaukee has developed a plant for converting a waste which was offensive and prejudicial to health into a real source of wealth.

REFERENCES

1 Wilson, *J. Am. Chem. Soc.*, **38**, 1982 (1916).

2 Beans and Eastlack, *Ibid.*, **37**, 2667 (1915).

3 Donnan, *Z. Elektrochem.*, **17**, 572 (1911).

4 Procter and Wilson, *J. Chem. Soc.*, **109**, 307 (1916). Wilson and Wilson, *J. Am. Chem. Soc.*, **40**, 886 (1918). Wilson, Chemistry of Leather Manufacture, New York, 1923; pp. 94–132.

5 Wilson and Heisig, *J. Ind. Eng. Chem.*, **13**, 406 (1921). Wilson, Copeland and Heisig, *J. Am. Water Works Assoc.*, **8**, 486 (1921). Wilson, Copeland and Heisig, *Ind. Eng. Chem.*, **14**, 128 (1922); **15**, 956 (1923). Wilson, *Proc. International Conference on Sanitary Engineering*, 110 (1924). Wilson, *Mechanical Engineer*, **47**, 837 (1925). Wilson, *Nation's Health*, **7**, 305 (1925). Wilson, Bogue's Colloidal Behavior, New York, 1924, chapter 34. Milwaukee Sewerage Commission, Annual Reports, 1920–25.

THE DIRECT MEASUREMENT OF OSMOTIC PRESSURE

BY

J. C. W. FRAZER
PROFESSOR OF CHEMISTRY
JOHNS HOPKINS UNIVERSITY

Lecture delivered at Columbia University in the special course on Contemporary Developments in Chemistry given in the Summer Session of 1926 on the occasion of the opening of the Chandler Chemical Laboratories

COLUMBIA UNIVERSITY PRESS
NEW YORK
1927

Printed from type. Published March, 1927

Printed in the United States of America

THE DIRECT MEASUREMENT OF OSMOTIC PRESSURE

By J. C. W. Frazer

For many years after its discovery the phenomenon of osmotic pressure was of little scientific interest and remained so for chemists until it was made the most striking feature of van't Hoff's theory of solutions. This theory dominated chemistry for many years and has been the inspiration for an enormous amount of experimental and theoretical work. In spite of the interest given it by van't Hoff's work and by its numerous applications by others, it is remarkable that the number of measurements of this force is very small. One can almost count on the fingers of one hand the number of solutes whose osmotic pressure has been measured directly. This lack of experimental material, however, is largely supplied by the enormous amount of data obtained by indirect methods.

The subject of Osmotic Pressure may be discussed from several standpoints. First, its theoretical interest as bearing on the behavior of substances in solution; secondly, the applications of the phenomenon, which carries us at once into the complicated field of biology and, lastly the experimental investigations upon which our knowledge of the phenomenon depends. This lecture is to deal almost exclusively with the direct experimental investigation of osmotic pressure, in which I hope to point out some of the experimental difficulties encountered in this work that account for the small amount of experimental material on record.

The existence of phases composed of the pure solvent in equilibrium with the solution enables one to measure the osmotic pressure by a number of indirect methods, for the application of thermodynamics to the conditions for these equilibria gives us the freezing point, boiling point and vapor pressure methods for studying solutions. The vapor pressure method is the only one of these capable of use at different temperatures and will be the only one of these indirect methods mentioned here, in spite of the fact that the freezing point method is a most convenient and excellent one for use in the neighborhood of the freezing point of the solvent. In spite of the difficulty of obtaining osmotic pressure measurements I believe the osmotic method is the only method which can give the desired data with the accuracy required by the more modern theories of Debye, Hückel and others.

It is not surprising that certain practical applications of the phenomenon of osmotic pressure are more evident than is its bearing on chemistry and so we find that the first interest was shown in these applications. Thus the first important work was done not by chemists but by botanists, since its bearing on the motion of liquids in animals and plants was evident. In fact the measurement of osmotic pressure was made possible by the work of two botanists, Traube and Pfeffer.

The experiments of Traube[1] were performed with the simplest possible apparatus. A straight open tube of small bore with a drop of a solution of one of the membrane-formers on one end was dipped into a solution of the other membrane-former and there was at once produced over the surface of the drop a thin layer of the precipitate which hung from the end of the tube as a small sac. By investigating the behavior of these precipitated membranes toward various solutes Traube was able to show that their permeability toward solutes varied, but for some of them at least certain of these precipitated membranes were, as far as he could tell, entirely impervious while still allowing the passage of the solvent. Thus, we have established for us the first important step in measurement of osmotic pressure, the possibility of forming artificially semipermeable membranes which up to that time were known only in living organisms. Much skill and patience were required to establish the semipermeable character of these delicate membranes, for it must be borne in mind that they were nothing more than very delicate sacs attached to the end of small tubes and suspended freely in the liquid beneath. If only a slight difference in osmotic pressure or mechanical pressure existed on the two sides such delicate membranes were ruptured at once.

The next step in the experimental investigation was taken by W. Pfeffer.[2] The outcome of these experiments of Pfeffer was to show how it was possible to support the membranes of Traube so that pressures of considerable magnitude could be measured. Pfeffer's first attempts were but little improvement on the experiments of Traube. His first efforts were made by forming the membrane so that it closed the end of an open capillary tube. After this had failed, attempts were made to support the membrane by forming it on closely woven silk or linen. In all these experiments the membrane was forced through the pores by very small pressure differences and appeared beneath the supporting material. Parchment paper and animal membranes were also found to be unsatisfactory supports and finally porous clay vessels such as are used for batteries were investigated. After long experimentation with this means of support, suitable porous vessels were obtained and satisfactory methods devised for depositing the membrane within its pores. This, however, was accomplished only after much care even though, to aid in the solution of these difficulties, he had an entire pottery at his disposal. Pfeffer gives no details for the production of these cells though he gives in considerable detail the method used for the successful deposition of the membrane. The cell was treated first with a solution of caustic potash and then with a solution of hydrochloric acid (3 per cent) to remove iron and other materials which seemed to act injuriously. After thorough washing with water the cell was dried. The air was then pumped from the pores of the cell with an air pump and the cell immersed in water thus filling the interior spaces of the cell wall with water. In this condition it was immersed in 3 per cent copper sulphate solution for some hours. The interior was then washed out rapidly several times with distilled water, dried with strips of filter paper and left for some time in the air until it felt barely moist. It was then filled with a 3 per cent solution of potassium ferrocyanide solution. After standing for 24–48 hours it was completely filled with this solution and closed. It then developed slowly a certain pressure. After standing for another 24–48 hours in this condition it was opened and filled with some of the same ferrocyanide solution to which 1½ per cent of potassium nitrate was added. Under these conditions a pressure of about 3 atmospheres was produced. The gradual production of pressure in this way and slow building up of the membrane was found necessary for the production of useful cells. Only by careful operation in this way could successful membranes be obtained. When the pressure was allowed to rise rapidly the membrane was ruptured and frequently the copper ferro-cyanide appeared on the outside of the cell. The membrane could be deposited in the interior of the cell wall by allowing the membrane-formers to diffuse into the cell wall from opposite sides. When broken the membrane appeared as a thin line throughout the cell wall. Much better success was had by depositing the membrane in the pores immediately within the interior surface.

Any one who has attempted to repeat these osmotic pressure measurements of Pfeffer must admire his skill and regret that he does not give the method used for producing the clay cells. For years no one was able to repeat them even though careful directions were given by Pfeffer for every step except the production of the cell. This last really is one of the most important and difficult parts of the procedure. To judge Pfeffer's success we have his statement that of twenty cells he had practically no failures.

Pfeffer studied many other membranes than copper ferrocyanide, such for instance as Berlin blue, calcium phosphate, ferric hydroxide and ferric phosphate; though copper ferrocyanide was found to be the best. His results show clearly that, for the range of pressures covered, his apparatus was very trustworthy and the mechanical weakness of the apparatus alone limited the range of concentration covered. His membranes appear to be truly semipermeable for such substances as sucrose and dextrin as no leakage of solute was detectable in twelve days. Pfeffer's study of the effect of concentration and temperature on the osmotic pressure especially of sucrose solutions was of fundamental importance in establishing van't Hoff's theoretical views and will always remain a model of experimentation.

Many attempts were made to repeat Pfeffer's experiments but without success. These failures were almost certainly due to the lack of cells of suitable texture as Pfeffer's description of the method of depositing the membrane is given in considerable detail.

The next improvement in experimentation was the electrolytic deposition of the membrane. H. N. Morse while engaged in the electrolytic production of permanganic acid noticed that at times the porous cell separating the electrodes exhibited osmotic activity after manganese dioxide had deposited in its pores. This suggested the idea of forming osmotic membranes electrolytically in porous cells and this was successfully demonstrated by Morse and Horn.[3]

A careful investigation has been made to determine the desirable qualities which an osmotic cell must possess. A microscopic study of cells good, bad, and indifferent has shown the necessity of having a fine uniform texture of cell wall. A cell that lacks these qualities gives imperfect support to the membrane as had been previously noted by Pfeffer in certain of the supporting materials he first used. To secure the desired texture we have used only carefully bolted clays and a mixture of two clays chosen which required the addition of no binder. This material was molded into cylindrical form under high pressure and the cell finally turned from this on the lathe. Fig. 1 shows a photograph of a section of a potter's cell (A) and one made by this process (B). The difference is quite evident. Such cells have been useful for the study of solutions of solutes of non-electrolytes of high molecular weight.

The experiments of Berkeley and Hartley[4] were made in a porcelain tube of Pukal mixture open at both ends and glazed at its two extremities. The membrane of copper ferrocyanide was formed near the exterior surface by a method somewhat similar to that described by Pfeffer and finished by the electrolytic method. A gun metal jacket was then fixed around the tube with provision for making water tight joints at the glazed ends of the tube. The glazed ends extended outside the metal jacket, and one end of the tube was closed by a glass tap and the other carried a glass tube of small bore. The solution was placed inside the metal jacket around the porcelain tube and water filled the interior of the porcelain tube and partially filled the capillary glass tube mentioned above. Mechanical pressure

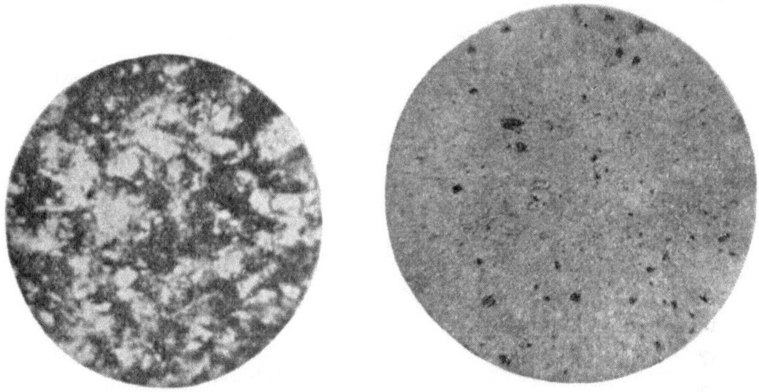

A B

FIG. 1.

was applied to the solution and gradually increased until the motion of the meniscus in the capillary showed the osmotic pressure had been exceeded. By altering the pressure and noting the motion of the liquid in the capillary the true osmotic pressure was located between two slightly differing pressures and the osmotic pressure calculated from the motion of the liquid in the capillary at the two pressures, one slightly in excess of, the other slightly below, the true osmotic pressure.

It is to be noted that this method may have certain advantages. As mechanical pressure was used it would seem that results may be obtained by this method with a less perfect cell and membrane than by Pfeffer's method of allowing the pressure to come to equilibrium pressure solely by the passage of water through the membrane into the solution.

The following results have been obtained by this method.

Osmotic Pressure of Cane Sugar Solutions at 0°

(Proc. Roy. Soc. 1916, A, **92**, 477)

Concentration (Grams of Sucrose in 100 Grams of Water)	Osmotic Pressure, Atmospheres
3.3927	2.23
10.184	6.85
20.525	14.21
30.608	21.87
33.945	24.55
81.262	67.74
111.731	100.13
141.107	134.84

Osmotic Pressure of Glucose Solutions at 0°

(Phil. Trans., 1906, A, **206**, 486)

Concentration (Grams of Glucose in 1 Litre of Solution)	Osmotic Pressure Observed, Atmospheres	Osmotic Pressure Calculated According to van't Hoff Atmospheres
99.8	13.21	12.42
199.5	29.17	24.82
319.2	53.19	39.72
448.6	87.87	55.83
548.6	121.18	68.27

Osmotic Pressure of Galactose Solutions at 0°

(Phil. Trans., 1906, A, **206**, 486)

Concentration (Grams of Galactose in 1 Litre Solution)	Osmotic Pressure Observed, Atmospheres	Osmotic Pressure Calculated According to van't Hoff Atmospheres
250	35.5	31.1
380	62.8	47.29
500	95.8	62.22

Osmotic Pressure of Mannitol Solutions at 0°

(Phil. Trans., 1906, A, **206**, 486)

Concentration (Grams of Mannitol in 1 Litre Solution)	Osmotic Pressure, Observed, Atmospheres
100	13.1
110	14.6
125	16.7

OSMOTIC PRESSURE OF SOLUTIONS OF α-METHYL GLUCOSIDE AT 0°

(Proc. Roy. Soc., 1916, A, 92, 477)

Concentration Grams of Solute in 100 Grams of Water	Osmotic Pressure, Atmospheres
22.873	29.87
34.767	47.72
44.928	63.84
54.869	80.78
63.845	95.97
75.709	117.19

OSMOTIC PRESSURE OF SOLUTIONS OF ISODULCILI. AT 0°

(Proc. Roy. Soc., 1916, A, 92, 477)

Concentration Grams of Solute in 100 Grams of Water)	Osmotic Pressure, Atmospheres
31.07	45.98
31.11	46.53
31.15	47.01
35.67	53.62

OSMOTIC PRESSURE OF SOLUTIONS OF α-TETRAMETHYL FERROCYANIDE AT 0°

(Proc. Roy. Soc., 1916, A, 92, 477)

Concentration (Grams of Solute in 100 Grams of Water)	Osmotic Pressure, Atmospheres
7.304	5.96
58.524	52.32

The measurements given above are not all equally trustworthy. Cane sugar solutions are most easily measured but in the case of mannitol solution the membranes used were not perfectly semi-permeable. The same is true of the measurements of α-methyl glucoside.

In all of the measurements that have been made in the Johns Hopkins laboratory the cell is required to develop pressure by the intake of water through the membrane as was done by Pfeffer. This imposes much more rigid requirements on the cell than the method used by Tamman and Berkeley and Hartley. By the former method a cell can be relied on to develop the maximum pressure of a solution only when in excellent condition. To develop a cell so that reliable measurements can be obtained, often requires continuous attention for periods of six months or longer. This involves subjecting the cell to the membrane-forming process several times each week and frequently allowing it to develop considerable osmotic pressure in order to perfect the membrane. During all of this development the cell is kept at constant temperature. It is frequently necessary to keep the solution in the cell for three weeks or more to be sure the maximum pressure has been developed. The amount of water the cell is required to take in during an experiment to compress the air in the manometer is very small as the manometers used were about 0.5 mm. diameter and approximately 400 mm. long. Under these conditions it is easy to see that an extremely small leak would prevent the development of the maximum pressure of the solution, and give unreliable results. For measuring the osmotic pressure of more concentrated solutions a different form of apparatus has been used which in its final form is admirable for such experiments.[5]

The details of construction are shown in the accompanying diagram (Fig. 2) in actual size. *M* is the clay cell with the membrane of copper ferrocyanide *K* deposited on its exterior. This extends only as far as indicated by the heavy line *K*, the neck and shoulder of the cell being glazed to ensure tight joints. *J* is the bronze cylinder into which the cell is fastened by means of the plug *P*. The latter has three legs upon which the apparatus stands and which serve as a means of screwing the plug into the cylinder. *N* is a thick rubber washer and *O* is a piece of commercial "Rainbow Packing," about a millimeter and a half thick. This packing gives perfect satisfaction, while ordinary rubber will not withstand higher pressures. At all other joints, *N*, *G*, *H*, the packing is surrounded on all sides so it can not get away, hence in these places the softer rubber washers answer very well. In fact, it is desirable to have *N* and *G* thick (about 4 mm.) and soft so that initial pressure may be secured by their compression.

R is a glass U-tube held in the mouth of the cell by means of the rubber stopper *Q*. This is simply to keep the solvent *L* up in the cell and yet leave it open to the atmosphere. *I* is the solution to be measured. *E* is the manometer attachment and is fastened onto the cylinder by means of nut *F*, *G* being a thick rubber washer described above. *A* is a piece of capillary glass tubing to which the manometer is sealed. The square shoulder at the lower end of *A* is secured by sealing on to a piece of tubing, of the size desired for *A*, another piece, having somewhat larger external dimensions. The joint is kept soft and the glass allowed to flow together until a considerable enlargement is secured, the internal diameter being kept of original size by careful blowing. After cooling, the larger piece of tubing is cut off close to the enlarged joint. *A* is then mounted in the lathe and the enlarged end ground to the desired size and shape. The grinding

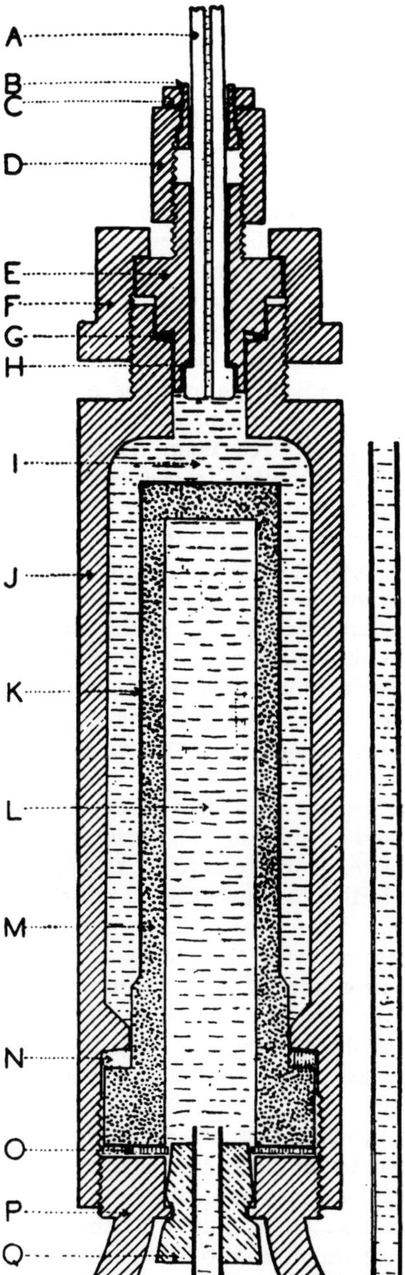

Fig. 2.

is done with carborundum dental wheels mounted on the lathe carriage and rotated at a high speed by a small electric motor. In this manner the lower end of *A* is made to fit accurately into the lower end of *E*, as shown in the diagram. To hold *A* in position, *B* is sealed onto *A* by means of sealing wax (Khotinsky) and after putting *D* into position, *C* is screwed onto *B*. Now it will be seen that by turning *D* to the left, *A* is raised and held tightly in position in *E* against the washer *H*. By turning *D* to the right, *A* is lowered and excess solution is allowed to escape. This method of joining glass and metal is of general application and has proven most satisfactory. Such a joint will hold tight under any pressure which will not actually break the glass.

Several points of advantage over the old type of cell deserve especial mention. In the first place the clay cell will withstand far greater pressure exerted upon its exterior than upon its interior, as would be expected from a consideration of the principles of arch construction. Secondly, the membrane itself is made stronger, for the pressure tends to force it farther into the pores and so compresses it, making it more firm and compact. With the membrane on the inside of the cell the pressure of the inclosed solution tends to expand the membrane out into the pores of the cell thus stretching the particles of the membrane and so making ruptures more likely to occur.

The cells used have the membrane deposited on the outer surface greatly increasing the area, hence equilibrium is much more readily established.

This form of apparatus fulfilled all expectations, proving convenient, strong, and of quick action, as the following results show. Equilibrium was reach in from 10 to 12 hours and in even less time when the initial pressure was as much as half the equilibrium pressure. The so-called "Thermometer effects"[6] which were so troublesome with the old type of apparatus, proved negligible in the new form of cell. Solutions whose osmotic pressures are not over 100 atmospheres can be measured with ease and expediency in the apparatus as described. It was impossible, however, to get glass manometers which would withstand greater pressure than this.

For pressures greater than 100 atmospheres the glass manometer was replaced by an electric resistance gage[7] developed and used by Johnston and Adams at the Geophysical Laboratory of Washington and later an adaptation of the water interferometer as shown in Fig. 3 was found to be much superior as it is devoid of hysteresis effects.[8]

The instrument is a modification of the Zeiss water interferometer. The cell ordinarily used in this instrument is replaced by a bronze block as shown in Fig. 3. Two longitudinal holes P_1', P_2', drilled parallel to each other are the two water compartments, which are closed at each end by heavy plate glass windows capable of withstanding the pressure. The pressure is communicated to one of these chambers through a mercury trap as shown while the other is open to atmospheric pressure.

On calibration, the pressure limit of adjustment of the interferometer was found to be approximately 60 atmospheres. It was therefore found necessary to devise some means by which the range of the instrument could be increased. This was accomplished by inserting a thin glass plate in the path of the beam passing through the non-pressure chamber, the glass plate being of such thickness that, when inserted, it would compensate for

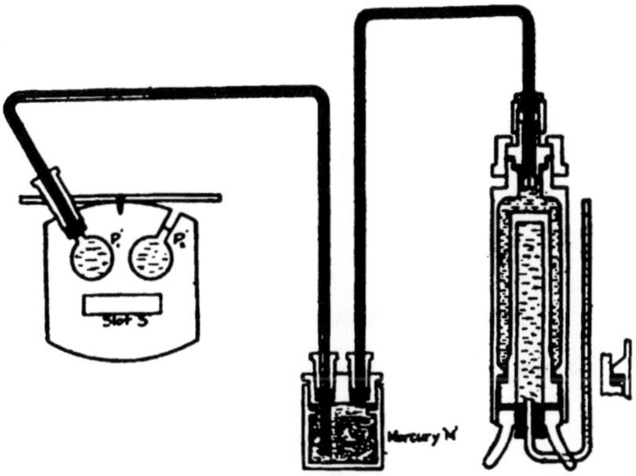

FIG. 3.

approximately 60 atmospheres. With this plate inserted the range of the instrument was from 60 to 120 atmospheres, approximately. The final adjustment for reading is accomplished by the usual method of tilting the glass plate of the interferometer. Thus, by making up several slides with plates of increasing thickness we were able to extend the range of the apparatus to any pressure required.

The glass plates mentioned above were attached firmly by Canada balsam to brass frames which were machined to fit into slides attached to the water bath of the interferometer.

We found that thin microscope-cover plates were quite regular in thickness and proved easily adaptable to the problem at hand. The plates used varied in thickness from 0.01016 to 0.3556 mm.

The method of mounting these thin glass plates was such that they could be replaced in exactly the same position and thus any error that would arise from any slight irregularity in the plate was avoided.

The apparatus was calibrated against an absolute pressure gage in the Geophysical Laboratory. The calibration was made at both 20° and 30°.

The following tables contain some of the experimental data obtained by the use of these methods.

Osmotic Pressure of Glucose

(Morse, Carnegie Institution Publication No. 198, p. 158)

Concentration	22°–25°	0.06°–0.38°	10°–10.40°
0.1	2.40	2.40	2.39
0.2	4.77	4.67	4.76
0.3	7.15	7.04	7.11
0.4	9.68	9.34	9.52
0.5	10.04	11.69	11.91
0.6	14.39	14.12	14.31
0.7	16.84	16.43	16.70
0.8	19.23	18.86	19.05
0.9	21.59	21.39	21.39
1.0	24.05	23.75	23.80

Osmotic Pressure of Cane Sugar

(Carnegie Institution, Publication No. 198, p. 184)

Concentration	0°	5°	10°	15°	20°	25°	30°
0.1	(2.462)	2.452	2.498	2.540	2.590	2.634	2.474
0.2	4.723	4.819	4.893	4.985	5.064	5.148	5.044
0.3	7.085	7.198	7.335	7.476	7.605	7.729	7.647
0.4	9.443	9.608	9.790	9.949	10.137	10.296	10.295
0.5	11.895	12.100	12.297	12.549	12.748	12.943	12.978
0.6	14.381	14.605	14.855	15.144	15.388	15.625	15.713
0.7	16.886	17.206	17.503	17.815	18.128	18.435	18.499
0.8	19.476	19.822	20.161	20.535	20.905	21.254	21.375
0.9	22.118	22.477	22.884	23.305	23.717	24.126	24.226
1.0	24.826	25.280	25.693	26.189	26.638	27.053	27.223

Osmotic Pressure of Mannitol

(Carnegie Institution, Publication No. 198, p. 207)

Concentration	10°	20°	30°	40°
0.1	2.314	2.395	2.467	2.557
0.2	4.609	4.781	4.943	5.107
0.3	6.940	7.181	7.430	7.664
0.4	9.209	9.570	9.881	10.216
0.5	11.613	11.960	12.345	12.804
0.6	....	....	....	15.315

OSMOTIC PRESSURE OF SUCROSE

(Lotz and Frazer, J. Am. Chem. Soc. 43, 2504 (1921))

Experiment	Sugar/1000 Cc. of Solution, Grams	Sugar/1000 G. of Water, Grams	Equilibrium Time, Minutes	Osmotic Pressure, Atmospheres
	30°			
1	478.3	680	90	57.5
2	472.0	665	12	56.6
3	597.0	958	90	87.2
4	605.4	980	150	90.4
5	608.5	990	165	92.0
6	700.2	1260	195	129.5
7	695.0	1242	225	127.4
8	781.4	1549	180	169.1
9	781.4	1549	225	168.6
10	777.3	1533	90	164.1
11*	831.5	1758	300	198.2
12	826.2	1737	195	200.2
13	839.8	1796	165	206.1
	55.7°			
14	477.2	674	22	61.0
15	481.4	685	20	63.1
16	610.6	996	35	97.4
17	612.7	1000	16	98.7
18	702.3	1270	30	132.4
19	706.4	1284	55	133.5
20	782.5	1556	75	170.6
21	791.9	1590	90	178.7
22	856.5	1877	120	222.0
23	842.9	1810	135	213.8
24	900.2	2112	165	259.3
25	910.6	2190	135	265.6
26	no analysis		135	273.0

*Unseasoned cell.

OSMOTIC PRESSURE OF PHENOL SOLUTIONS AT 30°

(Grollman and Frazer, J. Am. Chem. Soc., 45, 1707 (1923))

Concentration	Osmotic Pressure Calculated from Formula	Osmotic Pressure Observed	Calc. O. P. = Obs. O. P. Degree of a Association
0.1	2.46	1.46	1.68
.2	4.92	2.84	1.73
.3	7.38	3.93	1.88
.4	9.83	5.12	1.92
.5	12.28	6.40	1.92
.6	14.73	7.62	1.93
.7	17.16	8.82	1.95
.8	19.59	10.05	1.95
.9	22.02	11.28	1.95

Osmotic Pressure of Sucrose at 30°

(Frazer and Myrick, J. Am. Chem. Soc. 38, 1916 (1916))

Experiment	Grams Sugar, 1000 Grams Water	Grams Sugar, 100 Grams Solute	Final Concentration Grams Sugar, 1000 Grams Water	Time at which Equilibrium was Reached	Osmotic Pressure in Atmospheres
1	....	....	33.96	3–15 days	2.474
1	339.6	25.351	339.60	3–15 days	27.22
1	679.2	40.448	676.96	29 hours	57.70
2	679.2	40.448	668.86	1 hour	57.47
6	679.2	40.448	664.98	18 hours	57.16
3	1018.8	50.464	957.35	19 hours	88.10
4	1018.8	50.464	997.24	20 hours	90.04
19	1018.8	50.464	1018.80	5 hours	94.75
20	1018.8	50.464	1007.29	1½ hours	94.20
8	1358.4	57.598	1302.34	37 hours	130.80
9	1358.4	57.598	1289.30	6 hours	130.20
10	1358.4	57.598	1326.70	3½ hours	134.00
11	1358.4	57.598	1327.68	6½ hours	133.90
12	1358.4	57.598	1300.97	3 hours	131.50
15	1698.0	62.932	1597.00	16 hours	174.30
22	1698.0	62.932	1618.74	3 hours	175.80
17	2037.6	67.079	1942.56	7 hours	220.40
21	2037.6	67.079	1917.65	6 hours	216.10

Mention has been made of the fact that progressive improvement of the cells has been secured by increasing the fineness and uniformity of texture of the cell wall. Recently efforts have been made to secure still further improvement in this respect by depositing within the pores of the cell such substances as magnesium silicate before depositing the copper ferrocyanide membrane. This has resulted in more perfect cells and has eliminated the time required to get a cell into condition for giving reliable measurements. When such "filled" cells are dried out there is great danger that the cell wall will be disrupted by contraction of the gelatinous precipitation. The cell wall may be "filled" to much better advantage in the following way. The cell is first thoroughly dried in an air bath at an elevated temperature and when cold the cell wall is filled with such substances as the esters of orthosilicic acid. The cell is then exposed to water vapor and the ester hydrolysed in place leaving the porous silica in the interstices. As this silica occupies almost the same volume as the ester there is only a small contraction of the silica when the cell is dried out and much less danger of cracking the cell. This treatment may be repeated, this time allowing the ester to penetrate only the portion of the cell wall where the membrane is to be located. In this way a cell wall is obtained in which the support for the membrane is silica gel of fine, uniform porosity.

Only qualitative experiments have been made with these cells, but they seem to show that still further improvement has been secured. It is not necessary that the original cell, if improved by the method mentioned above, should be prepared with the great care that has been taken in the past. (The formation under pressure, etc.)

The vapor pressure method of studying solutions is perhaps the most desirable of all. The vapor space above a solution of a nonvolatile solute is an ideal semipermeable membrane as it enables the separation of the pure solvent from the solution. The measurement of the difference of the vapor pressure of solutions and the pure solvent at the same temperature is quite as important as the determination of the difference in pressure (the osmotic pressure) that must be put on solution and solvent to place them in equilibrium. The vapor pressure method as applied to the study of aqueous solutions is mentioned briefly below. At this point I wish to call attention to a somewhat different method of obtaining osmotic data and the object in view in the experiments that are under way.

In the ordinary method of determining osmotic pressure the solution is brought into equilibrium with the solvent by subjecting the solution to a certain excess pressure, this excess is the osmotic pressure of the solution. Solution and solvent can be brought into equilibrium by exerting a tension on the solvent, the amount necessary to bring about equilibrium is again the osmotic pressure of the solution. Thus if pure solvent evaporates from a porous surface which closes a vessel filled with the pure liquid solvent, distillation may be expected to continue from the pure solvent to the solution until the solvent is under sufficient tension to lower its vapor pressure to that of the solution and, if measured by a mercury column, would give the osmotic pressure of the solution. It will not be possible to use this method for pressures much above one atmosphere. The method will therefore be limited to quite dilute solutions but the results should be quite accurate. The only difficulty is to get the solvent to sustain the necessary tension. It has been shown that this can be done for higher tensions than mentioned above even when the solvent contains dissolved air. The importance of the method will be to obtain accurate data on solutions of electrolytes at great dilution.

The importance of vapor pressure measurements as a means of calculating osmotic pressure has already been mentioned. The two principal methods used to measure the vapor pressure of aqueous solutions are (1) the dynamic method and (2) the static method.

Berkeley, Hartley and Burton [9] have brought the dynamic method to its highest state of perfection and have recorded many accurate measurements of the vapor pressure of solutions of sucrose and a few other solutes. These results are in excellent agreement with their own measurements of the osmotic pressure of these same solutions. They are also in quite good agreement with the osmotic pressures of Frazer and Myrick as shown by the table below.

The following table is found in the second edition of Millard's Physical Chemistry for Colleges. In this table the author has compared the results of Frazer and Myrick on cane sugar solutions with the osmotic pressures calculated from the vapor pressure data of Berkeley and Hartley (loc. cit.) by the use of the equation.

$$\pi = \frac{RT}{V_1} ln \frac{p_0}{p}.$$

CANE SUGAR SOLUTIONS AT 30°

(Millard's Physical Chemistry for Colleges, 2d Ed., p. 162)

Molal Concentration	Measured Osmotic Pressure (Frazer and Myrick), Atmospheres	Calculated Osmotic Pressure (Berkeley, Hartley and Burton)	Percentage Error
0.10	2.47	2.47	
1.00	27.22	27.0	1.0
2.00	58.37	58.4	
3.00	95.16	96.2	
4.00	138.96	138.3	0.5
5.00	187.3	182.5	2.5
6.00	232.3	230.9	0.6

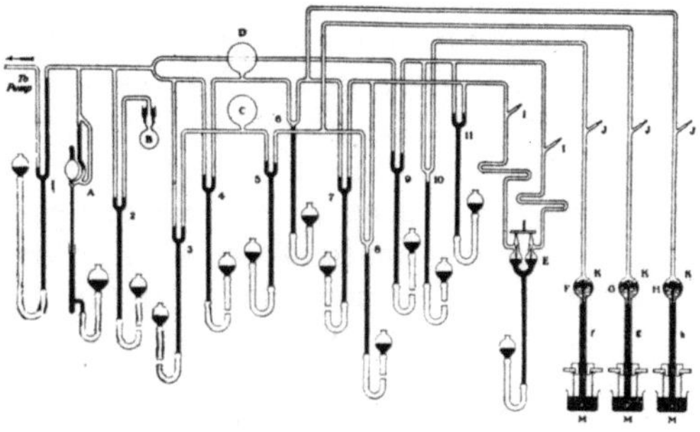

FIG. 4.

The static method for studying the vapor pressure of solutions has been used in all the experiments at Johns Hopkins.

As described in various articles this method has been sufficiently perfected so that it is believed the quantity measured, which is the vapor pressure difference between solvent and solution, is correct to something less than 0.001 mm. mercury.[10]

Fig. 4 shows the apparatus for the use of the static method, Fig. 5 the Rayleigh manometer with which the pressure differences are measured and Fig. 6, the construction of the constant

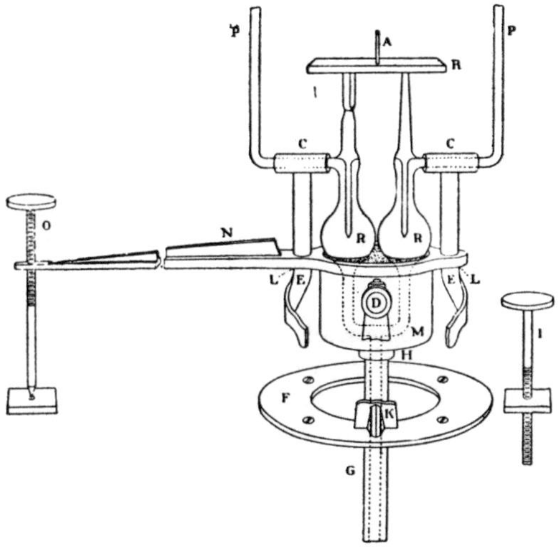

FIG. 5.

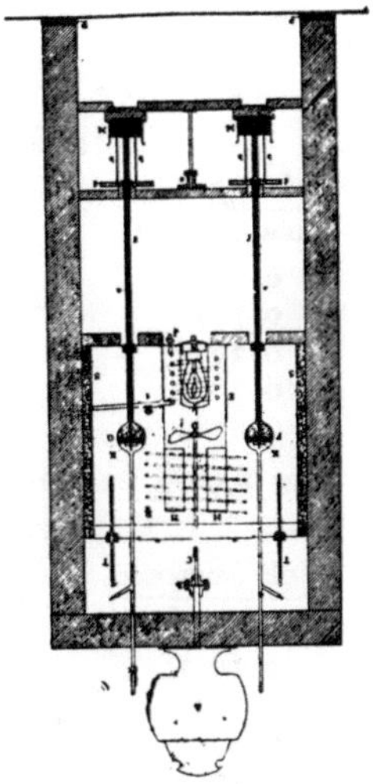

FIG. 6.

temperature bath with two of the bulbs for solution and solvent and stirrers in position.

Below are given some tables of data obtained by this method from which the osmotic pressure may be easily calculated.

THE VAPOR PRESSURE OF AQUEOUS SOLUTIONS OF MANNITE AT 20° C.

T. H. Rogers, J. Am. Chem. Soc., **42**, 1801 (1920)

Mols Mannite / 1000 g. H_2O	$p_0 - p$	Mols Mannite / 1000 g. H_2O	$p_0 - p$
0.0984	0.0307	0.5958	0.1863
0.1977	0.0614	0.6934	0.2162
0.2962	0.0922	0.7927	0.2478
0.3945	0.1227	0.8913	0.2791
0.4938	0.1536	0.8922	0.2792
0.5944	0.1860	0.9908	0.3096

THE VAPOR PRESSURE OF AQUEOUS SOLUTIONS OF CANE SUGAR AT 20° C.

H. K. Parker, Dissertation, Johns Hopkins University, 1921

Mols Cane Sugar / 1000 g. H_2O	$p_0 - p$	Mols Cane Sugar / 1000 g. H_2O	$p_0 - p$
0.0991	0.0312	0.9937	0.3379
0.1992	0.0627	2.0162	0.7432
0.4011	0.1276	2.9916	1.1734
0.6011	0.1984	4.0368	0.7432
0.6960	0.2301		0.9339
		4.9815	1.1734
			0.9750

THE VAPOR PRESSURE OF LITHIUM CHLORIDE SOLUTIONS AT 20° C.

W. H. Bahlke, J. Am. Chem. Soc., **45**, 2933 (1923)

Mols LiCl / 1000 g. H_2O	$p_0 - p$	Mols LiCl / 1000 g. H_2O	$p_0 - p$
0.0968	0.0559	0.5897	0.3566
0.1842	0.1071	0.7257	0.4428
0.2952	0.1730	0.7810	0.4782
0.3862	0.2290	0.9268	0.5749
0.4782	0.2858	1.0316	0.6489

THE VAPOR PRESSURE OF AQUEOUS SOLUTIONS OF CANE SUGAR AT 25° C.

L. C. Bear, Dissertation, Johns Hopkins University, 1922

Mols Sucrose / 1000 g. H_2O	$p_0 - p$	Mols Sucrose / 1000 g. H_2O	$p_0 - p$
0.1012	0.0381	0.7954	0.3487
0.1991	0.0791	1.0604	0.4812
0.3012	0.1195	1.9800	0.9762
0.4014	0.1637	2.9625	1.5566
0.5130	0.2224	2.7834	1.4543
0.5956	0.2622	3.9566	2.1994
0.6987	0.3027	4.7656	2.7411

THE VAPOR PRESSURE OF AQUEOUS SOLUTIONS OF SODIUM CHLORIDE AT 20° C.

W. R. Norris, Dissertation, Johns Hopkins University, 1922

Mols NaCl / 1000 g. H_2O	$p_0 - p$	Mols NaCl / 1000 g. H_2O	$p_0 - p$
0.0422	0.0247		0.5553
0.0510	0.0297	2.5707	1.0195
*0.0542	0.0301		———
*0.0953	0.0538		1.5748
0.1019	0.0590		
0.1976	0.1139		1.5748
*0.2090	0.1199	3.7891	0.8664
*0.3094	0.1781		———
*0.4233	0.2434		2.4412
*0.5217	0.2985		
0.5527	0.3173		2.4412
0.7072	0.4068	5.8290	1.6133
*0.8312	0.4771		———
0.9599	0.5553		4.0535
1.6133	0.9516		

* Read against Old Solvent.

THE VAPOR PRESSURE OF AQUEOUS SOLUTIONS OF SODIUM CHLORIDE AT 25° C.

S. S. Negus, Dissertation, Johns Hopkins University, 1923

Mols NaCl / 1000 g. H_2O	$p_0 - p$	Mols NaCl / 1000 g. H_2O	$p_0 - p$
0.0422	0.0338		0.1976
0.0709	0.0535	2.4165	1.7913
0.1015	0.0790		——
0.1976	0.1535		1.9889
0.3051	0.2366		
0.4112	0.3180		0.5689
0.5132	0.3967	3.0264	1.9889
0.6095	0.4714		——
0.7255	0.5631		2.5578
0.8230	0.6410		
0.9084	0.7097		1.1094
1.0284	0.8059	3.5795	1.9889
1.5642	1.2467		——
			3.0983
	0.9843		
2.0813	0.7097		0.8319
	——	4.3817	3.0983
	1.6940		——
			3.9302
	0.5446		
2.1916	1.2467		0.6135
	——	4.9607	3.9392
	1.7913		——
			4.5437

THE VAPOR PRESSURE OF AQUEOUS SOLUTIONS OF POTASSIUM NITRATE AT 20° C.

R. K. Taylor, Dissertation, Johns Hopkins University, 1923

Mols KNO_3 / 1000 g. H_2O	$p_0 - p$	Mols KNO_3 / 1000 g. H_2O	$p_0 - p$
0.0979	0.05448	1.2029	0.53795
0.1947	0.10524		(0.09388)
0.2913	0.15396		
0.3875	0.20040	1.4652	0.63183
0.4837	0.24570		
0.5415	0.27238	1.8098	0.74267
0.6261	0.30979		(0.15808)
0.7631	0.36802		
0.8205	0.39318	2.3504	0.90075
0.9114	0.42690		(0.26892)
		2.8873	1.03889
			(0.13814)

THE VAPOR PRESSURE LOWERING OF AQUEOUS SULFURIC ACID SOLUTIONS AT 25° C.

Arthur Grollman, J. Am. Chem. Soc., **47**, 714 (1925)

$\frac{\text{Mols } H_2SO_4}{1000 \text{ g. } H_2O}$	$p_0 - p$	$\frac{\text{Mols } H_2SO_4}{1000 \text{ g. } H_2O}$	$p_0 - p$
0.073	0.060	1.282	1.193
0.241	0.201	1.671	1.664
0.315	0.264	1.772	1.786
0.549	0.469	2.009	2.086
0.636	0.547	2.468	2.698
0.892	0.791	2.871	3.337
1.097	0.995		

VAPOR PRESSURE OF AQUEOUS SOLUTIONS OF HYDROCHLORIC ACID AT 25° C.

C. V. Holland, Dissertation, Johns Hopkins University, 1924

$\frac{\text{Mols HCl}}{1000 \text{ g. } H_2O}$	$p_0 - p$	$\frac{\text{Mols HCl}}{1000 \text{ g. } H_2O}$	$p_0 - p$
0.0421	0.0274	0.6959	0.5890
0.0745	0.0504	0.8187	0.7188
0.0997	0.0718	0.8837	0.7778
0.1444	0.1047	0.9691	0.8722
0.1817	0.1331	1.0716	0.9737
0.2205	0.1737	1.2748	1.2049
0.2477	0.1959	1.4588	1.4159
0.2665	0.2123	1.7033	1.7218
0.2972	0.2320	1.7597	1.7906
0.3089	0.2380	2.3646	2.6499
0.3384	0.2786	2.4051	2.7146
0.4326	0.3481	2.7074	3.1913
0.4909	0.4090	3.1460	3.9507
0.4936	0.4094	3.5511	4.7195
0.6023	0.5026		

REFERENCES

1 Archiv. f. Anatomie und Physiologie, 1867, 87 Botan. Zeitung, 1875, 56.
2 Osmotische Untersuchen, Leipzig, 1877.
3 Am. Chem. J. **26**, 80 (1901).
4 Phil. Trans. A, **206**, 486 (1906).
5 Frazer and Myrick, J. Am. Chem. Soc., **38**, 1907 (1916).
6 Publication No. 198. Carnegie Institution of Washington.
7 Lesell, Dissertation, Upsala, 1903; Lafay, Compt. rend. **149**, 566–569 (1909); Bridgeman, Proc. Am. Acad. Arts and Sci., **47**, 11, 355 (1911); **49**, 11, 640 (1913).
8 J. Am. Chem. Soc., **43**, 2498; 2501 (1921).
9 Phil. Trans., A, **218**, 295 (1919).
10 Frazer, Lovelace & Miller, J. Am. Soc., **38**, 515 (1916); Frazer, Lovelace & Rogers, *ibid.*, **42**, 1793 (1920); Frazer, Lovelace & Sease, *ibid.*, **43**, 102 (1921); Frazer Lovelace & Bahlke, *ibid.*, **45**, 2930 (1923); H. Parker, Dissertation, Johns Hopkins Univ., (1921); L. C. Beard, Dissertation, Johns Hopkins Univ. (1922); W. Norris, Dissertation, Johns Hopkins Univ. (1922); R. K. Taylor, Dissertation, Johns Hopkins Univ. (1923); S. S. Negus, Dissertation, Johns Hopkins Univ. (1923); Grollman, J. Am. Chem. Soc., **47**, 714 (1925); Holland, Dissertation, Johns Hopkins Univ. (1924).

CHEMISTRY OF BACTERIA

BY

TREAT B. JOHNSON

PROFESSOR OF ORGANIC CHEMISTRY

YALE UNIVERSITY

Lecture delivered at Columbia University in the special course on Contemporary Developments in Chemistry given in the Summer Session of 1926 on the occasion of the opening of the Chandler Chemical Laboratories

COLUMBIA UNIVERSITY PRESS

NEW YORK

1927

Printed from type. Published March, 1927

Printed in the United States of America

CHEMISTRY OF BACTERIA

TREAT B. JOHNSON

Professor of Organic Chemistry, Yale University

Introduction

The causation of disease by microorganisms is a subject of the very greatest scientific and economic importance. So promising of practical results is it as a field of research that large numbers of workers in various branches of science all over the world are devoting their time and skill to the study of *bacteria*. As a result of their activities we are accumulating most valuable scientific data, and acquiring, thereby, a knowledge of many of the bacterial diseases which is leading to discoveries of the most practical and far-reaching benefit to mankind.

A bacterium is a creature composed of a single cell exhibiting the activities characteristic of living things. He is a perfectly independent being doing everything for himself and he plays in nature a role which is an exceedingly important one. In fact, bacteria are not only curious and interesting as objects of study but they are indispensable to the continuance of the higher forms of life upon the earth. "They may well be called, in general, man's invisible friends; for without them the earth would soon be depopulated and lapse into what at first it was—a lifeless waste." *

We have good reasons for believing today that, when life first appeared on this earth, it showed itself in a primitive form and perhaps as is revealed to us in a living bacterium. In other words, bacteria are representatives of one of the first stages of an evolutionary, cellular change, revealing to us the original simplicity of living organisms. So small are bacteria and so simple are they in their activity and physical structure, that it has not been easy to decide from biological data whether they are plants or animals. It is now the prevailing opinion, which is supported by chemical evidence, that they are plants. They are characterized biologically by their speedy rate of reproduction. So rapid is this change that a single cell by its process of growth and sub-

* T. M. Prudden, "The Story of the Bacteria," G. P. Putnam's Sons, publishers.

division can give rise in some cases to more than 16,000,000 of similar organisms in twenty-four hours.

The bacteriologist has acquired a wonderful mastery over the life changes of these simple organisms. Not only has he learned how to grow them under both natural and artificial conditions, but he can modify their life processes at will by proper selection of stock and replanting. He has developed a technic, whereby characteristic capacities of these microorganisms can be profoundly altered and harmful bacteria can be rendered harmless, and the virulence of others can be increased. Armed with his microscope the bacteriologist can see his bacteria, observe the things they do and from the data thus obtained, trace out their complete biological history. Bacteriology was in its early infancy as late as 1890. It is still a highly undeveloped field of biological science, but out of its revelations in the last twenty-five years, there have been won treasures of new knowledge of the greatest practical beneficence to mankind. Our knowledge of the higher and more complex forms of living organisms will be advanced as we continue to study and investigate these lowest forms of life. In the future development of this field of science organic chemists will undoubtedly play an important role.

The Field of Chemical Bacteriology

The discovery of the microbe created the new biological science—bacteriology—and through its development medical ideas and procedures have been radically modified. It is a statement of fact that every advance made in the science of bacteriology in the past has resulted indirectly in a corresponding benefit to medicine. Historically speaking, three outstanding periods have been especially significant in the development of this branch of science. The first period had its origin in the pioneer work of Pasteur, who may be called the father of bacteriology, and through whose researches we first obtained a fundamental understanding of infectious diseases. The second period, which was really inspired by the work of Pasteur, led to the important discoveries by Lister, who introduced into medicine antiseptic methods that have revolutionized the practice of surgery. The third period, which is characterized by the formulation of a sound conception of a rational control of bacterial activities, took its rise after the establishment of the science of bacteriology on a sound basis by Koch, through his researches on tuberculosis, cholera and anthrax.

Notwithstanding the advances already made in bacteriology, the chemical aspects of this subject have not received the attention that they deserve. In the past, emphasis has been laid on the effects produced by bacteria, and a study of the conditions regulating or influencing their natural development. While there are scattered fragments of information regarding chemical changes that take place in the bacterial cell during its growth, for the most part all these facts remain today as isolated, and uncorrelated items of information, which are scattered through chemical, medical and physiological scientific publications. It is only recently that the bacteriologist has begun to realize the value of the service of the chemist and to appreciate the importance of attacking some of his problems by chemical methods. In fact, the special field of chemical-bacteriology lies unexplored. The writer ventures to make the prediction that it is only through a closer cooperation between chemist and bacteriologist in the attack on bacteria, with introduction of chemical methods, that new results will be obtained which will have an influence on medicine comparable with that of the discoveries of the past.

In the past, the term bacteria has not been one received with favor and has generally been associated with a meaning implying deadly germs infecting mankind, and operating as a force for evil. In fact, as a consequence of the destructive nature of pathogenic organisms, bacteria have gained a notoriety which has forced the chemist to avoid the trails that lead into the realm of bacteriology. As chemists, we have failed to realize what civilization owes to these lower forms of life. It is now recognized that these organisms are the chemists of the living earth which operate in nature's laboratory performing synthetic processes of the most complicated nature, but which are essential for the perpetuation of higher animal and plant life.

Bacteriology has finally reached that stage of development where it requires the aid and cooperation of the chemist in the solution of its problems. We have to deal in the study of bacteria with physico-chemical phenomena for which anatomy cannot account. While the microscope has played its part well and will always find application as an instrument of research, it is realized by our leading bacteriologists that progress cannot be limited to its use. It is not enough to know the anatomy of a cell to understand its function and normal changes. A correct definition of cell processes cannot be given without a knowledge of the chemical composition and structure of the cell.

The development of the gas motor, the dynamo, the X-ray, photography and wireless telephony came as a result of the discovery of new facts and laws of physics. A corresponding advance in bacteriology and indirectly medicine, is tied up with a more accurate knowledge of the chemical processes that take place in living bacteria. Here we are dealing with a complicated but fundamental mechanism for whose correct interpretation we need to develop a better understanding, as chemists, of the established laws regulating diffusion, dialysis, adsorption, interfacial changes and solution.

As long as the organic chemist is guided by the view-point or conclusion that the study of bacterial cells offers a barrier that is insurmountable, and that cells in consequence of their natural unstable condition cannot withstand the treatment necessary for chemical analysis, no progress in this field along chemical lines will ever be made. We are told by biologists that protoplasm is indefinite in composition, undergoing rapid changes during the process of growth, and as a consequence of its complicated nature is excluded from the ordinary chemical methods of investigation. Some even tell us that biochemistry is not yet concerned with the chemical composition of cells of any type, but only with their decomposition or metabolic products and the conditions regulating exchanges between cells and their surroundings. The writer cannot subscribe to such a point of view, as it seems most probable to him that a study of cell composition will throw light on the chemical nature of protoplasm, and also reveal to us a clearer picture of the mechanism of the fundamental chemical changes of cytoplasm regulating cell life.

It is the belief of many chemists, who have followed carefully the developments of recent years, that our knowledge of the science of immunity can be advanced by application of the chemical method of attack, and that a solution of many of the problems of immunity will be enlightened by the results to be obtained as a result of a thorough analytical study of the proteins and other constituents of bacterial cells. If an application of the chemical method of investigation will hasten a logical development of this newer science, simplify our knowledge of the composition of protoplasm, and introduce more accurate methods for handling cellular materials and advance the technic of bacteriology, then no one will deny that chemistry will again have demonstrated its usefulness as a fundamental science. The chemistry of immunity as well as the chemistry of the bacterial cell and of enzymes are all practically on the same plane and of equal importance.

It is very important that we increase our knowledge of the proteins of bacteria. It has been established that the chemistry of the proteins of blood, for example, is not always the same even with closely related species of animals and that some proteins may be quite identical in widely differentiated species. The specificity of blood proteins has been established by very accurate research, and after their purification by chemical methods it seems to have been quite definitely established that there is a very limited group of factors which determine specificity in a protein molecule, and that if two proteins are truly different chemically they will also show a quite distinct behavior immunologically. In other words, it seems to have been demonstrated almost conclusively that immunologic specificity depends on chemical composition and nature of chemical linkage rather than on biologic origin. In fact, we have today positive evidence that different species of bacteria may contain common antigenic proteins as well as specific antigens.

In the light of these recent and most interesting biological developments it is most important, therefore, that we increase our chemical knowledge of the composition and structure of these living cellular organisms. If it is true that biological specificity of characteristic antigenic properties, is solely dependent on chemical structure, and that immunological differences, which exist between different species of bacteria, are dependent likewise on differences in chemical composition of the proteins and other organic constituents entering into the composition of their respective cells, then it is high time that organic chemists began to interest themselves in this field of research. Just as the dividing line between organic and inorganic chemistry has grown less and less distinct as our knowledge of chemistry has advanced, so also may we expect to see, when we begin to unravel and explain the innermost secrets of the living cell, that we may soon have difficulty in distinguishing between biological and chemical changes.

The Biological Nature of Bacteria

It was not until a short time ago, when the writer began his researches on the chemical composition of the *tubercle bacilli,* that he began to appreciate how scanty is our knowledge of the chemistry of bacteria. When attempts were made to isolate definite organic combinations from these pathogenic organisms, we were confronted with a complicated mixture of chemical substances as to the nature of which we were able to obtain practically no help from the literature. Although many definite chemical substances have been isolated from animal and plant tissues, no comprehensive chemical study of the contents of any single cell or bacterium has ever been made. Until such knowledge is obtained and we can speak understandingly of the chemical constituents of bacteria how can the bacteriologist and chemist be expected to formulate jointly a rational conception of the chemical and physical processes occurring within the bacterium. In such a confusing mixture as is present in a bacterial cell the organic chemist finds it no easy task to develop a technic which will enable him to make out much in this jumble. He has, however, already surpassed nature in her power to synthesize, and it is our hope to look forward to the time when we may have learned some direct and practical means of dissecting chemically the bodies of these organisms. In order to accomplish this we must first gain a new vantage-ground and establish a base line from which to make our chemical attack with a finer precision than has been accomplished hitherto. It is with this aspect of this problem that we have been working in the Yale Laboratories during the past three years. This investigation has been generously supported by a grant from the National Tuberculosis Association, and an extensive program has been outlined for another year. Our work has now been developed to the point where we have succeeded in adopting a systematic, analytical procedure, that makes it possible to outline and direct a coordinated series of attacks on the different constituents of any bacteria, which can be produced in sufficient quantity for chemical research.

Let us consider for a moment the biological nature of the interesting problem we have selected for our study. In anatomy we learn that the body has a framework around which is grouped tissues and different organs. We study their shape, size, number, and their relations, and name the various parts functioning in the life processes of the body. Leaving anatomy, we then enter the more complicated field of histology. Here we travel into a domain where the naked eye does not further serve us, and with the aid of the microscope we find that these various parts of the body encasing the framework are composed of tiny and highly developed units which we call cells. These are masses of organized matter of very peculiar but definite chemical constitution which exhibit a complex of phenomena that we call life. In bacteria, we also have to deal with life, but it is revealed to us by a complex organization operating completely within a given cell while the life of the higher animals is the sum of the coordinated lives of the various cells which compose it.

Bacterial cells are filled with substances largely in a colloidal state, but we know very little regarding the nature of these colloids. They also contain organic and inorganic substances in true solution, but it is not until recently that we have begun to understand what these are, and what are their relations to the cell colloids. We are informed by the literature on our subject that bacteria contain proteins, fats, carbohydrates, lipoids, and other extractable organic substances, but with this limited knowledge we must admit that we are unable to formulate any definite conception today of the chemical processes involved in cell metabolism. More is actually known about the chemistry of cells of higher organisms than bacterial cells, but here also investigations in this more highly developed field are far from complete.

Since life would be impossible unless the contents of a cell were in a state of chemical instability it is unlikely that it will ever be possible to obtain in a laboratory beaker the same organic combinations or constructions as those that function in the living cytoplasm of the cell. The organic chemist believes, however, that he will be able to approach this goal more nearly than has yet been accomplished in the past.

From the very nature of the conditions under which bacteria grow and multiply by subdivision it would be predicted, from a chemical point of view, that one would find all stages of cellular development from the simplest combinations up to the most complicated ones in the growing bacterium. In some cases these changes are favored by aerobic and in others by anaerobic conditions. Almost any adsorbed reagent might rupture molecular combinations between cell constituents and furthermore, the activity of specific enzymes present in the cell might destroy very quickly stable organic complexes. The whole subject of localized antiseptic action leading to death and destruction of cells through adsorption of chemical substances is closely associated with these phenomena.

The velocity of chemical change in bacteria is very rapid. Complete synthesis of highly developed organic structures, such as proteins and aromatic combinations, are effected within a few hours as new cells develop. Therefore, one might expect to find evidence of every intermediate stage of organic development at all periods of cell growth from time of birth to death. In other words, a bacterium is a most remarkable illustration of a completely organized chemical system, undergoing fundamental changes, operating at very high speed, and with an accuracy and completeness that seems almost incomprehensible. It is a knowledge of this highly developed cell mechanism that we desire very much to obtain. Not until the organic chemist begins to operate with a bacterial cell like tubercle bacillus, for example, is he made to realize the complexity of a cellular biological process. If enzymes are active and function in the bacterium during the preparation of the material, which is to serve as the starting point for a chemical investigation, the chemist is faced at once with a mixture of normal constituents of the bacterium with secondary products of autolysis. In other words, he is not obtaining data which reveals the true picture of the cell composition or structure before its death. The writer often questions whether we will ever be able to solve this biological riddle. Living bacteria differ from dead bacteria in that they possess characteristic activities through which their life is expressed; they can maintain their individuality, reproduce and are characterized by specific function, with power of elaborating definite chemical products. The organic chemist, however, is very inquisitive and he desires to know how the living cell does these things. He realizes that the processes may be very primitive, or they may be very complicated, but the whole subject is very important and he is very anxious to know how the living cell achieves its end.

In the selection of the pathogenic organism—*tubercle bacillus*—for chemical study, we have seized upon an arch enemy which we desire very much to capture. We are conscious, when we encounter this organism that we are to battle with a known and comprehensible foe, and it is not a question of groping in the dark after a mysterious and unknown enemy. We have advanced beyond the stage of operating with the microscope. It is now possible to obtain this pathogenic organism in sufficient quantity to make possible a practical application of modern organic technic and determine its chemical composition. On account of the availability of the bacilli for research work, we now hopefully look forward to the time when we may have learned some direct or indirect means of destroying by chemical means the invading germs after entrance into the body, or neutralizing the poisonous toxins which they generate. But how can we accomplish this goal, learn about the poisons secreted by these arch enemies of man, learn how to neutralize them, and impede the growth and development of the bacteria in the body until we know more about the chemical processes which lead to the production of those poisonous substances. We have got to get behind the scene, develop a new line of attack and learn, if possible, how these cellular changes are brought about. This cannot be accomplished with the microscope alone, nor by confining our attention to the study of *physical forms.*

Application of Organic Chemistry to the Study of Bacteria

In our work on the chemistry of bacteria, we have devoted our attention chiefly, up to the present time, to the study of tubercle bacilli. Only one other bacterial organism—the *Bacterium lactis aerogenes*—has thus far been studied by us. This was selected particularly on account of its ease of growth by artificial means and because of its non-pathogenicity. It seemed of especial interest to compare by our own chemical procedure two bacteria—(one highly pathogenic and the other non-pathogenic)—and see whether this biological difference can be correlated with any peculiar differences in the chemical composition of the two organisms. From the results of our proximate analysis it is quite apparent that these two bacteria differ decidedly in their chemical composition, especially in their fat and protein contents, and also in the nature of their respective proteins. Whether similar differences will be observed in the case of other bacteria remains to be determined as our work progresses.

In order to obtain as accurate a picture as possible of the chemical composition of these bacteria, it has been necessary to reduce to a minimum all possibility of cell change due to autolysis while working with the cell. Such changes begin very soon after the death of the bacteria. Also it is known that certain constituents of bacterial cells, especially the contents of the lipoid fraction, are very susceptible to oxidation, and consequently erroneous results are obtained on analysis, if bacteria are exposed to the atmosphere for too long periods. This applies to plant cells as well as to bacterial cells. Another treatment that led to erroneous results was the common practice of killing bacteria by excessive heat treatment or autoclaving. We found that this treatment brought about very pronounced changes in the constitution of the bacterial cell. Heating above 75° leads to coagulation of highly soluble albumin constituents, thus leading to results in our protein analysis which do not represent the actual composition of the cell. Autoclaving also leads to other changes, which are revealed by our methods of cell analysis. In other words, experience has taught us that the usual practice of autoclaving bacteria leads to changes in cell composition and, therefore, such treatment must be avoided if one expects to obtain accurate analytical results revealing the true nature of the cell. In our work dealing with the study of bacterial proteins, our procedure has been to wash our bacilli, after filtering from the culture media, with cold water as quickly as possible under aseptic conditions, and then to dry the cells immediately in a vacuum of semi-commercial size at a temperature not higher than 37°. Undoubtedly some slight autolytic change takes place during this rapid drying process, but no pronounced alteration of the protein constituents is detectable by standard chemical tests. In fact, long drying of cells in a current of slightly warm air may lead in some cases to the formation of a considerable quantity of products of autolysis. It has been reported to me by Dr. Osborne of Yale University that while boiling water extracts from live yeast about 17 per cent of its solids, from yeast which has been dried in a current of air at 60° upwards of 36 per cent may be extracted. It is an unsettled question whether this increase in soluble products by extraction of yeast with hot water results from decomposition of cell complexes caused by heat, or is due to a physical change in the permeability of the walls of the yeast cells. From observations that have been made in our work on tubercle bacilli it seems very probable that cell permeability has much to do with this behavior. It seems very plausible that this striking increase is influenced by both these factors.

In cells and bacteria we have to deal with the principle of surface adsorption, which is illustrated in its very highest degree. Here we have a power of selection under perfect control in the living cell and organic molecules are rendered unusually active by being carried in colloidal suspension. Whether we will be able to differentiate chemically between loose molecular combinations held together by secondary or partial valency forces, and organic molecules adsorbed on colloidal surfaces remains to be determined. In bacteria we have no circulatory system or blood stream to transport the building units to local points for special synthesis. All the chemical transformations take place in a cytoplasmic mass, and, as a result, we have to deal with a condition which calls for a special organic technic in separating the cell constituents.

To extract tubercle bacilli, for example, with hot water would be a technic destructive of a part of the proteins in the cell due to coagulation. Our object, in our tuberculosis research, has been to separate if possible, the albumin or globulin fractions of the bacterium which are soluble in cold water, and to study the biological action of these substances when introduced into the body. It is important to determine whether any of these protein fractions are specific in their action, or exhibit the characteristic biological properties of tuberculin, which is manufactured today from the concentrated culture media after removal of the tubercle bacilli cells. The specificity of tuberculin is attributed, on the basis of prevalent conceptions in immunology, to the presence in tuberculin of protein unique for the tubercle bacillus. If the activity of tuberculin is due to the presence of a specific protein substance as many investigators in this field assume, it seems not improbable that the same protein construction may be one of the normal constituents of the bacterial cell. If this was found to be true and the specific protein was possible of isolation from the cell in a far more potent and purer form than is the tuberculin manufactured from the culture medium, the commercial manufacture of tuberculin preparations would be revolutionized.

Before applying any aqueous extraction process to dissolve water-soluble products from tubercle bacilli, it is necessary to remove first the fat and waxy envelope which encases this unique organism. This acid-fast bacterium is characterized by its high lipoid content. On removing this fatty fraction, it is very necessary, if one is to utilize the bacterial residue for extraction of labile proteins, to operate under experimental conditions that will alter as little as possible the primary composition of the cell leading to coagulation or denaturation of protein. In our practice we have found ordinary ether the most useful solvent for removing this fat. We allow this to act on the desiccated cell at ordinary temperature, and the maceration is continued for several days. We feel that we are working here under the most favorable conditions. There is no danger of putrefactive or hydrolytic change. We overcome any danger of body infection by destroying the virulence of the bacteria through contact with ether; remove the major proportion of the cell fat and bring about a plasmolysis of the cell without any appreciable autolysis and, as a result, increase the permeability of the cell walls to cold water. In this manner, we are able to accomplish very rapid and efficient extractions of protein under sterile conditions, and are actually beginning an investigation with cytoplasm which has not been subjected to rough chemical treatment.

Only a small percentage of the proteins of the bacteria are removed by cold water extraction, and this seems to be the conclusion of workers with other cellular organisms. This fraction represents about 2 per cent of the weight of the defatted bacteria in the case of tubercle bacilli. The protein is coagulated by heat and, according to the results of the biological tests which have been reported by other workers to our laboratory, it is the most potent tuberculin reagent that has thus far been separated from the tubercle bacillus cell. It is precipitated from aqueous solution at its isoelectric point by acetic acid, and dissolves in cold dilute alkali without immediate loss of its biological activity. Prolonged treatment with alkali, however, is injurious. This water-soluble protein can be obtained free from carbohydrates and nucleic acid, and behaves in every way like a true albumin. Attempts to separate a globulin from this protein by dissolving in sodium chloride solution and subsequent dialysis have been unsuccessful. If globulins are present in the cell of tubercle bacillus they exist in very small quantities only. We have also made a similar observation in the case of the *Bacterium lactis aerogenes.* The cold water extract carrying in solution this very active protein also contains a large quantity of chemical compounds about which very little is known. In fact, about one third of all the nitrogen of the tubercle bacillus is represented by the nitrogenous substances dissolved in this aqueous solution. Nucleic acid is present but not in sufficient quantity to account for the total amount in the bacterium.

We are dealing here with a chemical picture quite different from that presented by the products of hydrolysis of a protein molecule. We do not believe that these nitrogenous substances present in this aqueous extract result from the breaking down of protein complexes, but are the building blocks which the cell has synthesized and is carrying in reserve for construction of the more complicated parts of his cellular system. Whether they are intermediate products resulting by synthesis of simpler nitrogen combinations introduced into his culture media or are degradation products of his reserve cell protein we do not yet know. It is entirely probable that they originate from both these sources, and it is also to be expected that they very probably represent all stages of his synthetical processes. It is not improbable that organic combinations are present here which have not hitherto been revealed as existing in bacterial cells. The fruitful results obtained by Dr. Vickery in the laboratory of Dr. Osborne, in their work on extractive principles from yeast and plant cells, reveals the complexity and productivity of such cell solutions.

The bacterial residue remaining after removal of water soluble proteins from tubercle bacilli is a highly insoluble and inert substance. In fact, it represents, so to speak, the cellular framework of the bacterium. It constitutes the chief bulk of the tubercle bacillus outside of the fat and waxes, and the same applies also to *lactis aerogenes*. This residue is composed chiefly of a protein fraction, insoluble in water, but soluble in 0.5 per cent alkali solution. The protein is precipitated immediately from its dilute alkali solution by addition of acetic acid. Biologically this is not a potent substance. It exhibits very weak antigenic action when introduced into the body. It is practically devoid of tuberculin activity. Chemically it is entirely different from the protein extracted by cold water. This is revealed by comparison of the results obtained by determination of aminoacid nitrogen according to the Van Slyke method of analysis. The two proteins obtainable from the cell of tubercule bacilli are as distinct and characteristic substances as are the two proteins formed by the silk worm namely—*sericin* and *fibroin*. As we have no positive evidence that sericin is an oxidation or degradation product of fibroin, so we cannot say at present whether our water-soluble protein from our bacterial cells has any generic relationship to the alkali-soluble protein separated from the defatted cell residue by extraction with cold alkali.

The high phosphorus content of this alkali-soluble protein is quite significant. It is not, however, a representative of the nucleoproteins. All the tests for sugar were negative. When subjected to acid hydrolysis a waxy fraction is released which leads us to believe that we are dealing here with a lipoprotein combination. It is not at all improbable that we have here an interesting case of adsorption, and that protein and wax do not enter into chemical combination. These characteristic properties were not revealed by our work with the corresponding protein fractions from the bacterium lactic aerogenes. In other words, neither of the two proteins isolated by us from tubercle bacilli or lactis aerogenes are representatives of nucleoproteins. The most distinguishing biological difference in the case of the tubercle bacilli proteins is the important observation revealed by our research, that the water-soluble albumin protein of the cell is an active tuberculin, while the water-insoluble protein shows practically no tuberculin activity when tested with guinea pigs. We consider this discovery of protein specificity and high tuberculin activity as our most important contribution to the chemistry of the proteins of the tubercle bacilli. Whether either of our two proteins are mixtures and can be split up into simpler protein units remains to be determined. It is our belief that they are purer forms of bacterial proteins than have hitherto been described in the literature. As soon as material is available these two proteins will be subjected to hydrolysis and a study made of their aminoacid contents. The Van Slyke analyses already indicate proteins of entirely different composition.

In connection with our protein study we have also been interested in the question of sulphur linkages in our bacterial cells. This subject is one of special interest at the present time when we realize the necessity of correctly explaining the factors controlling oxidation-reduction mechanism in the cell of the tubercle-bacilli. Hopkins, by his isolation of glutathione, gave definiteness to this most important subject, and showed conclusively that cellular, oxidation-reduction phenomena are regulated by the presence of this definite mercapto compound in tissues. This is the most important autoxidizable constituent of the animal cell thus far discovered. We now find that both of our proteins isolated from tubercle bacilli contain sulphur, but in very small amounts. Thus far, we have obtained no evidence that glutathione is present in the cell. By using the same methods which, when applied to the tissues of the rat or pigeon showed the presence of this sulphur combination, workers in other laboratories have also concluded that glutathione does not occur in defatted cultures of tubercle bacilli. By hydrolysis of both of our proteins from this bacterial cell cystine is formed and can be detected in the protein hydrolysates. If it be true that the oxidation-reduction mechanism of this pathogenic organism—tubercle bacillus—is controlled by a different autoxidizable substance than glutathione, and that sulphur in mercapto form is not a factor to be considered in regulating oxidation-reduction phenomena, then an important new field of cell research will have been opened up for future workers. We are dealing here with one of the most important problems of biochemistry, and we are fortunate in having means at our command whereby the bacilli can be obtained and produced in such large quantities.

It is now possible to conclude from the results obtained in our researches on the chemistry of bacteria that the two organisms—tubercle bacillus and lactis aerogenes—differ radically in their chemical composition. Not only do they differ in their relative proportions of fats and carbohydrates but also in the character of their protein constituents. Tubercle bacillus contains an appreciable amount of protein of the albumin type soluble in cold water, while lactis aerogenes does not. This is a most important and significant difference, and indicates that high pathogenicity may be dependent on the presence of characteristic water-soluble proteins possessing toxic properties. The analytical determinations made for distribution of nitrogen on the major protein fractions of these two bacteria show decidedly that each organism is characterized by its own specific protein. The basic protein from tubercle bacillus contains about 25 per cent of arginine and about one-fourth as much lysine. The protein from lactis aerogenes contains more than twice as much amide nitrogen as the corresponding protein from tubercle bacillus.

CONTACT CATALYSIS

BY

WILDER D. BANCROFT

World War Memorial Professor of Physical Chemistry

Cornell University

Lecture delivered at Columbia University in the special course on Contemporary Developments in Chemistry given in the Summer Session of 1926 on the occasion of the opening of the Chandler Chemical Laboratories

COLUMBIA UNIVERSITY PRESS

NEW YORK

1927

Printed from type. Published March, 1927

Printed in the United States of America

CONTACT CATALYSIS

By Wilder D. Bancroft

Activation

The essential problem in contact catalysis is how and why the activation of the reacting molecules takes place. It is the fashion at present to speak of the molecules acquiring "critical increments of energy" and people calculate the values of these increments. While this treatment may be right, it is not adequate unless one postulates localized critical increments of energy. We can convert alcohol into acetaldehyde and hydrogen practically exclusively or practically exclusively into ethylene and water, or we can get nearly any desired ratio of these reaction products at will. I do not see how such a state of things can be accounted for solely on numerical values of critical increments of energy.

The inadequacy of the prevailing point of view is also shown by the fact that we can duplicate the action of most catalysts by means of light and we know that it is only the light which is absorbed which acts. The wave-lengths which cause the conversion of alcohol into acetaldehyde and hydrogen are not the same as those which convert it into ethylene and water. With light we are unquestionably dealing with localized increments of energy and consequently we must be doing the same thing with contact catalytic agents.

What do we mean by localized increments of energy? The absorption of light of any given wave-length by any substance may be an atomic or a molecular phenomenon. If it is an atomic phenomenon and if we use Bohr's terminology, the light is tending to drive an electron into an outer orbit and may do it under some circumstances. If the absorption of light is a molecular phenomenon and if we use the valence terminology, the light tends to weaken or break a regular valence bond or a so-called residual valence bond or contravalence bond. We know that all light which is absorbed tends to make the molecule more reactive. Whether a reaction takes place is a question of chemistry. A reaction will always take place in presence of a suitable depolarizer.

In the case of silver bromide we know that the absorbed light tends to decompose the salt into silver and bromine. In the case of alcohol we do not yet know what happens; but I think that we must assume that light of some wave-length tends to decompose alcohol into $CH_3CH_2O + H$ and light of another wave-length tends to decompose it into $CH_3CH_2 + OH$. In so far as the photochemical and the catalytic decompositions are the same, the contact catalytic agents must tend to decompose alcohol into $CH_3CH_2O + H$ in one case and into $CH_3CH_2 + OH$ in another case. This is Baly's theory of contact catalysis as I understand it.

Taylor considers that formic acid is adsorbed in the same way whether it decomposes into carbon monoxide and water or into carbon dioxide and hydrogen; but it seems to me more plausible to postulate a different adsorption in the two cases.

This point of view harmonizes with Langmuir's theory of oriented adsorption, which postulates that an adsorbed substance is held to the adsorbing agent by a regular valence bond or by a residual or contravalence bond. If we assume that the adsorption of hydrogen by platinum involves a regular bond, we must have monatomic hydrogen on the platinum and Taylor at Princeton seems to have shown that this does occur. If we assume that the adsorption is due to a residual or contravalence bond, we shall have $-H_2$ which will probably be more reactive than molecular hydrogen and less reactive than atomic hydrogen. In the ammonia synthesis we must have an activation of nitrogen because monatomic hydrogen will not react with molecular nitrogen. Taylor believes that active nitrogen is monatomic nitrogen; but he has not proved this to his satisfaction.

If activated hydrogen is atomic hydrogen and activated nitrogen is monatomic nitrogen, both atoms of hydrogen or both atoms of nitrogen must be adsorbed and we have what Burk calls multiple adsorption. There is no reason why we should not have CH_3CH_2O and H adsorbed separately by the same atom or by adjacent atoms of nickel and CH_3CH and OH adsorbed separately by the same molecule or adjacent molecules of alumina. It is of course possible that one radical may be adsorbed by aluminum and the other by the oxygen radical or by one of the oxygen atoms. Until we know why titania, for instance, gives mixed products, I do not see how we can tackle the very puzzling problem discovered by Adkins that the compound from which the alumina is prepared apparently has an effect on the ratio of two possible sets of reaction products obtained with an alumina catalyst. There is no obvious reason why alumina from aluminum nitrate should differ theoretically from alumina from aluminum ethylate or propionate. There must be some factor which has been overlooked by Adkins.

Multiple adsorption or preferential adsorption at two adjacent points might well account for the observed cases of actions at an interface. Pease and Taylor have shown that the reduction of copper oxide by hydrogen takes place at the interface between copper and copper oxide and that the reaction between hydrogen and oxygen in presence of copper and copper oxide takes place at the interface. The action of promoters may be in part a case of reaction at an interface, though one would then expect that a larger proportion of the promoter would be desirable.

While thoria can increase the apparent catalytic action of nickel by acting as a support, increasing the surface and preventing sintering, this should not be considered a case of promoter action any more than one would consider asbestos a promoter in the case of platinized asbestos. Actually the thoria does act as a promoter to nickel in methanation and in the hydrogenation of ethylene compounds. With ten per cent of thoria in a nickel catalyst on pumice, the increase in surface as shown by adsorption was about twenty per cent while the increase in reaction velocity was ten-fold.

Medsforth found that a concentration of 0.5 per cent ceria in the nickel catalyst causes a ten-fold increase in the reaction velocity. Taylor points out that this means one molecule of ceria to every 1130 atoms of nickel in the catalyst surface, if we assume a uniform distribution of promoter in the catalyst. This minute change in surface concentration brings about a ten-fold increase in catalytic activity. Taylor considers that this may mean that the active nickel atoms are of the order of 1 in 11300 atoms in the case of a catalyst containing no promoter.

A promoter may act by introducing an interface, by activating one of the other constituents of the reacting mixture, by increasing the number of active atoms of the catalyst, by changing the ratio of the adsorbed gases, or in some still unspecified way. We ought to study the effect of promoters on the dehydrogenation of alcohol by nickel, because that would eliminate two of the possibilities.

There is one difficulty in regard to the activation by oriented adsorption and that is that we apparently ought to get catalysis at any temperature. As a matter of fact, there is always a minimum temperature for any sample of catalyst, below which no appreciable reaction takes place in any reasonable time. This is not an absolute minimum temperature, as it varies considerably with differently made samples of the same catalyst. For the moment there seem to be only two possible explanations—neither of them satisfactory. The orientation may change with the temperature or the activated products may re-form the original substance. In the case of alcohol, there are two possible reactions:

$$CH_3CH_2O + H = CH_3CH_2OH$$
$$CH_3CH_2O + H = CH_3CHO + H_2.$$

If the first reaction becomes all-important at lower temperatures, everything is simple; but we cannot predict that it will and we do not know that it does. This difficulty does not exist for the people who talk about critical increments of energy because they say that the alcohol molecules do not receive sufficient increments of energy and let it go at that.

Rideal has discussed the question of induction for the particular case of hydrogen and ethylene. If equal volumes of hydrogen and ethylene are brought in contact with nickel, there is a long period of induction and then reaction takes place. If one puts the nickel in a vacuum and lets in first hydrogen and then ethylene, the period of induction is very small and can scarcely be measured. If the ethylene is put in first and then the hydrogen, there is a long period of induction. This apparently means that the nickel activates the hydrogen so that it will react with ordinary ethylene; but does not activate ethylene so that it will react with molecular hydrogen. It is therefore necessary to wait until the hydrogen has displaced some of the ethylene at the nickel surface.

With the present conception of activation, it becomes of less importance whether the intermediate stage is an adsorption or the formation of a definite chemical compound. It is desirable, however, to be clear on such points and that has not been the case in the past. The catalytic decomposition of hydrogen peroxide by mercury [1] seems to be a case where there is an intermediate formation of mercury peroxide, though this is now questioned by Hedges and Myers.[2] When acetone is formed by passing acetic acid vapor [3] over heated barium carbonate, it seems probable that there is an intermediate formation of barium acetate. The simplest explanation of the oxidation of carbon monoxide by oxidation in presence of the mixed oxides [4] of cobalt, manganese, etc., is that we have an alternating oxidation and reduction of these oxides. It is important, however, to prove in any case whether definite intermediate compounds are or are not formed. If we are dealing with a case of activation by adsorption, no definite intermediate compound is formed and it is quite possible that same cases, now assumed to involve definite intermediate compounds, may prove really to be activation by adsorption. In the catalytic oxidation of carbon monoxide, it is assumed that there is alternate or simultaneous reduction and oxidation of the catalyst. The oxygen carrier is supposed to oxidize the carbon monoxide and to be reoxidized itself by the oxygen of the air. Against this is the fact that the rate of oxidation of carbon monoxide is so low relatively that it may be a negligible factor in the reaction velocity. Bray believes that it is useless to try to decide whether, at the dynamic equilibrium, a molecule of oxygen at the surface of the catalyst actually changes some of a lower oxide to a higher (and the reverse change with carbon monoxide),[5] or whether the oxygen is merely held on the surface in an active condition ready to combine with carbon monoxide. In either case all the processes are taking place simultaneously.

It does not seem wise not to attempt to solve a problem like this and the first step is to formulate it. Suppose we have a hypothetical case in which copper activates oxygen. We may have an intermediate formation of cuprous oxide, cupric oxide, or even a peroxide, which then reacts,[6] or we may have copper or one of the oxides acting as an adsorbent, which we will call A. Our adsorbent may take up monatomic oxygen forming A–O–; or it may take up molecular oxygen, forming $A–O_2–$. Neither of these is a definite chemical compound in the ordinary sense. There must be some way in which one can distinguish between these different hypothetical cases.

The oxidation of alcohol by air in the presence of osmium tetroxide can easily be run in two stages, because osmium tetroxide will oxidize alcohol in the absence of air and air will oxidize the dioxide back to the tetroxide; but we do not know whether the single reaction velocities are sufficient to account for the rate of catalysis. If not, we shall have to postulate activation of oxygen independently of the definite chemical reaction. This might be a good case with which to start.

We know that there is practically no adsorption and no activation at a vitreous surface and that, at some temperature, there is a definite parallelism between adsorption and activation; but we are still pretty much at a loss as to the exact relation between the quality of the surface and the amount of adsorption and the degree of adsorption. It has always been believed that a porous substance was the best for adsorption and this is of course true for capillary condensation; but that is a different thing. The best charcoal for gas masks was apparently a charcoal with many, fine pores. When Langmuir advanced the doctrine of monomolecular adsorption on plane surfaces, he recognized explicitly the possibility of thicker films in porous substances.

All our good catalysts are porous or, at least, pulverulent.

Taylor[7] believes, however, that the catalytic action takes place on peaks and not in valleys, that the active atoms (or molecules) are the few, scattered over the surface, which are held to the mass of the catalyst by perhaps a single bond-valence.

"The X-ray examination of metallic hydrogenation catalysts has shown that these catalysts, even when prepared at low temperatures, possess the definite lattice structure of the crystalline material. A granule of such a catalyst must, therefore, possess, in part, the ordered arrangement of the atoms found in crystalline material. The method of preparation of active materials suggests, nevertheless, that the ordered arrangement of the atoms has not been completely attained and that, here and there, on the surface of a partially crystalline material there are groups of atoms in which the process of crystallization is not yet complete.

"The atoms in the plane surface of any face of the crystal, *e.g.*, a face-centered cubic crystal of nickel, will be practically saturated by the neighboring atoms in three dimensions, with the exception that there will be a certain degree of unsaturation towards the gas phase. In a sufficiently large plane surface of this type each atom will be identical in properties with the great majority of the others in the surface. There will be a certain variability in the surface fields of attraction with variation in the exposed face. The fields will be weakest in the closest packed faces, *i.e.*, the 111 faces, stronger in the 100 faces and strongest in the 110 faces. Atoms in the edges of such a crystal will be one degree less saturated than the atoms in the surface, by reason of the fact that they are to a less degree surrounded by nickel atoms. For this reason they will possess stronger attractive forces for impinging atoms. Evidence for this is available in a study by Volmer and Estermann[8] of the growth of mercury crystals from the vapor. The crystals grow in the form of thin hexagonal leaflets the thickness of which is of the order of 1×10^{-4} of the hexagonal face. From a quantitative analysis of the experimental observations Volmer concludes that molecules condensing on the face of the leaflets are held temporarily in the force field of the surface, retaining, however, their motion until they either (1) evaporate or (2) are attracted to, and held at, the edges of the surface or (3) through contact with other molecules form the nucleus of another lattice layer. This latter alternative only becomes important when the plates, by growth of the edges, have attained a length and breadth of 0.3 mm. This increased attractive force at an edge will be surpassed by that obtaining at a corner. In the incompletely ordered atoms the attractive force will progressively increase as the degree of saturation by neighboring nickel atoms becomes less and less, until, finally, in atoms which are only held to the granule by one attachment, *e.g.*, Ni-Ni, the unsaturation of the metal atom and the attractive force for impinging molecules will become a maximum. The methods of preparation of active catalysts and the activation of inactive masses all tend to produce arrangements of atoms in which these varying characteristics are multiplied.

"It is at once evident how this view of the composite nature of a catalyst surface is in harmony with the observations already made as to the adsorption characteristics of such catalysts. For, the ratio of faces to edges to corners and to extra-lattice atoms will vary with varying preparations and with heat treatment. The attractive force of a surface atom may be adequate to retain one gaseous molecule, whereas another gas would suffer a practically completely elastic collision. The less saturated atoms in the catalyst surface will be preferred positions of attachment of catalyst poisons. Hence, the varying ratio of adsorption between poisoned and unpoisoned catalyst. Hence, also a mechanism for progressive poisoning.

"The present concept of the composite catalyst surface is more embracing than that stressed by Langmuir [9] in dealing with the problem of catalyst activation. Langmuir emphasizes that the distribution of catalyst atoms more or less at random will 'vary the distances between adjacent adsorbed molecules over wide range and some of these distances will be exactly right for the reaction to occur at the highest possible speed.' Variation in the distances between catalyst atoms, otherwise identical, does not appear to be adequate to account for *all* the variations in adsorption characteristics of single gases recorded in preceding paragraphs. The present concept substitutes for varying atomic distances varying degrees of saturation of surface atoms, from those in a plane surface to those held to the surface by a single atom-atom linkage. With this varying degree of saturation in the lattice structure comes not only the varying attractions for impinging molecules but also a varying chemical reactivity. For, the outermost atoms of the catalyst granule, linked by a single bond to the solid mass, are the atoms of a solid body most closely resembling the gaseous molecule. Now, a gaseous nickel atom can attach to itself four molecules of carbon monoxide to form nickel carbonyl. It does not seem improbable, therefore, that the outermost singly-linked nickel atoms shall be able to take up three molecules of carbon monoxide, or three molecules the linkage of which to nickel is that obtaining between the metal and carbon monoxide. The possibility thus arises of both hydrogen and unsaturated body being attached to the same nickel atom in a single complex at a catalyst surface, thus ensuring a more intimate electronic association than could be secured by adsorption on adjacent catalyst atoms."

It is not certain that so radical a change in our point of view is desirable. If the active atoms or molecules were put in the pores of the catalyst, they would apparently function equally well and would tie in better with our previous views. Multiple adsorption seems a more unlikely phenomenon on a peak than in a valley. Also, one does not see why a support should be so desirable for a catalyst, if the active material is on a peak far from the support. The two essential features of Taylor's hypothesis are that only a small fraction of the catalytic surface is active and that the active atoms (or molecules) are the semi-detached. Either of these assumptions may be true without the other one being true.

Taylor gives some independent evidence for his theory. "The theory of a catalytic surface which has been put forward is thus far an *ad hoc* hypothesis designed to fit more of the experimental facts now known than is possible with preceding theories. Experimental evidence that unsaturation of the surface such as has been postulated does exist would strengthen considerably its probability. Such evidence appears to be available from a recent research. Garner and Blench [10] have recently published data on the heat of adsorption of oxygen on charcoal. They have obtained values ranging from 50,000 calories at ordinary temperatures to the astonishing value of 224,000 calories at low pressures and 450° C. The astonishing nature of this latter value is at once evident when it is remembered that the heat of combustion of solid carbon to gaseous carbon dioxide is only 94,000–97,000 calories. The heat of adsorption may therefore be more than twice as great as the heat of combustion of the solid body.

"It is at once evident that on the basis of the theory put forward in the preceding section this high value is to be expected. The high value recorded would represent the heat effect produced by the interaction of oxygen with the atoms of carbon most detached from the main mass of the solid carbon. The progress occurring will be intermediate between the interaction of gaseous carbon and oxygen in which no carbon linkages are broken and that of solid carbon and oxygen in which four carbon-carbon linkages are broken in the oxidation process. The combination of gaseous carbon and oxygen involves the abnormally high heat effect of 350,000 calories owing to the high heat of vaporization of carbon or alternately to the endothermic breaking of the carbon-carbon linkage.[11] The value obtained by Blench and Garner is intermediate between the values for the combustion of the solid and the gaseous forms, and constitutes experimental evidence for the existence of isolated extra-lattice atoms."

Poisons

One very striking characteristic of solid catalytic agents is the ease with which they become inactive or are poisoned. In technical catalytic processes, the great difficulty is to keep the catalyst active, as the presence of any one of a number of substances even in minute amount will poison the catalytic agent and render it inert. Up to a few years ago, this poisoning of the catalytic agent was considered a most mysterious phenomenon, but the theory of it is now quite satisfactory. Since the reaction takes place in or at the surface of the catalytic agent, any substance, gas, liquid, or solid, which decreases the rate at which the reacting substances reach the catalytic surface,[12] or which prevents them from reaching it,[13] will decrease the reaction velocity and may destroy the catalytic action completely.

Berliner[14] has shown that traces of fatty vapors from the air or from the grease on the stop-cocks will decrease the adsorption of hydrogen by palladium from about 900 volumes to nothing. Pollard[15] observed a similar decrease with platinized asbestos, from about 160 volumes to a negligible amount. Faraday[16] proved that traces of grease destroy the catalytic action of platinum black. Lunge and Harbeck[17] found that carbon monoxide inhibits practically completely the catalytic action of platinum on a mixture of ethylene and hydrogen. Taylor and Burns[18] have shown that this is because carbon monoxide decreases the adsorption of hydrogen by platinum black. Working under more favorable conditions, Pollard[19] has obtained a similar, but more striking, result. With platinized asbestos, kept very clean from grease, he obtained an adsorption of 160 volumes of hydrogen per volume of platinum. On introducing carbon monoxide the hydrogen adsorption dropped to about seven volumes, which was practically negligible under the conditions of the experiment. Taylor and Burns conclude from their experiments that, even when the pressure of carbon monoxide does not exceed a few centimeters, the platinum surface is probably covered so thoroughly with carbon monoxide that hydrogen and ethylene are unable to reach it. They also point out that platinum, which holds carbon monoxide so tenaciously, is not a good catalytic agent for the reduction of carbon monoxide to methane, whereas this reduction takes place readily at a palladium surface, from which carbon monoxide can be displaced easily by hydrogen at ordinary temperatures.

Schönbein[20] pointed out that the hydrides of sulphur, tellurium, selenium, phosphorus, arsenic, and antimony act very energetically in cutting down the action of platinum on mixtures of air with hydrogen or ether. Since he did not realize that an adsorbed gas film might keep out other gases, he decided that these hydrides must decompose and plate out a solid film on the platinum. This hypothesis is not necessary to account for the phenomenon; but Schönbein was right in at least one case. The most complete experimental study of poisons made so far is by Maxted,[21] who has shown that hydrogen sulphide is decomposed by platinum black with the evolution of hydrogen, and that the "sulphurized" platinum does not adsorb hydrogen. With varying amounts of hydrogen sulphide, both the adsorbing power and the catalytic action decrease linearly over a certain range of concentrations. When platinum is poisoned by lead, one milligram of lead poisons nearly nine milligrams of platinum, this figure applying only to platinum black prepared in a given way and, therefore, having a given ratio of active surface to mass.

Pease[22] has shown that the vapor from one cubic millimeter of liquid mercury introduced into one hundred grams of a reduced copper catalyst inhibits completely the reaction between hydrogen and ethylene at 0°. Furthermore, the reaction is still extremely slow at 100°. The mercury suppresses practically completely the adsorption of hydrogen by the copper, but has very little effect on the adsorption of ethylene.

Harned[23] has shown that the rate of adsorption of chloropicrin by a charcoal which has been cleaned by washing with chloropicrin is much greater than by a charcoal which has not been so cleaned, although the final equilibrium is apparently about the same in the two cases. This is analogous to the evaporation of water when covered by an oil film. The oil cuts down the rate of evaporation very much but has practically no effect on the partial pressure of water at equilibrium.

It is easy to see that a piling up of any of the reaction products on the surface of the catalyst will decrease the reaction velocity if this hinders or prevents the reacting substances from coming in contact with the catalytic agent. This has been observed in the contact sulphuric acid process.[24] The explanation that the decrease in the reaction velocity is due to a decreased adsorption of the reacting substances was first given by Fink,[25] who is the real pioneer in this line. Although the reaction between carbon monoxide and oxygen is practically irreversible, it occurred to Henry,[26] nearly ninety years ago, that the presence of the reaction product, carbon dioxide, might slow up the rate of reaction, and he proved his point by increasing the reaction velocity when he removed the carbon dioxide with caustic potash. Water vapor checks the catalytic dehydration of ether [27] and of alcohol [28] somewhat, and hydrogen cuts down the catalytic dehyrogenation of alcohol. In fact, nickel and copper tend to dehydrogenate substances in the absence of hydrogen and to hydrogenate them in its presence.

Since the poisoning of a catalytic agent is due to marked adsorption, which cuts down the adsorption or the rate of adsorption of the reacting substances, and since the presence of sulphur trioxide, the reaction product, tends to decrease the rate of reaction of sulphur dioxide and oxygen, it follows that an extremely strongly adsorbed reaction product will act as a catalytic poison. In such a case the extent to which the reaction will run will depend on the relative amount of catalytic agent present.[29] If a large amount of catalytic agent be added to a mixture which does not react perceptibly in finite time in the absence of the catalytic agent, the reaction will run to an end or to true equilibrium before the catalytic agent is poisoned completely. If there is only a small amount of catalytic agent, it will be poisoned very early in the course of the reaction and we shall have an apparent equilibrium, reached from only one side, which will vary with the amount of the catalytic agent. For any given small amount of catalytic agent we shall get an apparently definite end-point; but the value of the end-point will vary with the amount of the catalytic agent taken. This is called autotoxic catalysis.

While there seems to be no doubt but that a poison acts because it is adsorbed and because it decreases the rate at which the reacting substances, or some of them, reach the surface of the catalyst, the amount of substance which poisons the reaction practically completely is sometimes only a fraction [30] of that which would be necessary to cover the surface of the catalyst completely. Taylor [31] accounts for this by assuming that only a fraction of the surface of the catalyst has the power of catalyzing reaction.

If one uses colloidal platinum, for instance, as a catalyst, it may be necessary to keep it in suspension by some protecting colloid, which may have an effect on the adsorption and on the catalytic action. Gröh [32] showed that the stabilization of colloidal platinum by gelatin decreases the catalytic action on hydrogen peroxide, Table I.

All subsequent work has confirmed the generalization that we pay for stabilization by a protecting colloid through decrease in the catalytic action. This effect may be masked to some extent if the protecting colloid increases the dispersion and there-

TABLE I

EFFECT OF GELATIN ON CATALYTIC ACTION OF PLATINUM ON HYDROGEN PEROXIDE

Per Cent Gelatin	Time for Half Decomposition	Per Cent Gelatin	Time for Half Decomposition
0.00	100	0.05	620
0.001	437	0.10	983
0.01	460		

fore the surface of the catalytic agent,[33] or if the protecting colloid is itself a catalytic agent. Iredale [34] finds that the better a substance is as a protecting colloid the more it inhibits the catalytic action of platinum on hydrogen peroxide.

There are a couple of interesting cases of what appears to be poisoning of the catalytic agent.[35] Although the reaction between phosgene and water to give carbon dioxide and hydrochloric acid is not reversible, it apparently does not run to an end in the presence of concentrated hydrochloric acid. The most plausible explanation of this is that it is a wall reaction and that the adsorbed film of hydrochloric acid keeps the phosgene from coming in contact with the walls. Trichloromethylchloroformate, $ClCO_2CCl_3$, decomposes to carbon tetrachloride and carbon dioxide in presence of alumina and to phosgene in presence of ferric oxide, neither reaction being reversible to any appreciable extent. During the war some of this compound was placed together with ferric oxide in a glass tube connected with a closed manometer. There was a rapid decomposition at first, as shown by the increase in pressure; but the reaction soon came apparently to an end. On raising the temperature, the reaction went a little farther, and did not reverse when the temperature was brought back to its original value. This observation was not checked; but, if the facts are as stated, there must have been a poisoning of the ferric oxide, which was decreased by the rise of temperature. If this is the real explanation, it should apparently be possible to carry to an end the reaction between ethyl butyrate and water in presence of a small amount of lipase [36] by letting the temperature oscillate.

The poisoning of chemical reactions which take place rapidly in the absence of a catalytic agent is a phenomenon familiar to the chemist, though not usually classified under that head. Aluminum and nickel should corrode readily in the air; but do not, owing to the formation of a film of oxide or hydroxide which protects the underlying metal. So long as iron is coated with a coherent film of ferroferric oxide, the amount of rusting is negligible. Parkerized or phosphatized iron is protected in part by an oil film and is to that extent analogous to platinum black which is not a catalytic agent when covered with a film of grease. Lead and lead peroxide in sulphuric acid constitute the lead storage battery; but the peroxide plate is not a short-circuited cell because the peroxide coats the lead grid and keeps it from coming in contact with the acid.

Beneficial Poisoning

While the poisoning of a catalytic agent is usually a thing to be avoided, Rosenmund and Zetsche [37] have made use of what they call partial poisoning in order to increase the yield of certain desired products. If an acid chloride, benzoyl chloride for instance, is dissolved in pure benzene and is then treated with hydrogen in presence of colloidal palladium, practically no benzaldehyde is formed. The authors consider that the benzene poisons the palladium. If ordinary benzene is used, a good yield of benzaldehyde is obtained.[38] This seems to prove that some impurity in the benzene counteracts the poisoning. Instead of searching for the particular impurity in the benzene which had this effect, the authors studied the effect of adding known amounts of definite substances to the pure benzene or other aromatic hydrocarbon. With 2 g. of benzoyl chloride in 10 cc. of xylene and using 0.7 g. of palladium, addition of 1 mg. of thioquinanthrene brought the yield at 150° up 78 per cent, while addition of sulphurized quinoline (impure thioquinanthrene) brought the yield up nearly to 90 per cent. Addition of 0.1 mg. distilled quinoline [39] reduced the yield practically to zero because the reduction went further—to benzyl alcohol. Addition of xanthone accelerates the reaction between benzyl chloride and benzyl alcohol, forming benzyl benzoate. In toluene as solvent instead of xylene, benzyl benzoate is the chief product because the toluene checks the reaction between benzyl benzoate and hydrogen which gives rise to benzoic acid and toluene.

It has also been found that addition of quinoline to benzyl alcohol dissolved in cymene increases the yield of benzaldehyde when copper is added and air passed through the solution. A further addition of nitrobenzene increases the yield still more. In the absence of quinoline, copper, nickel, silver, magnesium, etc., behave about alike. Addition of quinoline makes copper, nickel, and silver much more effective than the other metals but in different ways, the copper increasing the oxidation of the alcohol, while silver and nickel increase the formation of benzyl ether and of resins.

Recent work by Adams and his co-workers indicates that permanent poisons on the catalyst surface may be used deliberately to direct the catalysis. With a platinum-iron catalyst, cinnamic aldehyde undergoes reduction to phenyl propyl alcohol. If the catalyst be poisoned suitably with zinc salts, the product is cinnamyl alcohol. The efficiency varies very much with the amount of zinc salts used to poison the catalyst.

Organic Chemistry

The present conception of activation has a bearing on organic chemistry which goes far beyond contact catalysis. Every reaction between so-called saturated compounds necessitates either a preliminary dissociation which means an opening of regular bonds, or a preliminary addition which means the opening of residual valence or contravalences. All reactions between saturated organic compounds therefore involve a preliminary activation of some sort.[40] The new organic chemistry will therefore be a chemistry of radicals and not of molecules. The organic chemistry of the present asks what compounds are formed and what are their constitutional formulas; the organic chemistry of the future will ask how the reaction takes place in the laboratory and in nature. The apparent difference between organic and inorganic reactions will disappear because we shall be dealing essentially in both cases with reactions of activated molecules. The ordinary reactions in aqueous solutions are reactions of activated molecules which we call ions. Many of the reactions in organic chemistry will involve activated molecules which we shall call radicals. An activated molecule which is electrically charged is an ion; an activated molecule which is not electrically charged is a radical. That is what we are led to by a consideration of contact catalysis from the viewpoint of the colloid chemist.

REFERENCES

1 Bredig and von Antropoff: Z. Elektrochem., 12, 581 (1906): von Antropoff: J. prakt. Chem. (2) 77, 273 (1908).
2 J. Chem. Soc., 125, 1288 (1924).
3 Squibb: J. Am. Chem. Soc., 17, 187 (1895).
4 Lamb, Bray and Frazer: J. Ind. Eng. Chem., 12, 217 (1920).
5 Benton considers that he has proved this, J. Am. Chem. Soc., 45, 900 (1923).
6 Cf. Sabatier-Reid: "Catalysis in Organic Chemistry," 50 (1922).
7 J. Phys. Chem., 30, 145 (1926).
8 Z. Physik, 5, 31, 188 (1921); 7, 1, 13 (1920); Z. physik. Chem., 102, 267 (1922).
9 Trans. Faraday Soc., 17, 618 (1922).
10 J. Chem. Soc., 125, 1288 (1924).
11 Fajans: Verh. physik. Ges., 14, 324 (1912).
12 Taylor. Trans. Am. Electrochem. Soc., 36, 149 (1919).
13 Bancroft: J. Phys. Chem., 21, 734 (1917).
14 Ann. Physik, (3) 35, 903 (1888).
15 J. Phys. Chem., 27, 356 (1923).
16 "Experimental Researches on Electricity," 1, 185 (1839).
17 Z. anorg. Chem., 16, 50 (1896).
18 J. Am. Chem. Soc., 42, 1285 (1921).
19 J. Phys. Chem., 27, 356 (1923).

20 J. prakt. Chem., **29**, 238 (1843).
21 J. Chem. Soc., **115**, 1050 (1919); **117**, 501 (1920); **119**, 225, 1286 (1921) **121**, 1760 (1922).
22 J. Am. Chem. Soc., **45**, 2299 (1924).
23 J. Am. Chem. Soc., **38**, 1145 (1916).
24 Bodländer and Koppen: Z. Elektrochem., **9**, 566 (1903); Berl: Z. anorg. Chem., **44**, 267 (1905).
25 Bodenstein and Fink: Z. physik. Chem., **60**, 61 (1907); cf. Bunsen: J. Chem. Soc., **26**, 736 (1873).
26 Phil. Mag., (3) **9**, 324 (1836).
27 Ipatioff: Ber., **37**, 2996 (1904).
28 Engelder: J. Phys. Chem., **21**, 676 (1917).
29 Bancroft: J. Phys. Chem., **22**, 22 (1918).
30 Armstong and Hilditch: Trans. Faraday Soc., **17**, 670 (1922).
31 J. Phys. Chem., **28**, 939 (1924).
32 Z. physik. Chem., **88**, 414 (1914).
33 Rideal: J. Am. Chem. Soc., **42**, 749 (1920).
34 J. Chem. Soc., **119**, 109 (1921).
35 Bancroft: Trans. Am. Electrochem. Soc., **36**, 140 (1919).
36 Kastle and Loevenhart: Am. Chem. J., **24**, 491 (1900); Bancroft: J. Phys. Chem., **22**, 40 (1918).
37 Ber., **54**, B, 435, 638, 1092 (1921).
38 Rosenmund: Ber., **51**, 585 (1918).
39 It is interesting to note that quinoline catalyzes the change from vitreous to metallic selenium, the action being more marked in the sunlight than in the dark. Saunders: J. Phys. Chem., **4**, 469 (1900).
40 Bancroft: J. Ind. Eng. Chem., **14**, 645 (1922); **16**, 270 (1924).

Water-Soluble Vitamins

BY

E. V. McCOLLUM

PROFESSOR OF BIOLOGICAL CHEMISTRY

JOHNS HOPKINS UNIVERSITY

Lecture delivered at Columbia University in the special course on Contemporary Developments in Chemistry given in the Summer Session of 1926 on the occasion of the opening of the Chandler Chemical Laboratories

COLUMBIA UNIVERSITY PRESS

NEW YORK

1927

Printed from type. Published March, 1927

Printed in the United States of America

WATER-SOLUBLE VITAMINS

E. V. McCollum

Johns Hopkins University, Baltimore, Md.

At present there is convincing evidence that there are five vitamins essential for the normal metabolism of most mammals. Two of these, viz., vitamins B and C, are soluble in water, whereas vitamins A, D, and E are soluble in fats. The latter three are especially abundant in certain fats and oils. Since the demonstration of the indispensableness of water-soluble material of a type first suggested by the experients of Eijkman and of Grijns, most investigators have, for simplicity in formulating a working hypothesis as a guide to the making of experimental diets, assumed that what has been called vitamin B represents but a single nutrient principle. The same may be said of water-soluble C, or more commonly vitamin C. The terms vitamin B and C were coined to designate preparations of greater potency for the prevention or cure of polyneuritis and scurvy respectively than were the food-stuffs from which they were prepared.

In this paper only vitamin B will receive attention. There has been for years a belief in the minds of several investigators that at least two indispensable nutrient principles are contained in all preparations which give the accepted tests for potency. The instability of vitamin C, and the experimental difficulties associated with testing for it with animals, have deterred investigators from making thorough-going studies of it. Accordingly we have no recent advances in the study of vitamin C, as we have of vitamin B. A brief catalogue of the principal events in the investigation of the latter will provide a setting in which to view effectively the latest experimental work in this field.

Eijkman (1897), who first observed the development of polyneuritis in fowls restricted to a diet of polished rice, likewise determined that not only the administration of the polishings of rice, but also that a water extract of the latter, would prevent or cure the disease. He pointed out the relation between polyneuritis gallinarum and human beri-beri. Grijns (1909) was the first to suggest that beri-beri developed as the result of a lack of certain substances which are of importance in the metabolism of the peripheral nervous system. Funk (1922) summarized the knowledge available in 1911 of the nature of the anti-neuritic principle as follows: "It is soluble in water, alcohol and acidified alcohol; it is dialyzable; and is destroyed by heating to 130° C."

Funk (1911, 1912, 1913) made the first elaborate effort to isolate and identify the antineuritic vitamin. He employed phosphotungstic acid, silver nitrate and barium hydroxide, and picric acid for the separation of the vitamin. He secured active material which contained nitrogen but no phosphorus, and suggested that it was a pyrimidine. He tested his preparations on polyneuritic pigeons. Suzuki, Shimamura and Odake (1912) employed precipitation methods, resulting in the preparation of a picrate which appeared to be a mixture of two crystal forms, described as flakes and needles, the former of which appeared to possess anti-neuritic properties. A number of other investigators made studies along similar lines to those just mentioned, but none of these advanced our knowledge beyond the suggestion that perhaps an unstable modification of nicotinic acid (Funk, 1913) is the anti-neuritic substance; that adenine and certain α-hydroxy pyridines exist in an unstable isomeric modification which may exist in a number of types of substances possessing the power to restore the paralyzed bird.

McCollum and Kennedy (1916), and McCollum, Simmonds and Parsons (1918), concluded that the experience then available warranted the belief that the substance which exhibits anti-neuritic properties with the pigeon is probably identical with the substance which, when added to the diet of a rat kept on a diet which is complete except for lack of what has been called vitamin B, induces growth and the maintenance of health. This assumption was supported by the fact that when lacking in the anti-neuritic substance (vitamin B) the rat diet permitted the development of polyneuritis in the rat as well as in the pigeon. It was believed that the failure of growth and decline in weight and vitality under these experimental conditions were but secondary manifestations of impairment of metabolism due to the progressive depletion of the body of the young rat in the anti-neuritic substance.

This view was not accepted by all inquirers in this field. In 1920 Emmett and Luros presented an elaborate series of experimental studies in which they employed both pigeons and rats, with a view to demonstrating whether the anti-neuritic and so-called growth-promoting water-soluble vitamins are the same. They introduced their discussion by stating that the object of their paper was to present data which they believed indicated that a more definite distinction should be made between the anti-neuritic and certain of the other water-soluble vitamins, particularly those that relate to growth stimulation. They interpreted their findings as warranting the tentative conclusion that the anti-neuritic substance as tested for with pigeons, and the water-soluble B (vitamin B) tested for with rats, were not the same, and that it would be better to consider them as being different until further proof was forthcoming. Mitchell (1919) published a critical review of the literature relating to these two terms and favored the view that each is a distinct substance. The belief in their identity rested on the following facts: 1. The distribution of the two substances in natural foods is very similar if not actually parallel. 2. The lack of one or both of these principles in the diet of different species of animals generally, if not always, produces symptoms of nerve degeneration in a considerable number of individuals experimented upon. 3. All preparations possessing growth-promoting properties likewise possessed anti-neuritic properties. No clear case was known where a preparation was anti-neuritic but not growth-promoting when used to supplement a diet which was complete except for what passed under the term "water-soluble B," now "vitamin B." 4. The solubilities of the principles having the distinct effects of inducing growth under certain restrictions of feeding, and of relieving polyneuritis appeared to be identical. 5. So far as data were available the two hypothetical substances had identical properties as respects their behavior with solvents, precipitants and adsorbents. 6. Their stabilities in the presence of acids, alkalies and elevated temperatures appeared to be similar or identical.

Mitchell took exception to certain of the above statements, although the assumed facts available for basing any conclusions as to the relative richness of any one or more of a considerable list of natural food-stuffs which had been experimented upon were meager, and in the view of anyone familiar with this type of experimentation required some credulity for acceptance. One of these was that fresh and desiccated cabbage, onion, and carrot are poor sources of the anti-neuritic vitamin as compared with whole wheat, the former being on a par with fresh meat. The latter is known to be a poor source of both the hypothetical growth-promoting and the anti-neuritic substances. Osborne and Mendel (1919) had reported that cabbage and spinach are rather rich in the growth-promoting substance as compared with whole cereal grains. The fact that the growth-promoting tests were all made with rats, while the anti-neuritic tests were made with pigeons, forms a weak link in the comparison of these two properties in any list of food-stuffs. The rats used were young and growing, the birds were grown. The size, presence of feathers, difference in body temperature, as well as species difference, make it very difficult to secure data with these two subjects which may be regarded as in any sense quantitatively comparable. Furthermore, the data on the two properties have almost always been derived from the work of different investigators, some of whom confined their studies to pigeons and others to rats. As an instance of the variability of vitamin content of a vegetable, it is reported that young carrots are a better source of vitamin C than are old carrots. Tomatoes ripened on the plant are rich in vitamin C, whereas it is said the fruit picked green and ripened by keeping contains little of it. It is not improbable that similar differences may exist in respect to different samples of vegetables and grains. Withal it has never been possible to establish a very convincing case, on the basis of the kinds of data discussed above, for the multiple nature of so-called vitamin B. Mitchell employed the term "water-soluble B" to denote the hypothetical growth-promoting vitamin as distinguished from the anti-neuritic vitamin.

Certain of the earlier observations on the solubility of the growth-promoting and of the anti-neuritic substances appear to be more significant from the standpoint of deciding whether they are distinct substances. Thus Osborne and Mendel (1917) and Drummond state that the water-soluble growth-promoting substance is insoluble in absolute alcohol, and McCollum and Simmonds (1918) found it only incompletely soluble in 95 per cent alcohol. Eijkman (1911) states that the anti-neuritic substance is extracted from yeast by strong alcohol. Funk (1911–12) and Fraser and Stanton (1910) report on several occasions that the anti-neuritic substance was extracted from rice polishings by absolute alcohol. Cooper (1912–1914) also asserted that the anti-neuritic substance is extracted from dried beans by absolute alcohol, and that the constituent of dried egg yolk which cures polyneuritis in pigeons is readily extracted by ether. McCollum and Simmonds have repeatedly shown that the growth-promoting substance (water-soluble B) is insoluble in ether.

In this connection may also be mentioned the reports of McCollum and Kennedy (1916) that acetone and benzene extract from wheat germ previously extracted with ether the substance which cures polyneuritis in pigeons. Such extracts have always been found of doubtful potency in inducing growth in young rats which are declining due to the absence of water-soluble vitamin (vitamin B) in the diet. Steenbock (1917) also reported that by means of neutral solvents he prepared a water-acetone-soluble fraction of egg yolk which in small doses introduced intraperitoneally cured polyneuritis in pigeons.

Funk and McCallum (1916) concluded, after an eleborate attempt to fractionally separate two vitamins of the kinds now considered, that "The results obtained so far clearly indicate that the growth-promoting substance is analogous to and possibly identical with the beri-beri vitamin. . . . However, it must be admitted that while it is uncertain whether these two substances are chemically different, the results obtained do not exclude such a possibility." Drummond (1917) has also reported studies, the results of which he interprets as supporting the view that the anti-neuritic and growth-promoting vitamins are identical.

It has already been stated that the early efforts to separate and identify the anti-neuritic vitamin by precipitation methods have led to no differentiation in properties which would lead one to suspect the presence of two vitamins where but one could be demonstrated. The similarity in the solubility of the two substances, if there are two, has also been dealt with, and here we have the suggestion in several studies that especially the anti-neuritic vitamin may have been concentrated by employing solvents which do not effectively extract the growth-promoting vitamin. We may next discuss the results of separating by adsorption the vitamin or vitamins hitherto included in the vitamin B fraction.

Chamberlain and Vedder (1911) were the first to attempt the separation of the anti-neuritic vitamin by adsorption. They found that by filtering a solution through charcoal the filtrate was rendered inactive for the cure of polyneuritic pigeons. Cooper (1913) could not altogether confirm this observation, but found that six filtrations through charcoal reduced the activity of the solution by about 30 per cent. Seidell (1916) found the anti-neuritic vitamin to be adsorbed by fuller's earth. This "activated solid" has repeatedly been shown to promote growth in rats which are declining for the want of the vitamin B (fraction ?). From this activated solid the active material is extracted by alkalies. No differences have been found either in the solution or fuller's earth which has been in contact with it which have led anyone to suspect the presence of two active principles.

Harden and Zilva (1918) also reported that the vitamin B is adsorbed from autolyzed yeast by means of dialyzed iron, and that it may be recovered from the precipitate by means of 95 per cent alcohol. Myers and Voegtlin (1920) reported that olive oil and oleic acid dissolve or adsorb the vitamin from yeast filtrate.

No evidence has come from either of these procedures which indicates the multiple nature of the vitamin B.

Funk and Dubin (1921) state that Funk and Macallum have tested the phosphotungstic precipitate obtained from yeast, and whereas it was strongly curative for avian beri-beri, it induced only moderate growth in rats. Funk and Dubin report that by fractional adsorption of yeast autolysate with fuller's earth or norit (birch wood charcoal) it is possible to make an almost quantitative separation of the anti-neuritic vitamin and of another substance which stimulates the growth of yeast (bios) which they call vitamin D. McCollum, Simmonds, Shipley and Park have designated the anti-rachitic vitamin of cod liver oil by the symbol D, and this term has become generally adopted in this sense. They state that when 1 liter of yeast autolysate is shaken with 50 grams of fuller's earth, the filtrate was found in the majority of cases to be inactive for avian beri-beri. This filtrate was then treated twice with 100 grams of fuller's earth, which carried down quantitatively the yeast growth-promoting substance called by them vitamin D (bios ?).

They found pigeons to require only vitamin B (anti-neuritic substance ?), whereas rats fed a purified diet supplemented with cod liver oil to supply fat-soluble vitamins, supplemented with their anti-neuritic preparation of fuller's earth, or with the earth carrying the yeast growth-promoting fraction (bios), grew but little and began to die after two months. Increasing the amount of either of these vitamin preparations alone did not have any appreciable effect, but the addition of the missing component in both cases caused growth. Rats which were given both of these principles plus the last filtrate which contained neither B (anti-neuritic) nor D (bios) showed normal growth and behavior. They suggest that rats require two vitamins of the B type for growth and well-being. Seidell has studied the removal of the anti-neuritic substance from activated fuller's earth by sodium hydroxide and later by barium hydroxide, and more recently he has again returned to the use of sodium hydroxide. From vitamin extracts he has described the preparation of a highly active picrate (anti-neuritic), and of a silver compound of the vitamin. He states that there are two kinds of crystals in his active picrates, and is inclined to believe that one is an unstable isomer which possesses the anti-neuritic property. Seidell has very recently described the preparation of a non-hygroscopic preparation from fuller's earth which is extremely potent in anti-neuritic properties. It appears that he has found this line more promising than a continuation of study of the isolation of the anti-neuritic vitamin by means of the picrate or the silver compound. His work suggests the existence of two active principles in the vitamin

B fraction. In his discussions, however, he apparently has had in mind the isolation of the anti-neuritic vitamin.

Laird, working in the author's laboratory, sought during the winter of 1924–25 to study the vitamin requirements of the rat and pigeon with respect to the water-soluble vitamin fraction. The following tables taken from his paper show that remarkable differences in the nutritive requirements of these two species were manifested in their reaction to several preparations. These results would seem to be explainable only on the assumption that there are two or more indispensable nutrients in the vitamin B fraction hitherto regarded generally as supplying not necessarily more than one. This follows from the fact that certain preparations were positive with one species and negative with the other, and also because combinations of several preparations, each alone negative as a source of "vitamin B" with the rat, were positive with this species. (Examples are AA/AB, and AB/BA, Table IV.)

TABLE III

Summary of Results of Feeding Tests

Extracts of wheat germ, using 0.5 per cent acid in organic solvent. Tests on single extracts

No.	Symbol	Acid	Solvent	Effect on Rats	Effect on Pigeons
1	AA	Acetic	Acetone	Negative	Negative
2	BA	"	Benzene	"	"
3		"	Chloroform	Fair	*
4		"	Ether	Negative	*
5	AB	Benzoic	Acetone	"	Negative
6	BB	"	Benzene	"	"
7		"	Chloroform	Fair	*
8		"	Ether	Negative	*
9	AC	Citric	Acetone	"	Good (slow)
10		"	Ether	"	*
11		Formic	Acetone	Good	Fair
12		"	Ether	Negative	*
13		Gallic	Acetone	Fair	Good
14		"	Ether	Fair–good	*
15		Lactic	Acetone	Fair	Fair
16		"	Ether	Negative	*
17		Malic	Acetone	Fair	*
18		"	Ether	*	*
19	AO	Oleic	Acetone	Negative	Negative
20		"	Benzene	Fair–poor	*
21		"	Chloroform	" "	Fair
22		"	Ether	Fair	*
23		Oxalic	Acetone	Negative	*
24		"	Ether	"	*
25		Salicylic	Acetone	Excellent	Negative
26		"	Benzene	Good	"
27		"	Chloroform	Fair–poor	*
28		"	Ether	Negative	*

* Not tested.

TABLE IV

SUMMARY OF RESULTS OF FEEDING TESTS

Extracts of wheat germ, using 0.5 per cent acid in organic solvent.
Tests on two extracts combined

No.	Symbol	Effect on Rats	Effect on Pigeons
29	AA/AB	Good	Negative—2 birds
30	AA/AC	Negative	Good
31	AA/AO	Fair–good	Excellent
32	AA/BA	Negative	Negative
33	AA/AB	Good	"
34	AB/AC	Poor–excellent	"
35	AB/AO	Negative	"
36	AB/BA	Excellent	"
37	AB/BB	Fair–good	" —2 birds
38	AC/AO	Negative	Good—2 birds
39	AC/BA	"	Fair
40	AC/BB	Fair–good	Negative—2 birds
41	AO/BA	Negative	Excellent
42	AO/BB	Fair–good	Negative
43	BA/BB	Negative	" —2 birds

Experiments which are significant in showing whether vitamin B is of multiple nature have been reported recently by Smith and Hendrick (1926). These investigators found that 14 per cent of rolled oats as the only source of vitamin B, supplemented with purified food-stuffs and vitamin A, failed to induce satisfactory growth in rats. The addition of 2 per cent of dried brewer's yeast gave a better growth curve. They sought to inquire what constituent of the yeast corrected the deficiency. They observed that a diet containing 14 per cent of protein derived entirely from rolled oats (92.5 per cent) supplemented with salts and butter fat to provide vitamin A, did not induce normal growth. The addition of 6 per cent of dried brewer's yeast greatly stimulated growth. The addition of protein either in the form of gelatin or casein did not stimulate growth as did the yeast. They inferred from this that it was not the protein of the yeast which supplied the nutrient principal lacking in the oat. A diet consisting of rolled oats 40, casein 10, autolyzed brewer's yeast (dried) 5, salts, butter fat and starch to 100, induced satisfactory growth in rats. The yeast was autoclaved at 15 pounds pressure for 6 hours. They showed by appropriate experiments that this treatment completely destroyed its power to induce a growth response under the conditions of the rat test for vitamin B. It has been repeatedly reported that autoclaving yeast destroys the anti-neuritic substance. Evidently the supplementary value of autoclaved yeast for oats, when the latter is adequately supplemented by casein, salts and vitamin A, resides in some nutrient principle other than the anti-neuritic substance, since this has been destroyed. They found that 40 per cent of rolled oats furnish sufficient of the anti-neuritic substance to support normal growth over the principal growth period in the rat, since their rats grew from 50 grams to 200 grams during 13 weeks on this diet. That the factor in the autoclaved yeast which withstands heat and supplements a deficiency in oats is not an amino-acid they make practically certain by demonstrating that the test ration (177) with 6 per cent of coagulable yeast protein replacing the 6 per cent of autoclaved yeast induced but a slow rate of growth.

Another type of experiment confirms the conclusion of these authors that the substance furnished by the autoclaved yeast which supplements rolled oats is a newly discovered vitamin. A diet of purified food substances containing cod liver oil to furnish vitamins A and D, and 2 per cent of a vitamin B picrate prepared by Seidell, failed to induce growth. The same diet without the vitamin B picrate, but containing 5 per cent of autoclaved yeast also failed to induce growth, and resulted in a slow loss in weight of the animals. When the two were added to the basal ration the animals responded with rapid gains in weight.

Almost simultaneously with Smith and Hendrick, Goldberger and associates (1926), reported experiments which demonstrate the same thing. Goldberger and associates demonstrated that 27 per cent of autoclaved yeast failed to serve as a source of what has been called vitamin B when employed in a purified diet supplemented with fat-soluble vitamins. A similar basal diet was not supplemented with respect to vitamin B when a cold alcoholic (85 per cent) percolate of ground corn equivalent to 40 per cent of corn was added to the diet. On the other hand the diet of purified food-stuffs plus the fat-soluble vitamins A and D, and containing the corn extract equivalent to 5 per cent of corn in the diet, together with 8 per cent of autoclaved yeast, induced essentially the normal rate of growth.

A similar demonstration was made using the purified diet supplemented with 20 per cent of dried fresh beef as the sole source of vitamin B in the first period. After an initial gain during two weeks the animals began to decline in weight. When they had declined for 4 or 5 weeks and had reached their weight at the beginning of the experiment, the addition of the corn extract equivalent to 5 per cent of corn in the diet caused a prompt resumption of growth in those animals which survived the preceding treatment. This is interpreted as meaning that fresh beef contains the same principal which is still present in autoclaved yeast which gives it its supplementary value with respect to the corn extract.

Both of these observations are in harmony with the findings that both an extract of dried yeast and fresh beef have preventive or curative action in pellagra when the amounts fed were insufficient to prevent the appearance in human subjects of incipient signs of beri-beri. The same substances which are effective in preventing pellagra have been found effective in the prevention of a disease known as black tongue in dogs, and which is believed to be the analogue of pellagra in man. The heat-stable substance in yeast, which is also found in lean meat, which acts as a nutrient supplement in the above described experiments, and which is shown to prevent pellagra is designated by Goldberger and his associates as P-P (pellagra-preventive).

One appears justified from these experiments in concluding that what has hitherto been called vitamin B preparations contain at least two principles, both indispensable to normal nutrition. One of these is the anti-neuritic substance for which Goldberger suggests the retention of the name vitamin B, and the other the so-called P-P.

BIBLIOGRAPHY

Eijkman, C.: Virchow's Archiv, 1897, cxliv, 187.
Grijns, G.: Geneeskundig Tijdschrift voor Ned. Indie, 1901.
Funk, C.: The Vitamines, Baltimore, 1922.
Jour. of Physiol., 1911, xliii, 395; *ibid.* 1912, xliv, 50; *ibid.* 1913, xlvi, 173.
Suzuki, U., Shimamura, T., and Odake, S.: Biochem. Zeitschr., 1912, xliii, 89.
McCollum, E. V., and Kennedy, C.: Jour. Biol. Chem., 1916, xxiv, 491.
McCollum, E. V., Simmonds, Nina, and Parsons, Helen T.: Jour. Biol. Chem., 1918, xxxiii, 411.
Emmett, A. D., and Luros, G. O.: Jour. Biol. Chem., 1920, xliii, 265.
Mitchell, H. H.: Jour. Biol. Chem., 1919, xl, 399.
Osborne, T. B., and Mendel, L. B.: Jour. Biol. Chem., 1919, xxxvii, 187.
Osborne and Mendel: Jour. Biol. Chem., 1917, xxxi, 158.
Drummond, J.: Biochem. Jour., 1917, xi, 261.
McCollum, E. V., and Simmonds, Nina: Jour. Biol. Chem., 1918, xxxiii, 62.
Eijkman, C.: Arch. Schiffs- u. Tropenhyg., 1911, xv, 698.
Funk, C.: Jour. Physiol., 1911–1912, xliii, 395.
Fraser, H., and Stanton, A. T.: Lancet, 1910, ii, 1755.
Cooper, E. A.: Jour. of Hyg., 1912, xii, 436; *ibid.* 1914, xiv, 12.
Steenbock, H.: Jour. Biol. Chem., 1917, xxix, p. xxvii.
Funk, C., and McCallum, A. B.: Jour. Biol. Chem., 1916, xxvii, 63.
Drummond, J.: Biochem. Jour., 1917, xi, 255.
Chamberlain, W. P., and Vedder, E. B.: Philippine Jour. of Sci., B., 1911, vi, 251, 395.
Cooper, E. A.: Biochem. Jour., 1913, vii, 268.
Seidell, A.: U. S. Public Health Reports, 1916, xxxi, 325, 361.
Harden, A., and Zilva, S. S.: Biochem. Jour., 1918, xii, 93.
Myers, C. N., and Voegtlin, C.: Jour. Biol. Chem., 1920, xlii, 199.
Funk, C., and Dubin, H.: Proc. Soc. Exper. Biol. and Med., 1921, xix, 15.
Laird, C. N.: Amer. Jour. of Hyg., 1926, vi, 201.
Smith, M. I., and Hendrick, E. G.: U. S. Public Health Reports, 1926, xli, 201.
Goldberger, J., Wheeler, G. A., Lillie, R. D., and Rogers, L. M.: U. S. Public Health Reports, 1926, xli, 297.

QUANTITATIVE RESEARCH
IN THE
CHEMISTRY OF NUTRITION

BY

HENRY C. SHERMAN

MITCHILL PROFESSOR OF CHEMISTRY
COLUMBIA UNIVERSITY

Lecture delivered at Columbia University in the special course on Contemporary Developments in Chemistry given in the Summer Session of 1926 on the occasion of the opening of the Chandler Chemical Laboratories

COLUMBIA UNIVERSITY PRESS
NEW YORK
1927

Printed from type. Published March, 1927

Printed in the United States of America

QUANTITATIVE RESEARCH IN THE CHEMISTRY OF NUTRITION

By Henry C. Sherman

In the chemistry of food and nutrition we recognize four major factors of essentially coordinate importance: (1) energy, usually expressed in terms of calories; (2) protein or amino acids; (3) the vitamins; (4) the inorganic or mineral elements or "ash constituents" of plants and animals.

The energy relations of nutrition have been studied with many species; but since in this field of research much may be gained through the cooperation of the subject, the highest development of accuracy has been attained in work with man and it is mainly to studies of the transformation of energy in the human body that the present discussion will refer, though the principles are doubtless equally valid for other species.

Quantitative study of the energy transformations in the human body was first satisfactorily developed, during the period of twenty-five to thirty years ago, in the chemical laboratory of Wesleyan University at Middletown, Conn., through the joint work of W. O. Atwater, professor of chemistry at Wesleyan and director of nutrition investigations for the United States Department of Agriculture; E. B. Rosa, then professor of physics at Wesleyan, later physicist of the United States Bureau of Standards; and F. G. Benedict, then professor of chemistry at Wesleyan, now director of the nutrition laboratory of the Carnegie Institution of Washington.

Atwater, Rosa, and Benedict constructed in the basement of the Wesleyan chemical laboratory a respiration calorimeter by means of which they could measure, simultaneously and quantitatively, the respiratory exchange and the transformation of energy in the human body; and through years of experimentation and invention they developed this apparatus until it became truly an instrument of precision—while large enough so that a man could live normally in it for days at a time under constant accurate observation and with continuous measurement of the energy which was being transformed in his body and of his oxygen consumption and carbon dioxide production as well as the more usual determinations of the intake and output of chemical elements and compounds.[1]

To the work of Atwater, Benedict and their colleagues with this respiration calorimeter and the other forms of apparatus later developed from it is due the main part of the quantitative basis of our modern knowledge of the energy relations, which underlies and must precede a clear understanding of the other nutritive requirements. For only when energy requirements are adequately covered can individual nutrients be advantageously used and conserved.

The influence of the age, size, and activity of the organism, and of many physical and chemical factors of its environment and internal condition have been studied by the same and other investigators but the results of these investigations, important as they are, cannot now be reviewed, for time suffices only for the barest mention here of the types of quantitative research which have been and are being developed in the different branches of the chemistry of nutrition, and does not permit of the presentation of any of the detailed data, or discussion of the interesting bearings of the results.[2]

From the quantitative point of view, to which this lecture is devoted, special significance attaches to the close agreement between the heat calculated from the chemical data of the respiration calorimeter experiments and the heat production directly observed.

In a compilation of the data of six series of experiments, four with men, one with cattle, and one with dogs, Armsby [3] has shown that this agreement was in each of the six series within less than one per cent; and that in the entire combined series covering 336 complete experimental days, the total calculated heat production was 1,441,691 Calories, and the observed heat production was 1,445,398 Calories—or the former was to the latter as 1.0000 to 1.0026.

It seems altogether probable that this very small difference is merely accidental; but if it were taken to mean that the body has other sources of energy than the fuel value of its food, then it would also follow that the magnitude of all such other sources of energy put together is to that of the fuel value of the food as about one to four hundred. In other words, that if there are other sources of energy they are quantitatively negligible and that for all practical purposes it may be said that the body derives the energy, both for its internal life processes and for its external muscular work, from the oxidation of the organic foodstuffs; and gets from them the same amount of energy as would be computed from the chemical data involved.

Thermodynamically it is natural enough that in so far as the products of oxidation are the same whether the foodstuffs are burned in a living body or in oxygen in the laboratory, the energy transformed should likewise be the same. But it is quite a different matter to say how, at the relatively low temperature of the body, the oxidation of the foodstuff can proceed fast enough to supply energy at the rate needed for the maintenance of all the life processes. And this latter question is plainly of great importance to our understanding of the chemical aspects of nutrition—the chemical processes of life.

The carbohydrates, fats and proteins of the food first undergo cleavage in the body into substances more readily oxidizable; but even so, at the temperature of the body, both cleavage and oxidation must be greatly catalyzed if the chemical reactions involved are to go on rapidly enough to support life. Hence these catalysts are absolutely essential. They are formed in the body; but the materials of which they are formed must come either directly or indirectly from the food. A part of the nutritive function of the food must be to furnish the substance or substances from which these particular catalysts are derived.

What, then, are the chemical natures and nutritional antecedents of the catalysts which expedite the utilization of the organic foodstuffs as fuel? Here we find an interesting interrelation of the energy and the protein metabolism; for all of these catalysts which have been sufficiently studied to yield clear light as to their chemical nature, appear to be derived in one way or another from protein. That is, they appear to be either of the nature of proteins themselves or derived from certain of the amino acids which the proteins yield in the body. Thus the typical oxidative hormones thyroxin [4] and epinephrine (adrenaline) [5] appear to be derivatives of the amino acid tyrosine; glutathione, the oxidative catalyst widely distributed and apparently always present in the tissues, has been found [6] to be a derivative of the two amino acids cysteine and glutamic acid; and the chemical nature of insulin, while not yet fully determined, seems also to be such as to indicate an amino acid derivation.[7, 8]

But before these hormones can come into play in the utilization of the foodstuffs as sources of energy, it is necessary, in most cases, that the foodstuff shall undergo a preliminary cleavage; and these cleavages proceed rapidly enough only when catalyzed by enzymes—those of the digestive juices and of the body tissues. The typical enzymes are doubtless of much more complex molecular structure than are the typical hormones. For this and other reasons (some of which will be mentioned in what follows), the chemical investigation of typical enzymes has proven exceedingly difficult—even an account of the results of such investigation can hardly be reduced to easy reading and at the same time to a form sufficiently full and explicit to be conclusive—and some writers continue the statement that the chemical nature of enzymes is unknown. The researches of recent years have, however, thrown much more light upon this subject than is yet (August 4, 1926) generally realized. A part of the present lecture is therefore devoted to this subject.

As the most abundant of the organic foodstuffs is starch, and the starch-splitting enzymes or *amylases* are believed to be typical in this respect, the chemical nature of two representative amylases has been much studied, and these studies furnish a rather striking example of the quantitative type of research which I have been asked to discuss in this lecture.

In the study of enzymes, the chemist (paradoxical as it may at first appear) is apt to feel embarrassed and perhaps estranged by the very fact that he must depend so completely upon chemistry. As the typical enzymes have been met only in colloidal condition,* the familiar physical criteria of melting and boiling points, crystalline form, etc., which play such a large part in the identification and testing for purity of ordinary simple substances, are not applicable here. On the other hand *the* characteristic and significant property of each enzyme is its catalytic effect upon some chemical reaction. Thus while most substances are characterized largely by their physical properties, enzymes are characterized by their chemical properties only.

The existence of such substances as we now call enzymes first became known through the observation that barley malt had the property of hydrolyzing starch. Since this must have been due to something previously unknown but contained in the malt, a concentration was later sought by fractioning the malt in different ways, testing the fractions, rejecting those which did not show the peculiar property, and fractioning still further those in which it appeared.

The principle is well illustrated in the now classical investigations which led to the isolation of radium.

Uranium and thorium were known to be radioactive when Madame Curie discovered that two minerals, pitchblende and chalcosite, showed *greater radioactivity* than pure thorium or uranium, suggesting the presence in these minerals of some new element more radioactive than any of the elements previously known. The Curies then set to work to separate pitchblende into its constituents, much as they would have begun a complete

* Sumner's crystallization of jack-bean urease was reported after the above had been written and is referred to again below.

chemical analysis of any mineral, but with the added feature that after each separation or fractionation all the products were examined for radioactivity, not only qualitatively but quantitatively, by means of the electroscope. In this way they could not only follow the fate of the radioactive material but also could judge of the *extent* to which they were succeeding in concentrating it. Also, at each step, the fraction which showed the radioactivity was examined spectroscopically and it was only when they had concentrated the radioactivity 50-fold that they were first able to see new lines in the spectrum, indicating the presence of a new element. Further fractionation, guided still by measurements of radioactivity to show the concentration of the radioactive material, brought them finally to the attainment of a constant maximum radioactivity which meant the isolation of radium in the form of a pure salt. (For the sake of brevity we must omit description of the isolation of radioactive actinium and polonium from other fractions of the pitchblende.)

Now the process of purifying a natural enzyme, or separating it from the other substances with which it is mixed in the plant or animal tissue or secretion in which it is found, is similar in principle to that by which radium was isolated from pitchblende. Just as the isolation of the radium was best guided and judged by quantitative measurements of its characteristic property of radioactivity, so the isolation of the enzyme is best guided and judged by quantitative measurements of its characteristic enzymic activity.

Attempts to isolate an enzyme, by fractioning the natural material which contains it and continuing the fractionation until the characteristic property is concentrated to a constant maximum by removal of all other substances, involve very great difficulties.

First of all, the enzyme itself is apt to be undergoing disintegration during the process. Another difficulty is that the fractionation may remove something which while itself having no enzymic activity is helpful or possibly essential to the full activity of the enzyme, so that, if tested without sufficient regard to this possibility, an enzyme after purification may be purer in the sense that the product is more nearly a single chemical individual and yet may show less of the characteristic enzymic activity than before. This difficulty has had the very unfortunate effect of leading many students of the question, and even some prominent investigators, to minimize or neglect the importance of the measurement of activity as a guide both in the development of methods and in the interpretation of results of research upon the purification of enzymes.

Since the enzymic activity is the essential characteristic of every enzyme, any failure to make the fullest use of quantitative measurements of activity in enzyme research must seriously restrict its value.

The only thoroughly scientific method of meeting this difficulty is to work out experimentally, separately and quantitatively for each enzyme to be investigated, a knowledge of all the environmental conditions such as the effective hydrogen ion concentration (pH) and the optimum concentrations of such other ions as may be essential to the best activity of the enzyme, as well as the best conditions of time, temperature, and kind and concentration of substrate to be used in the quantitative measurements of the activity of the enzyme preparation.

When through long and systematic experimentation one has acquired a sufficiently complete and accurate knowledge of the environmental conditions required, first, to conserve the enzyme as well as possible from deterioration during investigation, and second, to permit of its exerting the full catalytic effect of which it is capable in the quantitative determination of enzymic activity, then only is one in position to attempt the isolation of an enzyme from the material which contains it in nature according to the same scientific plan of procedure as was followed in the isolation of radium from pitchblende. Working according to this general plan, Osborne[9] purified malt amylase and reported it to be of protein nature. We have confirmed his conclusion and extended the study to the investigation of pancreatic amylase which has yielded results of the greatest interest from the point of view of quantitative research.

Employing several different combinations and modifications of the processes of extraction, dialysis and fractional precipitation, workers in this laboratory repeatedly have obtained preparations of purified pancreatic amylase of apparently uniform chemical and physical properties and of practically constant *maximum* enzymic activity; and this product, like the malt amylase previously prepared by Osborne, was of protein nature. Inasmuch as the varied fractional precipitations to which this product was subjected would presumably have changed its composition and properties if it had contained any large admixture of foreign material, the fact that the enzymic activity showed quantitative constancy at the maximum is strong direct evidence of the protein nature of the enzyme.

In 30 minutes at 40°, this material splits 20,000 times its weight of starch and forms 10,000 times its weight of maltose. Although subject to serious deterioration in solution, the activity of this material is such that when tested in longer experiments it digested 4,000,000 times its weight of starch and formed no less than 2,800,000 times its weight of maltose before it had all become inactivated.

This marked enzymic activity was exhibited by the preparation at a dilution of 1 : 100,000,000 parts of water. The most delicate tests for proteins are not valid at dilutions greater than about 1 : 100,000. The material reacted like typical protein to the usual protein tests, but its own enzymic activity constituted a test for its presence which was 1,000 times more delicate. Thus the failure of protein reactions in solutions enzymically active does not show that the enzyme is of other than protein nature in its chemical composition, although this negative conclusion has been erroneously drawn by some investigators and is repeated by many writers.

Here it is of interest to recall the experience of the Curies in the concentration of radium from pitchblende as described above. It was only after they had intensified the property of radioactivity 50-fold by concentration of the radioactive element that its presence was revealed analytically even by so delicate a test as the use of the spectroscope. Delicate as is the spectroscopic test, the property of radioactivity constituted a test at least 50 times more delicate. When now in our work with pancreatic amylase we find in an analogous way that the property of enzymic activity is a test for this enzyme which is 1,000 times more delicate than any analytical test for the protein of which the enzyme preparation is materially composed, we see how easy it can be for an investigator to be misled as to the chemical nature of an enzyme by obtaining something which shows enzymic activity, but does not respond to analytical tests for protein.

In addition to quantitative elementary analyses and qualitative reactions in all of which our pancreatic and malt amylase preparations show results typical of protein substances, we have submitted both of these enzyme preparations to quantitative analysis with reference to the eight forms of nitrogen determinable by the Van Slyke method with results which show that the enzyme preparations yield all these products of hydrolysis in proportions within the range of variation shown by such typical protein materials as casein, edestin, hair, and hemoglobin. Time does not permit a fuller presentation of the direct evidence that the protein material which plainly constitutes almost the entire dry weights of these preparations is in the main identical with or essential to the enzyme itself.

This view of the chemical nature of the enzyme is supported by evidence of other kinds which, while indirect, is none the less instructive.

The investigations of this laboratory* upon pancreatic amylase have shown [10] that this enzyme loses its activity in aqueous solutions as already mentioned, that this loss proceeds more rapidly with increase of temperature, and that it is undoubtedly connected with hydrolytic changes taking place in protein material which constitutes either the enzyme molecule itself or an essential part of it. Extensive quantitative studies of the conditions influencing the rate of hydrolytic inactivations of the enzyme, especially as affected by added amino acids, give additional evidence of the protein nature of the enzyme.

This is further confirmed by the quantitative finding that the antiseptics which act chemically upon proteins are enormously more destructive of the enzyme than are the antiseptics of the lipoid-dissolving type.

The amylase of malt has yielded similar results; and it has also been possible in working with this more stable enzyme to demonstrate that it has a definite isoelectric point or zone and that the material which shows the enzymic activity is an ampholyte, presumably protein. Certainly it is quite improbable that, if the enzyme were a substance of some other chemical nature, it would be so inseparable from the particular protein with which associated as to migrate with it under the influence of the electric current in either direction according to the hydrogen ion concentration of the solution.

Thus we now have experimental evidence of four different kinds all pointing clearly and consistently to the protein nature of the enzyme; † and any question left open by any one of the four methods of research would seem to be adequately answered by at least one of the others. Largely because of its quantitative character, the newer evidence is more convincing the more fully and critically it is studied.

Hence it appears that the protein of the food must furnish material not only for the building and upkeep of the body tissues in the ordinary sense, but also for the production in the body of the enzymes and hormones which catalyze the cleavage and oxidation of the organic foodstuffs and thus make these chemical reactions run fast enough to furnish energy at the rate needed to support the processes of life.

Thus it is clear that some amino acids, in serving as material for the formation of hormones and doubtless also of enzymes

* We are greatly indebted to the Carnegie Institution of Washington for grants in aid of these investigations.

† Since the above was written, the protein nature of at least one enzyme has been beautifully confirmed by the work of Sumner, who has succeeded in obtaining the enzyme known as jack-bean urease in the form of well-crystallized protein, and has even recovered it unchanged upon recrystallization. (Journal of Biological Chemistry, August and September, 1926.)

in the body, have essential *specific* functions in addition to the general tissue-building function of the amino acids as a group. And from this it necessarily follows that different consequences may result from protein starvation, or inadequacy, according to which amino acid is first (or most markedly) reduced below normal amount or concentration.

The *average* protein requirement of normal adult maintenance in human nutrition has been fairly well established by the accumulation of over one hundred quantitative experiments bearing upon the problem.[11, 12] The causes and extent of variations from the average, and the relation of the requirements of growth to those of maintenance are still promising fields for quantitative research.

Osborne and Mendel's studies of individual proteins, and the interpretation of their nutritional relationships in terms of their amino acid make-up, introduced a new era in this field of research.[13] Osborne having carried the study of proteins about as far as it could be carried *in vitro*, they began feeding experiments with isolated proteins and, for reasons which time does not permit us to recite but which have been amply justified by subsequent experience, they selected the rat as their chief experimental animal.

So general has the use of such animals in chemical laboratories now become, and so greatly has it contributed to the recent development of chemistry that in a course of this kind a special lecture might well be given upon the subject of "rats as reagents"; but under our present need for brevity it must suffice to let a few examples speak for themselves. And in accordance with the general plan of this course, the lecturer has been requested and must be expected to draw his illustrations mainly from researches of his own laboratory.

The determination of the sulfur-containing amino acid, cystine, in proteins (especially those like casein in which it occurs in relatively small amounts) is a matter of importance to the chemistry of nutrition, but for which no satisfactory analytical method has been found, although much careful investigation has been devoted to the problem.

Osborne and Mendel had shown that when a rat is kept upon a diet poor in an amino acid such as lysine or cystine, its growth is retarded, and in such a case a marked increase in the rate of growth may be induced by the addition to the diet of the one amino acid which is its growth-limiting factor. Feeding experiments based on this fact have recently been made a method for the quantitative determination of cystine in casein, a problem which, as already noted, had baffled repeated attempts at solution by various analytical procedures. Working with rats as reagents, Dr. Woods found [14] that so long as the cystine was the sole growth-limiting factor in the experimental diet, the rate of growth responded quantitatively to the amount of cystine added to the food—whether it were added as pure cystine or in the form of casein, the protein whose cystine content it was desired to determine. Thus by means of feeding experiments carried out in a strictly quantitative way, and in large enough numbers to ensure average results of high accuracy notwithstanding natural variability of individual animals, it was found that casein was equivalent as a source of cystine to one fortieth of its weight of pure cystine. This indicates for casein a cystine content of about 2.5 per cent, which appears to be nearer the truth than any of the results which had been obtained in attempts to determine the cystine content of casein *in vitro*.

If chemistry *in vivo* can thus afford more accurate measurements than can chemistry *in vitro* in the case of such a well-known substance as cystine, we may feel reassurance in the use of the feeding method for quantitative research in connection with the vitamins.

The water-soluble vitamin now commonly called vitamin B was reported, on the basis of work done, of necessity hurriedly, to meet a practical problem arising in connection with the provisioning of British troops during the World War, to be "stable on heating at 100° but decomposed at 130°." This statement gained wide currency and led not only to practical applications of dubious validity but also to the view that the reaction through which the vitamin was destroyed on heating must take place suddenly or at a greatly accelerated rate at some particular temperature between 100° and 130°.

Dr. Spohn,[15] studying the behavior of vitamin B as it occurs in skimmed milk powder, found that when this was heated dry at 100° the destruction of vitamin B was not appreciable; but that it became plainly demonstrable at this same temperature when heated in solution. In this connection, through hundreds of careful comparative feeding experiments, she worked out a good basal ration for use in such investigations—one which furnishes the experimental animal with approximately optimal proportions of all needed nutrients except that it is devoid of the vitamin in question.

Using this basal ration, Dr. Grose [16] studied quantitatively the rate of destruction of vitamin B at each of the four temperatures 100°, 110°, 120°, and 130°; and his data are here cited to illustrate this type of research.

Using tomato juice as the source of vitamin B, he found

that systematically graded allowances to the experimental animals resulted in regular increments of the body weight curve as seen in Fig. 1. Such feedings were then repeated in parallel with others in which were used portions of the same tomato juice which, however, had been heated for four hours at 100°, 110°, 120°, or 130°. In each case the feedings were systematically varied from experiment to experiment until through long series

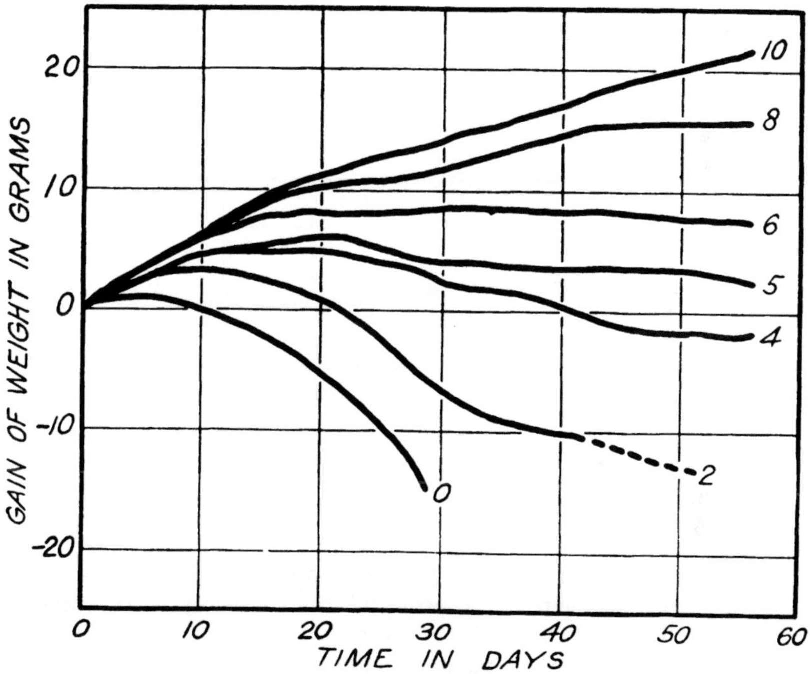

FIG. 1.—Weight curves of young rats showing the effect of graduated doses of vitamin B in the form of tomato juice. Rats shown in curve marked O received the basal diet only and therefore no vitamin B, while those whose averages are shown by the other curves received this ration supplemented by daily doses of 2 cc. to 10 cc. of the tomato juice as indicated on the curve. Each curve is based upon the average results of from 6 to 25 experiments, and it will be seen that with such numbers the average results can safely be interpreted to show differences in amounts of vitamin of the order of magnitude of five to ten per cent.

of experiments it was found just what were the relative proportions of the heated and unheated juices required to furnish the same amount of vitamin B* and thus to induce the same amount

* While the term vitamin B as hitherto used may prove to have covered more than one substance, it will be noted that the experiments here described are so designed as to afford quantitative measures of that growth essential which is here the actual growth-limiting factor.

of growth, as shown, for example, in Fig. 2. The quantitative significance of such experiments depends upon the selection of a level of vitamin intake at which either an increased or decreased amount of the vitamin will materially influence the result. The results as a whole very definitely cleared up the misapprehensions which had existed as to a marked acceleration of the rate of destruction between 100° and 130°; for these quantitative data show no departure from the orderly course of a chemical reaction under the accelerating influence of heat. Not only was there no sudden or exceptional acceleration at any point within the range of 100° to 130°, but the temperature coefficient throughout this range was below rather than above the general average for ordinary chemical reactions.

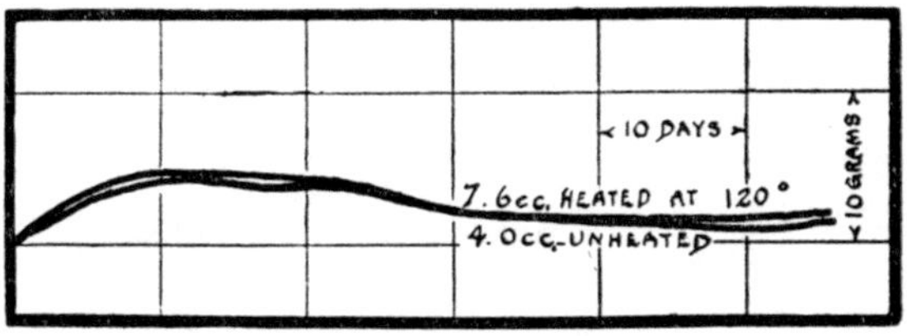

FIG. 2.—Weight curves of rats receiving 4 cc. of unheated tomato juice in comparison with those receiving an average dose of 7.6 cc. heated for 4 hours at 120°. As 4 cc. of the unheated and 7.6 cc. of the heated furnished the same amount of vitamin B it was shown that the four hours of heating at 120°C. had destroyed about 47 per cent of the vitamin B originally present. (Courtesy of the Journal of the American Chemical Society.)

Dr. Burton [17] employed the same method to study the influence of hydrogen ion concentration or activity (pH) upon the rate of destruction of the vitamin, and found that the rate of destruction was regularly increased as the reaction of the solution was shifted from acid toward alkaline—from pH 4.28 (the natural acidity of tomato juice) to 5.2; to 7.9; to 9.2; to 10.9. At pH 5.2, about 10 per cent; at pH 7.9, 30–40 per cent; at pH 9.2, 60–70 per cent; at pH 10.9, 90–100 per cent, of the vitamin B was destroyed upon heating for one hour at 100°. At pH 4.28 the destruction was less than at pH 5.2 by an amount too small to be measured by present methods, except by averaging the results of a prohibitively large number of experiments. The difference, however, became measurable when the time of heating was extended to four hours.

The fat-soluble vitamin, vitamin A, can be studied in the same general way as has just been described for vitamin B; but as the body may store relatively much more of the fat-soluble than of the water-soluble vitamin, it becomes important here to deplete the body store before beginning the feeding of graded doses of the material to be tested. When this has been done the average weight curves of sufficient numbers of experimental animals are found to respond quite regularly to systematic increments of the vitamin fed as was shown in the work of Dr. Munsell.[18] Dr. Quinn [19] studied quantitatively the destruction of vitamin A in tomato juice when heated four hours at 100° at its natural acidity as in the experiments of Dr. Grose upon vitamin B, and showed that, with vitamin A as with vitamin B, the destruction under these conditions was approximately 20 per cent.

The experiments of Dr. F. L. MacLeod [20] revealed the great importance of quantitative differences in the vitamin A content of the food even among cases which appear adequate as judged by growth experiments of the usual type. For example, twin sisters (rats) on diets differing in vitamin A content made almost equally good growth, but the one receiving the lower allowance of vitamin A failed utterly in reproduction whereas the sister receiving a more liberal allowance of this vitamin reproduced normally, and her offspring and descendants thrived vigorously. These twin sisters were mated to twin brothers so that hereditary factors were the same in the two families; and all factors of the environment except the food were likewise kept uniform. The experiment was then repeated until seventeen pairs of twin sisters had been thus compared.

The smaller amount of vitamin A proved sufficient for normal growth up to nearly average adult size, but not for successful reproduction, and rarely did it support satisfactory longevity. The parallel animals receiving the more liberal allowance of vitamin A grew to full average adult size, were successful in reproduction and the rearing of young and lived on the average about twice as long as those on the diet equally good in all other respects but lower in vitamin A.

These experiments show strikingly that a proportion of vitamin A in the food sufficient to support normal growth and maintain every appearance of good health, for a long time at least, may still be insufficient to meet the added nutritive demands of successful reproduction and lactation.

Among those receiving the lower allowance of vitamin A there usually also appeared in early adult life an increased susceptibility to infection and particularly a tendency to break down with lung disease at an age corresponding to that at which pulmonary tuberculosis so often develops in young men and women. The bacillus involved is different; but the close parallelism of increased susceptibility of the lung to infection at this stage of the life history appears very significant, especially in view of the fact shown by other experiments (cited below) that the vitamin-A content of lung tissue varies with that of the food.

Thus by quantitative research it was clearly shown that vitamin A is an even more important factor in the chemistry of food and nutrition than had previously been appreciated, for it must be supplied in liberal proportion not only during growth but in the food of the adult as well, if a good condition of nutrition and a high degree of health and vigor are to be maintained.

In Dr. Cammack's experiments,[21] some of the average results of which are summarized in Figs. 3 and 4, it was shown that the length of time an animal can survive the deprivation of

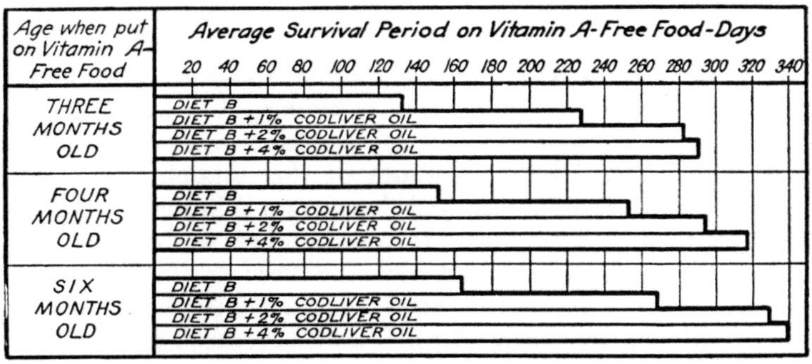

FIG. 3. This chart shows for experimental animals of different ages, the effect of different levels of feeding with vitamin A upon the duration of life when subsequently subjected to food otherwise adequate but devoid of this vitamin. The diets indicated in the horizontal boxes which represent the respective survival periods upon vitamin-A-free diet are those which the animals had received before being subjected to this vitamin deficiency. (Courtesy of the Journal of Biological Chemistry.)

vitamin A in its food can, within very wide limits, be experimentally determined and controlled by the extent of the opportunity which the previous feeding has afforded for the storage of vitamin A in the body. This furnishes a relatively clearcut and quantitative example of the actual nutritional control and predetermination of the life process through chemical research.

In the work of Dr. Boynton,[22] the distribution of stored vitamin A in the animal body was studied quantitatively by means of feeding experiments in which the tissues to be studied served as sole sources of vitamin A for young experimental animals. In the tissues of well nourished rats thus studied, the lung was found to be (weight for weight) over 40 times; and the liver between 200 and 400 times, as rich in vitamin A as was the muscle. Omitting the adipose tissue and skin, about nine-tenths of the vitamin A stored in the body of a typical adult rat was found in the liver and the remaining tenth about equally divided between the muscles as a whole, the blood, the kidneys, and the lungs. Diet had little effect on the vitamin A content of the muscles, but the richer the food in this vitamin the richer was found to be not only the liver (the chief place of storage for surplus) but the lung tissue as well. The probable significance of this latter finding has already been mentioned.

Quantitative research in the chemistry of nutrition has also in recent years been extended to some of the so-called mineral elements, particularly calcium and phosphorus. The average maintenance requirement for each of these elements in adult human nutrition has been established with a fairly satisfactory

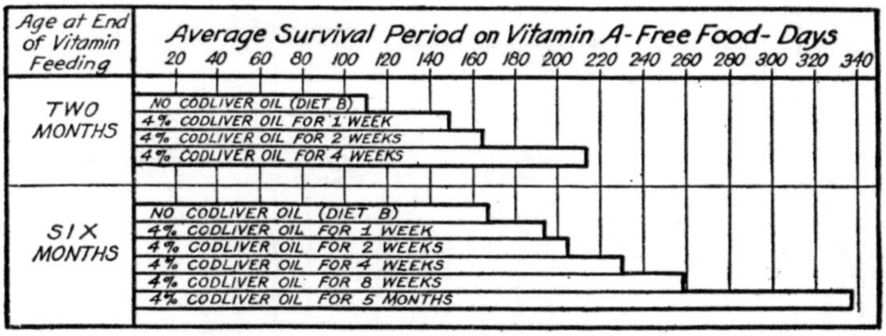

FIG. 4.—This chart shows the average results of experiments which, like those in Fig. 3, relate to the general question of the effect of supplying different amounts of vitamin A through the food, upon the relative amount stored in the body as indicated by the survival period or duration of subsequent life when the experimental animal is placed upon food good in all other respects but devoid of vitamin A. In the former series of experiments (Fig. 3) the codliver oil (used as a rich source of vitamin A) was fed at different levels, while in the experiments summarized here (Fig. 4) it was fed at the same high level but for different lengths of time. In both cases the length of life of the animal when subjected to vitamin-A-free food was found to be determined by the extent of the previous opportunity to acquire a bodily store of vitamin A from the food. (Courtesy of the Journal of Biological Chemistry.)

degree of precision,[23] and a long series of experiments with children of different ages has shown the relatively large amount of calcium required for optimum storage of this element in the body of the growing child, and hence presumably for the optimum development of the bones and teeth.[24]

The large increases not only in the amounts but also in the percentages of these elements which accompany the growth and development of the young are connected with heavy demands made by reproduction and lactation upon the calcium, and to a less extent the phosphorus, of the mother's body. Some indication of the great importance of both these factors may be conveyed by the accompanying chart (Fig. 5) which summarizes the results of hundreds of quantitative determinations, by Dr. MacLeod and by Dr. Quinn, respectively, of calcium and phosphorus in the bodies of rats of different ages.[25, 26]

As might be judged from the chart here shown (Fig. 5), a nutritional deficiency may often be divided between the mother and her young; and there may be cases in which nutrition seems normal and the food supply therefore adequate, but in which a better condition of nutrition may still be induced if the chemical factors are subjected to careful quantitative investigation.

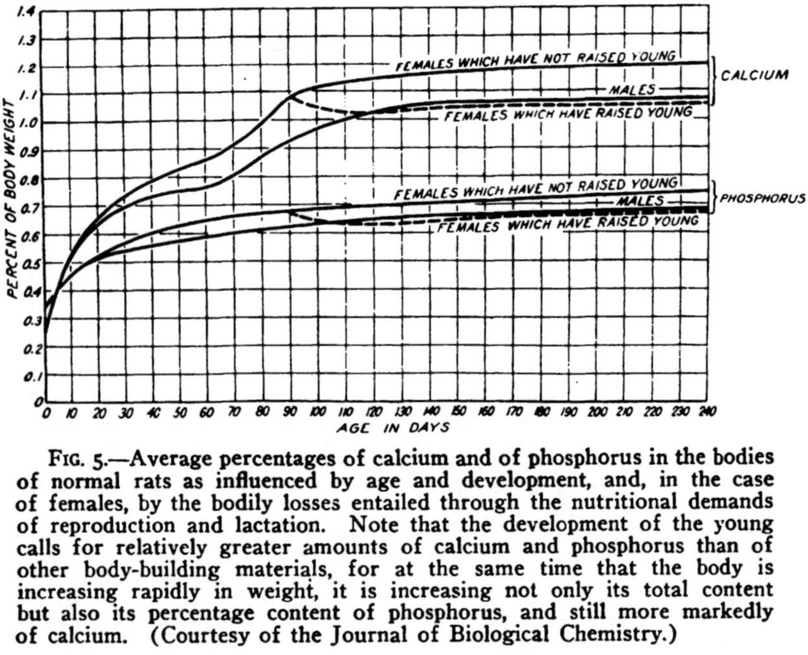

FIG. 5.—Average percentages of calcium and of phosphorus in the bodies of normal rats as influenced by age and development, and, in the case of females, by the bodily losses entailed through the nutritional demands of reproduction and lactation. Note that the development of the young calls for relatively greater amounts of calcium and phosphorus than of other body-building materials, for at the same time that the body is increasing rapidly in weight, it is increasing not only its total content but also its percentage content of phosphorus, and still more markedly of calcium. (Courtesy of the Journal of Biological Chemistry.)

And in the thorough and well-rounded scientific investigation of problems in the chemistry of nutrition there may frequently be need for feeding experiments of two types: (1) those in which the variable is a single chemical factor; (2) those in which the original experimental variables are the articles of food on which people and other animals subsist in actual life. In studies of the latter type, when significant scientific results are found, further experiments can be devised for the purpose of resolving the nutritional problem into its ultimate chemical terms.

Naturally in this lecture we have spoken chiefly of experiments of the type in which the variable is a single chemical factor in each case. The few remaining minutes may now be given to a brief (and necessarily sketchy and incomplete) account of a quantitative research in the chemistry of nutrition which has been developed upon the second plan.[27]

It is now well known that grains alone do not suffice for the normal nutrition of mammals such as human beings, cattle, swine, or rats. Excellent results can, however, be obtained by the feeding of the same grains adequately supplemented by milk. In view of the place of each of these types of food in nature, it is of much scientific (as well as economic) interest to know in what quantitative proportion must milk be consumed along with grain in order that the mixture may constitute a food supply adequate to meet all the needs of animal nutrition.

Experimentally it was found that a mixture of one-sixth dried whole milk and five-sixths ground whole wheat (Diet A) was *adequate,* for it supported growth and reproduction generation after generation in the experimental animals (rats) fed exclusively upon it, which is a rigorous test of the adequacy of a food supply; but a mixture of the same articles of food with the milk in a higher proportion, one-third dried whole milk and two-thirds ground whole wheat (Diet B), proved to be *better* in that it induced improved nutrition and a higher degree of health.

This was manifested in many ways: Growth was more rapid on Diet B than on Diet A when the two diets were fed to parallel groups of young rats of the same initial age, size, health, and heritage. Growth was also more economical in that there was greater gain in body weight per 1000 Calories of food consumed. Those on Diet B maintained a somewhat larger size at all ages, and, what is doubtless more significant of a higher degree of health, they showed both earlier maturity and a later onset of senility so that the prime of life was extended in both directions and became very materially longer. Further evidence of increased vitality was shown in the greater success of the females in the rearing of their young, as they were able to raise both larger numbers and a larger percentage of the young born; the percentage of families dying out was greatly decreased, and the young made a more rapid growth when the mothers were on the better diet, even though the number of young being reared was larger. Also, notwithstanding the greater number of young borne and reared, the average length of life was materially greater on Diet B than on Diet A, and the death rate was lower both in early and in adult life.

Each of these evidences of better nutrition and a higher degree of health and vitality, induced through improvement of a diet which was already adequate, was even more distinct in the second generation than in the first.

Each of the differences has been determined by careful quantitative observation and measurement upon large numbers of cases in which inheritance and all the factors of the environment except the food were exactly the same. And these large numbers of measurements have been subjected to statistical analysis with results which establish with entire certainty the fact that all of these differences are real and are due to the change in the food—none of them accidental or due to chance or physiological variability.

There is no reason to doubt and ample reason to believe that similar improvement in nutrition with resulting increase of vitality and higher degree of health, with longer life and especially a longer duration of the prime of life, can be realized in human experience by a like improvement of the food even though the original food supply is already adequate according to current standards.

In the experiments under discussion the difference between the two diets was, in the terms in which diet is usually considered, simply the one factor of a shift in the quantitative proportions of milk and wheat; while in chemical terms the change involved two significant factors, calcium and vitamin A, and it is possible that the lesser changes in the protein and antirachitic vitamin factors may also have had some slight influence. The resolution of the problem as to just how far the effects found are due to each of these chemical factors and in how far the factors are interlocking in relation to each other, is still in progress; but we already have a complete demonstration of the important fact that differences between adequate and optimal nutrition can not only be detected but measured. Not only is the problem amenable to quantitative study but it is only through quantitative research that findings such as those which we have just been discussing could have been established, for qualitatively the results of the two diets were alike adequate. And since the results on both diets are adequate in the average and variable in individual cases, a comparison might easily yield a misleading result if made on anything less than large numbers and in anything less than the most quantitative way both as to the actual measurements and the statistical interpretation of the data.

Exact scientific research, however, makes it clear that between the adequate and the optimal in nutrition there lies a broad field for chemical investigation which promises, when cultivated by rigorously quantitative methods—careful quantitative chemical experiments continued over long periods upon large numbers of laboratory animals—to prove fruitful of results important alike to our chemical understanding of life processes and our ability to control and improve them.

REFERENCES

1 Atwater and Benedict, A Respiration Calorimeter with Appliances for the Direct Determination of Oxygen, Publication No. 42, Carnegie Institution of Washington (1905); Atwater, Benedict, et al., Respiration Calorimeter Experiments, Bulletins 44, 63, 69, 109, 136, 175, Office of Experiment Stations, United States Department of Agriculture.

2 Lusk, Science of Nutrition; DuBois, Basal Metabolism in Health and Disease; Benedict, Measurement and Significance of Basal Metabolism, Mayo Foundation Lectures on Nutrition; also numerous publications of the Carnegie Institution of Washington.

3 Armsby, Food as Body Fuel, Bulletin 126, Pennsylvania Agricultural Experiment Station.

4 Harrington, Biochemical Journal, Vol. 20, pages 293–313.

5 Barger, The Simpler Natural Bases.

6 Hopkins, Biochemical Journal, Vol. 15, page 286.

7 Taylor, Braun and Scott, American Journal of Physiology, Vol. 74, page 539.

8 Abel, Proceedings of the National Academy of Sciences, Vol. 12, page 132; Abel and Geiling, Journal of Pharmacology and Experimental Therapeutics, Vol. 25, page 423.

9 Osborne, Journal of the American Chemical Society, Vol. 17, page 587.

10 Sherman, Proceedings of the National Academy of Sciences, Vol. 9, pages 81–86 and references there given; Sherman, Caldwell and Naylor, Journal of the American Chemical Society, Vol. 47, page 1702; Sherman, Thomas and Caldwell, Ibid., Vol. 46, page 1711.

11 Chittenden, Physiological Economy in Nutrition and Nutrition of Man.

12 Sherman, Journal of Biological Chemistry, Vol. 41, page 97.

13 Osborne and Mendel, Feeding Experiments with Isolated Food Substances, Carnegie Institution of Washington, Publication No. 156, Parts I and II, and a series of papers on proteins and amino acids in the Journal of Biological Chemistry.

14 Woods, Dissertation, Columbia University, 1924, and Sherman and Woods, Journal of Biological Chemistry, Vol. 66, page 29.

15 Spohn, Dissertation, Columbia University, 1922; Sherman and Spohn, Journal of the American Chemical Society, Vol. 45, page 2719.

16 Grose, Dissertation, Columbia University, 1922; Sherman and Grose, Journal of the American Chemical Society, Vol. 45, page 2728.

17 Burton, Dissertation, Columbia University, 1925; Sherman and Burton, Journal of Biological Chemistry, Vol. 70, page 639.

18 Munsell, Dissertation, Columbia University, 1924; Sherman and Munsell, Journal of the American Chemical Society, Vol. 47, page 1639.

19 Quinn, Dissertation, Columbia University, 1925, and subsequent experiments not yet published.

20 MacLeod (F. L.), Dissertation, Columbia University, 1924; Sherman and MacLeod, Journal of the American Chemical Society, Vol. 47, page 1658.

21 Cammack, Dissertation, Columbia University, 1925; Sherman and Cammack, Journal of Biological Chemistry, Vol. 68, page 69.

22 Boynton, Dissertation, Columbia University, 1924; Sherman and Boynton, Journal of the American Chemical Society, Vol. 47, page 1646.

23 Sherman, Journal of Biological Chemistry, Vol. 41, page 173, and Vol. 44, page 21.

24 Sherman and Hawley, Ibid., Vol. 53, page 375.

25 Sherman and MacLeod, Ibid., Vol. 64, page 429.

26 Sherman and Quinn, Ibid., Vol. 67, page 667.

27 Sherman and Campbell, Ibid., Vol 60, page 5, and much additional experimental work not yet published.

The Theory of Velocity of Ionic Reactions

BY

J. N. BRÖNSTED

PROFESSOR OF PHYSICAL CHEMISTRY
ROYAL POLYTECHNIC INSTITUTE, COPENHAGEN

Lecture delivered at Columbia University in the special course on Contemporary Developments in Chemistry given in the Summer Session of 1926 on the occasion of the opening of the Chandler Chemical Laboratories

COLUMBIA UNIVERSITY PRESS
NEW YORK
1927

Printed from type. Published March, 1927

Printed in the United States of America

THE THEORY OF VELOCITY OF IONIC REACTIONS

By J. N. Brönsted

According to the classical theory of chemical reaction velocity as developed by Wilhelmy, Harcourt and Esson, and van't Hoff, each molecule entering in the kinetic scheme of a reaction influences the velocity proportionally to its concentration. In spite of this law having been verified by a great number of investigations it does not seem generally valid since in many other cases apparently very marked deviations from it have appeared. It has been found on a closer examination of these *kinetic anomalies* that they are decidedly of systematic nature and so doubt has arisen as to the correctness of the fundamental kinetic principles, hitherto in use.

As to the character of these systematic deviations from the classical velocity laws it is now of great importance to note that kinetic anomalies in dilute aqueous solutions always appear when the reaction studied proceeds between ions, while in the case of reactions between an ion and uncharged molecules or between uncharged molecules only, the kinetics seem to be normal. From these facts it is natural to conclude that the deviations from the normal kinetics are in some way related to the anomalies known to exist generally in the static chemistry of ions, i.e., that to some extent the same reasons are to be made responsible for both static and kinetic anomalies of salt solutions.

Accordingly it will be expedient to outline quite briefly the chief feature in the theory of electrolytic solutions. In the first place I have to call to your attention that the modern theory of dilute salt solutions does not make use of the idea of a partial dissociation of strong electrolytes in order to explain the falling off of the osmotic effect, the molecular conductivity, etc., with increasing salt concentration. On the contrary, the modern theory looks upon a dilute solution of sodium chloride and similar strong electrolytes as practically completely dissociated in ions. On the basis of this assumption the theory of electrolytes in recent years has been the subject of a very fundamental and very fruitful innovation. While in the older theory the difficulties in studying the properties of the ions were greatly enhanced by the difficulty of first determining—by uncertain and unreliable methods—their true concentration in the solution, the corresponding question in the new theory vanishes in advance through the fact that ionic and stoichiometric concentrations are identical. The anomalies of the ions have nothing to do with their partial disappearance as undissociated molecules.

What is the reason then that the effect of say a chloride ion in a sodium chloride solution shows an abnormal dependence on concentration of the sodium chloride solution?

The new electrolytic theory explains this as an effect of the attracting and repelling forces operating between the ions on account of their electric charges. The influence of these interionic forces is accessible to a theoretical calculation, which shows in fact that the magnitude of their influence is quite sufficient for a quantitative explanation of the ionic anomalies, thus leaving nothing to be explained by the assumption of incomplete dissociation or other source of disturbance.

The influence of the electric forces is to *diminish* the thermodynamic activity of the ions, and this influence increases with increasing salt concentration. The activity as defined by G. N. Lewis [1] is a quantity, which in *ideal* solutions is proportional to, and actually can be made equal to concentration. In the case of *salt* solutions the activity is equal to the concentration only in the region of extreme dilution when the interionic forces can be neglected. Otherwise the activity is but a certain fraction f of the concentration, decreasing with increase in the latter. If c denotes concentration and a activity, we can therefore write

$$a = cf \qquad \text{(1)}$$

where f is unity in dilute solutions of non-electrolytes and in salt solutions of extreme dilution. f is called the activity coefficient. For instance in a 0.1 N. sodium chloride solution the activity is about 75 per cent of the concentration. Hence for the sodium or chloride ion we can write:

$$c = 0.1$$
$$f = 0.75$$
$$a = 0.075.$$

Fig. 1 gives a picture of the general dependence of the activity coefficient on the concentration of a sodium chloride or other uni-univalent salt solution in which the ion considered is present in small amount. The influence of increasing salt concentration is clearly a function of the valence, i.e., the electric charge of the ion considered. In extreme dilution the value of f in all cases is unity. In 0.1 N sodium chloride f_1 for a univalent ion has gone down to 0.7, f_2 for a bivalent ion to less than 0.4 and f_3 for a trivalent ion to about 0.1. It should be mentioned, however, that at such high concentration as 0.1 N not only the valence type, but also to some extent the individual nature of both solute and solvent salt influences the value of f.

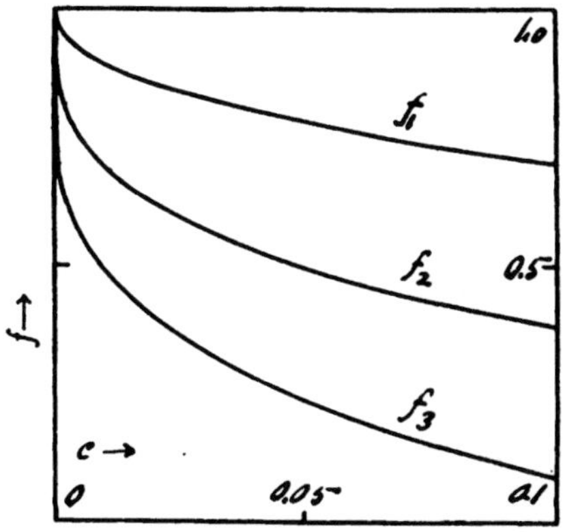

FIG. 1.—The activity coefficient (f) of an univalent (f_1), bivalent (f_2) or trivalent (f_3) ion when present in small amount in a solution of sodium chloride of the concentration C.

The great significance of the conception of activity lies in the fact that in all cases where the ordinary laws of equilibrium formulated by means of concentration, break down—as they do in salt solutions—introduction of activity instead of concentration makes them generally valid. So in the case of a saturated solution of a binary salt the ordinary solubility product law

$$c_A c_B = \text{constant}$$

breaks down except for the region of extreme dilution. By addition of a foreign salt, for instance, the interionic forces will lessen the activity in correspondence to the decrease of the product $f_A f_B$ and equilibrium is not reestablished until the solubility has increased so much as to impart to the product

$$c_A c_B f_A f_B = a_A a_B = \text{constant}$$

its original value. Quite similarly, when considering the following equilibrium in homogeneous solution:

$$\mathrm{A} + \mathrm{B} \rightleftarrows \mathrm{C} + \mathrm{D}$$

the ordinary mass action expression

$$\frac{c_C c_D}{c_A c_B} = K_c$$

is not generally valid, i.e., K_c is not generally a constant. However, introducing again activity instead of concentration:

$$\frac{a_C a_D}{a_A a_B} = \frac{c_C c_D}{c_A c_B} \cdot \frac{f_C f_D}{f_A f_B} = K_c \frac{f_C f_D}{f_A f_B} = K_a \quad . \quad . \quad . \quad . \quad (2)$$

a new mass action expression is reached, which is of universal validity.

If, as thus demonstrated, all the laws of chemical equilibria —except for infinitely dilute systems—necessarily suffer a fundamental transformation by introducing the conception of activity where hitherto concentration was used, then on account of the close relationship between velocity and chemical equilibrium the question obviously arises if a similar transformation of all *kinetic* equations is not bound to ensue.

This conclusion has been drawn indeed in recent time by several authors, who formulate the kinetic mass action equation, for instance in the case of a bimolecular reaction thus:

$$h = k a_A a_B, \quad . \quad . \quad . \quad . \quad . \quad . \quad . \quad . \quad (3)$$

instead of the usual expression:

$$h = k c_A c_B, \quad . \quad . \quad . \quad . \quad . \quad . \quad . \quad . \quad . \quad (4)$$

i.e., using the activity of the reacting molecules instead of their concentration. In the case of a reversible reaction:

$$A + B \rightleftarrows C + D$$

application of Equation (3) gives

$$h_1 = k_1 a_A a_B,$$
$$h_2 = k_2 a_C a_D,$$

which again in the case of equilibrium being reached leads to

$$\frac{a_C a_D}{a_A a_B} = \frac{k_1}{k_2} = K_a$$

identical with the thermodynamic mass action expression,[2] already advanced as universally valid.

This theory [2, 3] is known as the *activity rate theory* and is viewed by its adherents as strongly borne out by the derivability of the thermodynamic mass action law from its equations. The correlation between the kinetic expressions and the mass action expression undoubtedly shows the significance of the conception of activity in the theory of reaction velocity. It has, however, to be distinctly emphasized that even though Equation (2) is derivable from (3) the inverse argument is by no means justified. Thermodynamics tolerates an equation like (3) but it can on no account be made responsible for it. Moreover, in testing this theory experimentally by measuring both velocity and activity of the reacting or catalyzing substances the defenders of the theory have used in most cases strong salt solutions, where changes of the solvent due to the great concentration of salt have obscured the simple activity effect. Such experiments are hardly interpretable in favour of any simple velocity theory. If one declines to benefit by the existing general information and abundant evidence regarding the activity coefficient of salts in the simple cases of dilute salt solutions one will certainly fail to notice to what extent the theory in question is contradicted by fact.

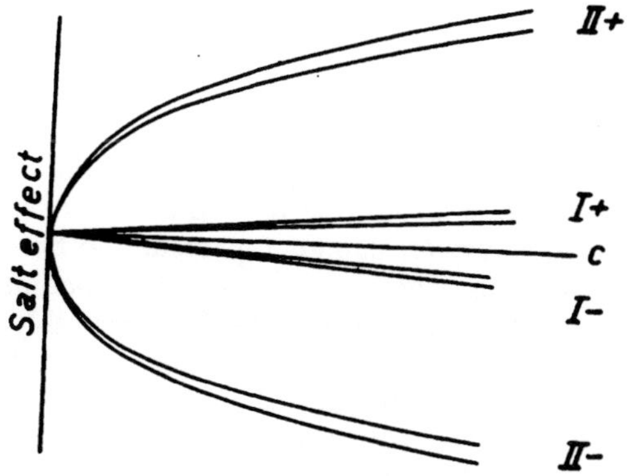

FIG. 2.—The effect of concentration upon the kinetic salt effect, illustrating positive and negative catalysis and the division into linear and exponential types.

When—in order to get a clear picture of the applicability, or inapplicability, of the activity rate theory—we consider the experimental evidence regarding kinetic salt effect, we are faced with the very remarkable fact that the phenomena of salt effect are divisible into two distinctly different groups. In one of these, denoted by *I* in Fig. 2, the salt effect is *linear,* i.e., proportional to concentration. In the other group, denoted by *II,* the salt effect is *exponential,* i.e., the curves showing the salt effect exhibit a marked curvature, particularly at lower salt concentrations. Furthermore, as a fact of no less importance, the salt effect may, as shown by the figure, in both cases be positive as well as negative, i.e., salts may accelerate or retard the reactions linearly as well as exponentially.

Confronting the activity rate theory with these experimental facts we find it in excellent agreement with case II (—) since the curves of the activity coefficient of ions, as we have seen above, are just characterized by a shape similar to this. The cases I (+), I (—) and II (+), however, the theory is quite unable to explain. The strong positive effects as well as the linear effects actually existing in a great number of reactions are not compatible with the activity rate theory.

The rigidity of this argument forces us to enter more thoroughly into the question of the mechanism of the chemical reaction in order to be able to point out the breach in the reasoning underlying the activity rate theory and Equations (3) and to replace it by a theory better adapted to the experimental facts.

For this purpose consider a homogeneous chemical reaction of measurable velocity in which a chemical system A_1 is transformed to another system A_2. The ordinary reaction scheme of such a process:

$$A_1 \rightarrow A_2$$

simply indicates the chemical nature of the initial and final systems and may indicate also physical conditions. A stoichiometric scheme like this, however, cannot represent the true mechanism of the transformation, since in that case the reaction, possessing a great tendency to occur, would be immeasurably rapid. As a matter of fact in between the systems of relative stability A_1 and A_2 there exists a number of intermediate unstable states through which the transformation proceeds before a molecule of the final system A_2 can be formed. Particularly it has to pass through what we shall call the *critical state,* in which the minimum energy required for the reaction is contained. It seems a cogent conclusion, that the formation of these critical products, which possess a high amount of energy, and are therefore of high instability, is the process that accounts for the slowness, i.e., determines the velocity of the reaction which we actually measure.

From the point of view here presented the complete picture of the reaction would be

$$A_1 \rightarrow [X] \rightarrow A_2.$$

It seems obvious therefore that the activity rate theory has not adequately considered the effect of changing activity coefficients in the reacting system. Instead of the expressions of the classical theory, viz.,

$$h = kc_A \quad \text{and} \quad h = kc_A c_B,$$

for uni- and bimolecular reactions respectively, the activity rate theory writes

$$h = kc_A f_A \quad \text{and} \quad h = kc_A c_B f_A f_B,$$

i.e., it only takes into consideration that the activity coefficient of the initial molecules is changed. Corresponding to the introduction of the factor f for the initial molecules one should deem it necessary, however, for the *final* molecule of the *true* kinetic reaction, i.e., the critical complex, for which the symbol X is used, to introduce the factor $\frac{1}{f_X}$. When doing so the expression becomes

$$h = kc_A \frac{f_A}{f_X}, \qquad (5)$$

and

$$h = kc_A c_B \frac{f_A f_B}{f_X}, \qquad (6)$$

which are the expressions of the new velocity theory.[4]

The actual basis of this transformation of the classical laws is the following hypothesis: (1) *The probability of a molecule or a system of molecules passing from ordinary conditions into a state of great improbability varies proportionally to the ratio of activity coefficients in the ordinary and improbable state.* For thermodynamic reasons this hypothesis is identical with the following: (2) *The probability for a molecule in a very improbable state to pass into ordinary conditions is independent of its activity coefficients in the two states.*

It is easy to see that both these hypotheses agree with the thermodynamic mass action law. From the first of these expressions (5) and (6) are directly derivable. A critical complex, namely, is a molecule in a very improbable state. Accordingly we have to introduce in the ordinary velocity equations the factors:

$$F = \frac{f_A}{f_X} \quad \text{or} \quad F = \frac{f_A f_B}{f_X},$$

expressing the changing probability of reaction for uni- and bimolecular processes respectively. F is termed the ***kinetic activity factor.***

The new velocity equations can be derived in a somewhat different way [5] by considering for instance in the case of a bimolecular reaction the equilibrium between the ordinary molecules A and B and a complex S of great improbability:

$$A + B \rightleftarrows S.$$

Applying to this equilibrium the thermodynamic mass action law:

$$\frac{a_S}{a_A a_B} = \frac{c_S}{c_A c_B}\frac{f_S}{f_A f_B} = K_a,$$

the concentration c_S is found to be:

$$c_S = K_a c_A c_B \frac{f_A f_B}{f_S}. \qquad (7)$$

Now ascribing to S the possibility of reacting, i.e., making S a critical complex X, removal of X from the equilibrium means formation of molecules of the final system C instead of its returning to A + B. This, however, on account of the very small concentration of X will make no change in the equilibrium. Introduction of hypothesis (2), i.e., putting the velocity of transformation proportional to c_S then immediately leads from Equation (7) to Equation (6), i.e., yields the velocity expression of the new theory.

This expression obviously differs from the expression of the classical theory by the factor

$$\frac{f_A f_B}{f_X},$$

and from the expression of the activity theory by the factor

$$\frac{1}{f_X},$$

and accordingly, in addition to the experimental evidence, we have also theoretical reasons to believe that neither of these theories can be correct.

The essential point in the new theory is the introduction of the coefficient f_X. This is the activity coefficient of the critical complex, formed by collision of two molecules. And the fundamental result, which enables us to evaluate the theory particularly in the domain of ionic reactions, is that the electric charge of this complex is computable as the sum of the charges of the colliding molecules.

If the reaction occurs between two univalent ions of opposite signs, we derive from the general formula

$$h = kc_A c_B \frac{f_A f_B}{f_X},$$

that the velocity will be given by

$$h = kc_A c_B \frac{f_1^2}{f_0}, \quad \text{. . . . . (8)}$$

if by the subscripts is indicated the number of charges of the ions in question. Since the activity coefficient f_0 of an electrically neutral molecule is but little and linearly affected by salts, this relation shows that there will be a strong *negative* salt effect.

In the case of reaction between two univalent ions of the *same* sign, the expression becomes

$$h = kc_A c_B \frac{f_1^2}{f_2}, \quad \text{. . . . . (9)}$$

and, since f_2 is much smaller than f_1^2, in dilute solutions, the salt effect will be strong and *positive.*

Finally, if a neutral molecule react with a univalent ion, the expression is

$$h = kc_A c_B \frac{f_1 f_0}{f_1} = kc_A c_B f_0, \quad \text{. . . . . (10)}$$

and there will be a very *small* effect, proportional to concentration.

These results, which can easily be extended to other reaction types, can be summarized as follows: In all cases where reaction occurs between neutral molecules or between a neutral molecule and an ion the kinetic activity factor F is approximately unity, and the normal velocity laws will hold good.

If, on the other hand, the reaction occurs between ions of the same sign, there will be a strong positive salt effect, i.e., velocity constant will increase with the concentration of the reacting ions or when neutral salts are added.

Finally in the case of reaction between ions of opposite sign the salt effect will be negative. The velocity constant will decrease when the concentration increases, or neutral salts are added. In the two last mentioned cases the salt effect is particularly pronounced in dilute solution—just as is the change of the activity coefficient of an ion. The effect will be what has been called above an *exponential effect.*

When the great amount of experimental work on reaction velocity is considered from the point of view of the theory here stated, we must admit that most of the difficulties and confusion hitherto prevailing in this region *disappear.* Application of the chemical laws has certainly been successful in a number of processes, namely—as mentioned already in the beginning—those belonging to the group first considered, in which the normal behavior is required also by the present theory. For instance, the cane sugar inversion:

$$C_{12}H_{22}O_{11} + H_3O^+ \rightarrow$$

belongs to this group. Otherwise you find the experimental evidence to be anything but satisfactory to the classical theory. In studying the reactions between ions very few authors have been able to prove strictly the validity of the kinetic equations. Particularly the device of varying the initial concentration to make the order of reaction appear clearly from the change of velocity involved is fatal for the interpretation of the experiment, because thereby a strong salt effect is introduced, causing the velocity constant to alter. It is a very common case met with in the chapter of velocity determinations that the order of reaction, as found by experiment on the basis of the classical laws is not expressible as an integral number.

The new theory is able to explain such difficulties. Indeed, it is on the whole, as we shall see from a consideration of available experimental data, in excellent conformity with the experimental facts.

Let us consider first reactions between an ion and a neutral molecule. Most reactions studied of this kind are of a catalytic nature. It is commonly claimed, for instance, in text books dealing with the subject of catalysis, that neutral salts exert a strong accelerating effect upon reactions catalyzed by hydrogen ions. This statement, however, is rather misleading if by "strong accelerating effect" is meant anything beyond what we have to expect from the general and well-known alteration of the solvent as a consequence of the addition of salts. When the activity coefficient of a nonelectrolyte is changed by 40 per cent adding 1 N sodium chloride to its aqueous solution, there is no support for the assumption that a similar kinetic effect may not arise from the same change. In fact, 0.1 N sodium chloride changes the velocity of inversion by a few per cent, and from the available information on this question the effect is linear, i.e., proportional to concentration. Such a linear salt effect exists in many other reactions of this type. Recently we have studied in my laboratory the effect of salts upon the diazoacetic ester catalysis, which as known proceeds proportionally to the hydrogen ion concentration. The effect obtained was a linear and positive effect just as in the case of cane sugar only numerically greater. We also examined the catalytic decomposition of nitrosotriacetoneamin by hydroxyl ions and found a linear effect, which in this case proved to be negative. All these phenomena agree with the above result, derived on the basis of the theory, that for this particular type of reaction the classical laws are approximately true, i.e., the salt effect is linear and relatively small in magnitude.

The small and linear effect of salts like sodium chloride on the catalysis by hydrochloric acid in dilute solution shows very decidedly that the activity of the hydrogen ion is not the determining factor. At the same time it is an argument of great value for the theory of complete dissociation of the strong electrolytes.

When now we turn to consider reactions in which the molecules partaking are both ions, the theory requires a strong, exponential salt effect, the sign of which as explained above may be positive *or* negative according to the type. The chemical literature provides a great many examples in conformity with this requirement of the theory.

T. S. Price[6] in 1898 examined the reaction between potassium iodide and potassium persulphate. This reaction takes place stoichiometrically according to the scheme:

$$S_2O_8^{--} + 2I^- \rightarrow 2SO_4^{--} + I_2$$

but kinetically is bimolecular. The velocity constant k_0 corresponding to the total equivalent concentration c, in experiments in which equivalent amounts of persulphate and iodide were taken, is given in the following table.

TABLE I

VELOCITY CONSTANT IN THE REACTION $S_2O_8^{--} + I^- \rightarrow$

c	k_0
0.025	50
0.028	53
0.033	58
0.040	61
0.050	66
0.067	76

The table and diagram evidently show a very marked positive salt effect. Since the process is between ions of the same sign, this behavior was to be predicted from the theory.

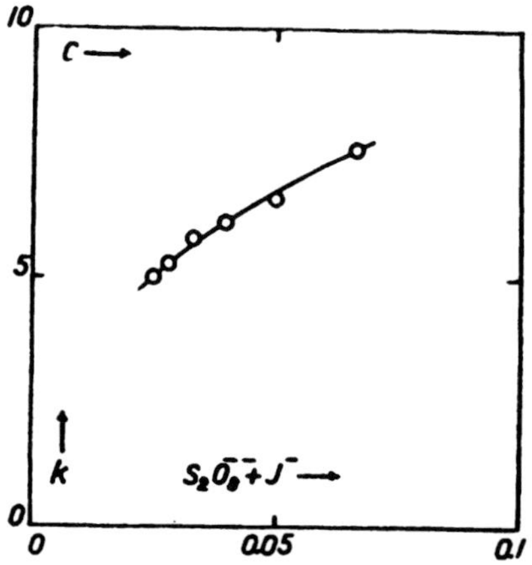

FIG. 3.—The effect of the total equivalent concentration upon the velocity constants for the reaction $S_2O_8^{--} + 2I^- \rightarrow 2SO_4^{--} + I_2$.

Holmberg,[7] in a series of papers, has studied the kinetics of the reactions between bromosubstituted anions and hydroxyl ions and similar processes, for instance:

$$\begin{matrix} CHBrCOO^- \\ | \\ CHBrCOO^- \end{matrix} + OH^- \rightarrow \begin{matrix} CHOHCOO^- \\ | \\ CHBrCOO^- \end{matrix} + Br^-$$

or

$$CH_2ICOO^- + SCN^- \rightarrow CH_2SCNCOO^- + I^-$$

The strong deviations from the classical laws demonstrated by such processes are grouped by Holmberg under the heading of "cation catalysis," since particularly the nature of the cation of the salts employed seemed to influence the variations of the effects. The phenomena agree with the proposed theory in the fact that also here a marked positive effect is present. The stronger influence of the cations is due to the reactions proceeding between negative ions, and forms an example of the general rule that activity coefficients are particularly sensitive to changes in the nature of ions with opposite sign of their own.

When two ions of opposite sign partake in the reaction the salt effect according to the theory should be negative. This is shown also by a number of investigations. If diazoacetic ester is decomposed by hydrochloric or nitric acid not only the simple catalysis by hydrogen ions:

$$N_2CH\cdot COOC_2H_5 + H_2O \rightarrow CH_2OH\cdot COOC_2H_5 + N_2$$

but also a side reaction [8]:

$$N_2CH\cdot COOC_2H_5 + H^+ + Cl^- \rightarrow CH_2ClCOOC_2H_5 + N_2$$

takes place. The first of these reactions, as mentioned, is normal. With rising concentration of HCl or KCl the second reaction becomes more and more predominating, the ratio of the two velocities h_2/h_1, however, does not show an increase proportional to the concentration of the chloride ion as required by the classical velocity laws, but increases much more slowly than this. This means that the velocity constant of the second reaction decreases with rising concentration and that again is tantamount to stating that there is a negative salt effect.

Of the same type is the transformation of chloroaceto aminobenzene into p. chloro acetanilide:

$$C_6H_5N{\genfrac{}{}{0pt}{}{Cl}{C_2H_3O}} \rightarrow Cl\cdot C_6H_4N{\genfrac{}{}{0pt}{}{H}{C_2H_3O}}$$

studied in a number of investigations. According to the measurements of Harned and Seltz [9] the velocity constant shows here also an exponential decrease, as predicted by the theory. Similar results are also obtained in the investigations of Bray and Livingston [10] on the decomposition of hydrogen peroxide in the presence of bromine and hydrobromic acid. The negative salt effect found by the authors agrees with the fact that the reaction takes place between a hydrogen peroxide molecule and both hydrogen and bromide ion. We shall consider this reaction in more detail later on.

Since, according to the theory all the effects here mentioned appear as a result of the changes in the ionic activity coefficients owing to the interionic forces changing with the salt concentration it should be possible to eliminate this effect and so turn the abnormal reaction into a normal one corresponding to the classical velocity theory, simply by using as a solvent salt solutions sufficiently strong to eliminate the effect of changes in concentration of the reacting ions themselves. As a matter of fact the classical reaction between stannous and ferric ions

$$2Fe^{+++} + Sn^{++} \rightarrow 2Fe^{++} + Sn^{++++}$$

studied by A. A. Noyes [11] and usually quoted as an example of a trimolecular reaction, but actually quite abnormal in its course, shows when examined in a strong salt solution, as recently shown by Timofeew, Muchin and Gurewitsch,[12] normal behavior as a second order reaction.

On the whole it is very probable that a reexamination of the numerous cases which are still unexplained on account of kinetic anomalies would allow of definite decision as to the true nature and order of reaction on the basis of the point of view of kinetics here adduced.

The examples mentioned above have been shown to agree qualitatively with the new theory of reaction velocity. When now proceeding to establish a more quantitative test of the theory we have to examine the possibility of determining accurately the different factors in the velocity expression of the theory. In this respect it will be sufficient to consider the kinetic activity factor in the case of a bi- and trimolecular reaction:

$$F = \frac{f_A f_B}{f_X} \quad \text{and} \quad F = \frac{f_A f_B f_C}{f_X},$$

respectively.

The activity coefficients of the reacting molecules f_A, f_B . . . are with more or less accuracy open to an experimental determination and are furthermore in sufficiently dilute solution theoretically computable. As to the activity coefficient f_X of the shortliving critical complex this of course is experimentally inaccessible, but since the sign and magnitude of the electric charge of the complex is known as the algebraic sum of the charges of the initial molecules we can make use of the general laws also to calculate f_X if the solution is sufficiently dilute.

The problem of the activity coefficients of ions has in recent years been subjected to very extended theoretical as well as experimental studies. It has been attempted to calculate the magnitude of the effect of the interionic electric forces in diminishing the activity of the ions and the same problem has been studied in several ways by accurate experimental methods. The theory culminates in the investigations of Debye and Hückel,[13] who by an ingenious method seem to have brought the problem of electrolytes in the very dilute range a considerable step further.

It would be out of place here to give a detailed account of the Debye-Hückel theory. It will suffice for our purpose to state the formula to which this theory finally leads. The simplest form of the activity coefficient equation to which the Debye-Hückel equation reduces is for the range of extreme dilution[14]:

$$-\log f = 0.5\, z^2\sqrt{\mu}, \quad . \quad . \quad . \quad . \quad . \quad . \quad (11)$$

where z is the valence of the ion having the activity coefficient f and μ the ionic strength of the solution:

$$\mu = \Sigma \tfrac{1}{2} m z^2 = \Sigma \tfrac{1}{2} c z,$$

m being the molal and c the equivalent concentration of an ion. This equation shows that f is independent of the nature of the ion, only depending upon its valence. In the case that the solution is not extremely dilute the activity coefficient formula reads:

$$-\log f = 0.5\, z^2 \frac{\sqrt{\mu}}{1 + 3.27.10^7 a\sqrt{\mu}} = 0.5\, z^2 \frac{\sqrt{\mu}}{1 + \alpha\sqrt{\mu}},$$

where a is the diameter of the ion, conceived of as a sphere, and this formula again for sufficiently dilute solution is practically identical with the equation:

$$-\log f = 0.5\, z^2\sqrt{\mu} + \beta\mu, \quad . \quad . \quad . \quad . \quad (12)$$

where β is a constant, depending upon the nature of the salt in question. This equation, also reached on experimental basis, shows that the deviations from the ideal law at higher concentration are linear. The Debye-Hückel theory has been verified already in a great many cases in the region of great dilution. Beyond this region, however, it loses its rigid character, since it rests upon several suppositions regarding the structure of ions, which are clearly fictitious. We shall regard the equations of the theory here as expressing a limiting law to which activity coefficients approach asymptotically when concentration approaches zero.

In order to test now the velocity theory for a bimolecular reaction in the range of great dilution we have to introduce the above equations for the activity coefficient in the kinetic activity factor

$$F = \frac{f_A f_B}{f_X}.$$

Application of Equation (11) yields:

$$\log \frac{f_A f_B}{f_X} = -0.5(z_A^2 + z_B^2 - z_X^2)\sqrt{\mu}$$

or since $z_X = z_A + z_B$

$$\log \frac{f_A f_B}{f_X} = z_A z_B \sqrt{\mu}. \quad \ldots \ldots \ldots \quad (13)$$

Hence:

$$F_2 = \frac{f_A f_B}{f_X} = 10^{z_A z_B \sqrt{\mu}}, \quad \ldots \ldots \quad (14)$$

and therefore:

$$h = k c_A c_B 10^{z_A z_B \sqrt{\mu}}. \quad \ldots \ldots \quad (15)$$

This very simple equation expresses the limiting law of ionic reaction velocity in the case of a bimolecular reaction. It shows in conformity with the above statement that the salt effect is positive, when z_A and z_B have the same sign, and negative when they have different sign. It shows also that the magnitude of the effect depends merely upon the magnitude of the charges but not upon their sign. Furthermore the great influence of the valence appears clearly from the form of the equation.

If the reaction is trimolecular the corresponding expression is:

$$\log \frac{f_A f_B f_C}{f_X} = (z_A z_B + z_A z_C + z_B z_C)\sqrt{\mu},$$

which in the most important case when $z_C = 0$, i.e., one of the reacting molecules is uncharged, leads to the same kinetic activity factor as given by Equation (13) for bimolecular reactions.

In order to test the equations here arrived at the best way would be to plot the logarithm of the ordinary velocity constant k_0 given by the equation

$$h = k_0 c_A c_B$$

against $\sqrt{\mu}$. Since the ordinary velocity constant is proportional to F, the plot of $\log k_0$, against $\sqrt{\mu}$, in the very dilute region should come out as a straight line having the slope $z_A z_B$, and the plot at higher concentration should show an asymptotical approach to this slope.

Four reactions, two with positive and two with negative salt effect, should be quoted as well fit to support quantitatively the theory in correspondence with the above calculations.

The first of these is the reaction, already mentioned, between hydrogen peroxide and hydrobromic acid, which has been studied by Bray and Livingston.[10] These authors showed that H_2O_2 not only oxidizes bromide to bromine, but also reduces bromine to bromide, from which observation the mechanism of the catalytic decomposition is entirely elucidated. The kinetic process can be taken as:

$$H_2O_2 + H^+ + Br^- \rightarrow$$

i.e., it seems to be trimolecular, but since one of the molecules is uncharged, the simple Equation (15) should be applicable. The following table shows the results in solutions containing only H_2O_2 and HBr. c, the concentration of HBr, is identical with μ.

TABLE II

Reaction between Hydrogen Peroxide and Hydrobromic Acid

c	$\sqrt{\mu}$	k_0	$2 + \log k_0$
0.0217	0.147	0.0348	0.542
.0338	.184	.0337	.528
.0432	.208	.0318	.502
.0677	.260	.0297	.473
.0945	.307	.0289	.461
.146	.382	.0279	.446
.206	.452	.0268	.428

In Fig. 4 these data are plotted, $\sqrt{\mu}$ being taken as abscissa and $\log k_0$ as ordinate. The heavy line represents the theoretical slope $_A z_B$ which according to Equation (13) is -1 for these cases. This slope is very well conformed to by the experimental points even though the ionic concentration is too high actually to reach the theoretical value. The relatively great deviations from the straight line at higher concentrations agree with the fact that the activity coefficient of the hydrogen ion does not decrease in this range of concentration to the same extent as most other univalent ions.

As the next example the saponification of an electrically charged estermolecule should be mentioned. While the saponification process by hydroxyl ions is a normal reaction with only little salt effect so long as the ester molecule is electrically neutral, an exponential salt effect is to be expected when the molecule is charged. When saponifying an ester of a bibasic acid as for instance succinic acid the second step of the reaction should therefore follow laws quite different from those of the first step.

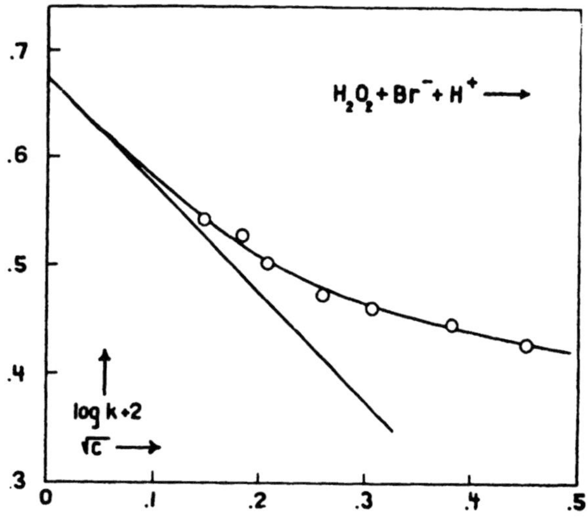

FIG. 4.—The velocity constants for the bromide ion catalysis of hydrogen peroxide as a function of the square root of the ionic strength, $\sqrt{\mu}$. The straight line represents the limiting values predicted by equation (13).

The particular substance studied from this point of view is the potassium nitrourethanate, which in aqueous solution dissociates completely according to:

$$\begin{array}{l} NO_2K \\ \| \\ N \\ | \\ CO_2C_2H_5 \end{array} \rightarrow \begin{array}{l} NO_2^- \\ \| \\ N \\ | \\ CO_2C_2H_5 \end{array} + K^+ . \quad . \quad . \quad . \quad (1)$$

The anion of this salt undergoes saponification with a measurable speed:

$$\begin{array}{l} NO_2^- \\ \| \\ N \\ | \\ CO_2C_2H_5 \end{array} + OH^- \rightarrow \begin{array}{l} NO_2^- \\ \| \\ N \\ | \\ CO_2^- \end{array} + C_2H_5OH . \quad . \quad . \quad (2)$$

As soon as this saponification has taken place, however, the doubly charged nitrocarbamate ion thus formed decomposes instantaneously forming carbonate ion and nitrous oxide:

$$\begin{array}{l} NO_2^- \\ \| \\ N \\ | \\ CO_2^- \end{array} \rightarrow CO_3^{--} + N_2O. \quad . \quad . \quad . \quad . \qquad (3)$$

One can take advantage of the gas evolution in the last reaction to follow the slow reaction (2) by providing an arrangement for measuring the gas pressure increasing proportionally to the progress of the reaction.

The following table contains the results obtained by Mrs. Delbanco [15] in my laboratory in this way, using various concentrations c_N of the nitrourethane ion and c_{OH^-} of the hydroxyl ion. In some experiments the change in salt concentration was effected by adding potassium chloride. The table gives the total salt concentration and the values of the bimolecular velocity constant k_0.

SAPONIFICATION OF THE NITROURETHANE ION

c_N	c_{OH^-}	c_{KCl}	c_{tot}	k_0	$\sqrt{c_{tot}}$	$\log k_0$
0.00581	0.0152	0.000	0.0218	0.177	0.145	0.248
0.00581	0.0250	0.000	0.0308	0.196	0.175	0.292
0.0200	0.0304	0.000	0.0504	0.205	0.224	0.312
0.00581	0.0150	0.010	0.0308	0.192	0.175	0.283
0.00581	0.0150	0.030	0.0508	0.201	0.225	0.303
0.00581	0.0150	0.080	0.1008	0.225	0.317	0.352
0.0200	0.0304	0.025	0.0755	0.227	0.275	0.356
0.0200	0.0304	0.050	0.1004	0.232	0.316	0.365
0.0200	0.0304	0.100	0.1500	0.251	0.387	0.400

The salt effect in this series is positive. In Fig. 5, $\log k_0$ is plotted against $\sqrt{c_{tot.}}$, which in solutions containing only univalent salts is identical with $\sqrt{\mu}$. The heavy line shows the theoretical slope $z_A z_B = +1$, and the curve drawn through the experimental points exhibits an asymptotic approximation to this line in agreement with the theory.

As further examples may be mentioned two reactions in the group of the cobaltammines. When bromopentammine cobaltic salts are mixed with strong bases the bromine atom is replaced by hydroxyl according to the following scheme:

$$\left[Co^{Br}_{(NH_3)_5}\right]^{++} + OH^- \rightarrow \left[Co^{OH}_{(NH_3)_5}\right]^{++} + Br^-$$

This reaction occurs with a measurable speed and under ordinary conditions goes practically to completion. When mixing the same pentammine ion with mercuric ions the strong affinity between the mercuric ions and the bromide ions makes itself

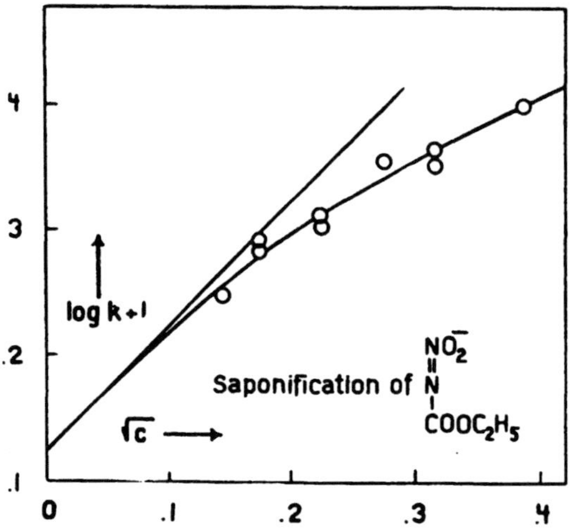

Fig. 5.

noticed in the formation of undissociated mercuric bromide according to the following stoichiometric scheme:

$$2\left[Co^{Br}_{(NH_3)_5}\right]^{++} + Hg^{++} + 2H_2O \rightarrow 2\left[Co^{H_2O}_{(NH_3)_5}\right]^{+++} + HgBr_2$$

Kinetically, however, this reaction is bimolecular as follows:

$$\left[Co^{Br}_{(NH_3)_5}\right]^{++} + Hg^{++} \rightarrow$$

Thus in either case the bromoion is decomposed, forming hydroxo- and aquopentamminions respectively. The two reactions are both of a high valence type, they are however quite different in the fact that in the first case the double charged positive cobaltion reacts with a singly charged negative and in the second case with a doubly charged positive ion. According to the theory we should expect therefore very marked salt effects in opposite direction.

Both of these reactions have been studied in my laboratory by Dr. Livingston by a colorimetric method. The bromoion is blue-violet, the hydroxoion red and the roseoion is pink. Although the color change as the reaction proceeds is marked enough, especially in the last of the two cases, to allow of a considerable accuracy in the determinations, the results seem to a certain extent to have been marred by some other disturbing influence. By the great number of experiments, however, a rather definite and unequivocal information as to the salt effect in this series is provided.

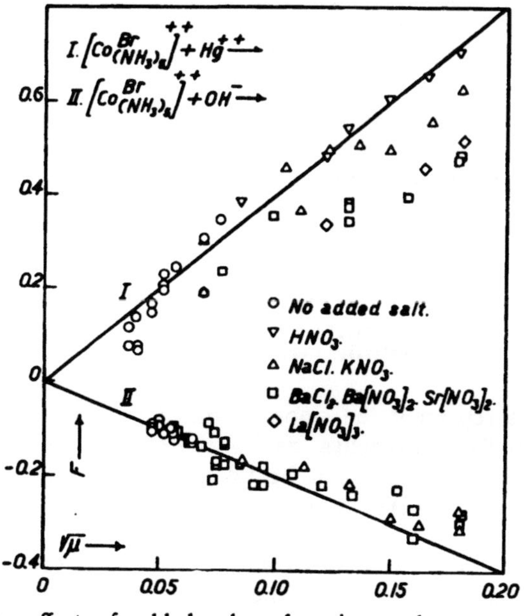

FIG. 6.—The effect of added salts of various valence types upon the kinetic activity factor F as a function of the square root of the ionic strength. The heavy lines represent the theoretical limiting slopes for the two reactions.

The results are given here only by means of a graph, the velocity constant at infinite dilution for both reactions being taken as unity. The heavy straight lines in Fig. 6 represent the theoretical limiting slopes. The various points of the figure correspond to experiments in some of which only the reacting ions were present while in others neutral salts or acid were added to change the ionic strength. The position of the points shows that the experimental accuracy of the method as mentioned is only moderate. Of the general character of the experiments, however, no doubt can exist. The approach of the data in the first reaction to the theoretical slope is very pronounced and mainly independent of the nature of the solution. In the second reaction the salt effect is more dependent on the type of the added salt. The effect seems to be more uniform when in the solution of the various salt types (KCl, $BaCl_2$, etc.) equivalent concentrations instead of ionic strength are compared. Likewise in this reaction, however, there is no question that the velocity constant varies approximately in concordance with the requirements of the theory developed.

This is the more remarkable as the change of the velocity constant in these experiments in spite of the great dilution is very pronounced corresponding to the high valence type of the reactions. In the reaction between the two positive ions the velocity constant changes fourfold when the ionic strength changes only from 0.0016 to 0.032.

In the 4 examples here mentioned for testing the theory the values of the valence product of the reacting ions:

Reaction	$z_A z_B$
$H_2O_2 + H^+ + Br^-$	-1
$N_2O_2CO_2^- + OH^-$	$+1$
$CoBr(NH_3)_5^{++} + OH^-$	-2
$CoBr(NH_3)_5^{++} + Hg^{++}$	$+4$

in all cases represent the slope to which the experimental $\log k_0$ curve actually approaches as concentration approaches zero. These results therefore can be taken as a strong proof of the validity of the new theory of chemical reaction velocity.

The salt effect so far dealt with is called *primary kinetic salt effect* because it is a purely kinetic effect in solutions of strong electrolytes, where no influence ascribed to changes in concentration of the reacting molecules is possible. Several phenomena suggest, however, that the effect of salts upon reaction velocity may be quite different in origin. So for instance Arrhenius [16] showed in 1899 that when cane sugar inversion is catalyzed by a weak acid such as acetic acid the acceleration caused by neutral salts is much greater than when the catalyst is a strong acid. A closer examination of this phenomenon discloses the presence of an exponential effect quite foreign to the theory given above for the simple case of inversion by hydrogen ions, and also showed that similar phenomena are generally met with when the molecules partaking in the reaction enter at the same time in the equilibrium scheme of the dissociation of a weak electrolyte.

As an explanation of the phenomenon demonstrated by him Arrhenius drew the conclusion that the salt caused an increase in the dissociation of the weak acid thus augmenting the concentration of the catalytically active hydrogen ion. He was not in a position, however, to explain the reasons of this increased dissociation. From the view points of the new electrolytic theory the phenomenon is easily understood, as a consequence of an equilibrium displacement owing to the interionic forces. And from this point of view the method of determining hydrogen ions in weak acid and buffer solutions is ranged alongside of the colorimetric and similar methods, in which the presence of neutral salts is known to give rise to similar disturbances.

To clarify the phenomenon here in question, to which the term *secondary kinetic salt effect* is used, since it is not here a question of a direct kinetic influence, let us consider the dissociation of a weak electrolyte like acetic acid (EH)[17].

$$EH \rightarrow E^- + H^+$$

When adding a salt such as sodium chloride the activity coefficients f_1 of the ions will decrease while the activity coefficient f_0 of the acid is practically unity also in the salt solution. Accordingly the equilibrium is disturbed and to make up for the diminution in the activity coefficients of the ions their concentration must increase, i.e., the acid will undergo dissociation until equilibrium is reestablished.

A quantitative calculation of the equilibrium displacement can be made by means of the equation:

$$K_a = \frac{a_{E^-} a_{H^+}}{a_{EH}} = \frac{c_{E^-} c_{H^+}}{c_{EH}} \cdot \frac{f_{E^-} f_{H^+}}{f_{EH}} = K_c \frac{f_1^2}{f_0},$$

or:

$$K_c = K_a \frac{f_0}{f_1^2}. \qquad \text{(16)}$$

In a 0.1 N sodium chloride solution, f_1 is about 0.7 and addition of this amount of salt to a dilute acetic acid solution therefore increases the dissociation constant K_c of the acid to double of its normal value and the hydrogen ion concentration by 40 per cent. This is not far from the value actually found by experiment. Furthermore Equation (16) shows that the effect must be exponential.

As obvious from this example the secondary salt effect here under discussion originates from the changes in the activity coefficient factor in the thermodynamic mass action expression. It is of great interest to note, however, that this factor does not bring about the same effect in all cases, but largely depends upon the type of weak electrolyte considered.

Representing the dissociation of a weak acid by the following scheme:

$$A \rightleftarrows B + H^+,$$

in which A functions as an acid and B as a base the relation between K_c and K_a, the ordinary and the thermodynamic dissociation constant is:

$$K_c = K_a \frac{f_A}{f_B f_{H^+}}, \quad . \quad . \quad . \quad . \quad . \quad . \quad . \quad . \quad (17)$$

and the secondary kinetic factor in question is:

$$\frac{f_A}{f_B f_{H^+}},$$

which in the case of acetic acid already considered reduces to

$$\frac{f_0}{f_1^2}.$$

The acid molecule, however, may be positively or negatively charged, and in that case the value of the activity coefficient factor will be quite different.

It is known that a chromic chloride solution for instance shows acid reaction. This in accordance with the above general scheme may most simply be explained as a dissociation:

$$[Cr(H_2O)_6]^{+++} \rightleftarrows \left[Cr{\,}^{OH}_{(H_2O)_5}\right]^{++} + H^+$$

and the kinetic factor for this case therefore is

$$\frac{f_A}{f_B f_{H^+}} = \frac{f_3}{f_2 f_1},$$

where the subscripts of the f indicate the charge of the corresponding ion. Contrary to the factor in the case of acetic acid the factor

$$\frac{f_3}{f_2 f_1},$$

however decreases with increasing salt concentration and correspondingly the salt effect must be negative.

The problem can be attacked generally for dilute solutions by means of the Debye–Hückel theory. In order to calculate the value of $f_A\, f_B f_{H^+}$, we introduce the equation already made use of above

$$-\log f = 0.5\, z^2\sqrt{\mu}. \qquad (11)$$

Hence

$$\log \frac{f_A}{f_B f_{H^+}} = 0.5(z_B{}^2 + 1 - z_A{}^2)\sqrt{\mu},$$

which equation since $z_A = z_B + 1$ reduces to:

$$\log \frac{f_A}{f_B f_{H^+}} = -\, z_B\sqrt{\mu}, \qquad (18)$$

or

$$\frac{f_A}{f_B f_{H^+}} = 10^{-z_B\sqrt{\mu}}. \qquad (19)$$

By means of this very simple expression for the kinetic factor, it is easy to predict generally the sign of the salt effect in hydrogen ion catalysis and to calculate it quantitatively in sufficiently dilute solutions. In fact the sign of the effect is simply opposite of that of the basic molecule corresponding to the catalyzing acid.

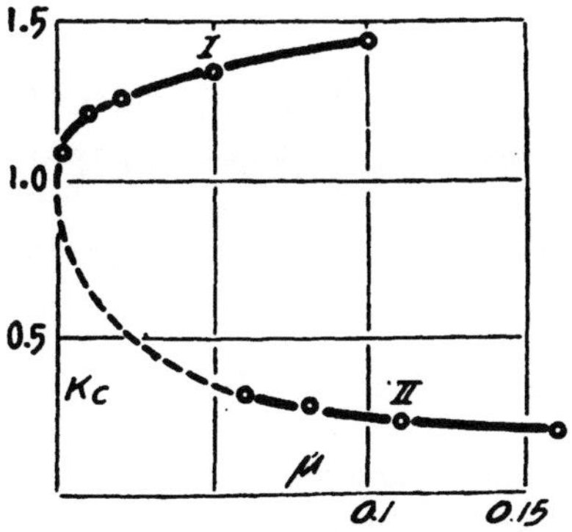

Fig. 7.—Secondary salt effects for the decomposition of diazoacetic ester.

Fig. 7 shows the secondary salt effect upon the decomposition of diazoacetic ester by acetic acid (curve I) and chromic nitrate (curve II) in the presence of neutral salt of concentration *c*, the velocity constant at $c = 0$ being taken as unity in both cases.[18] The positive effect in curve I and the negative effect in curve II is quite in agreement with the theory. The same reaction using chromic perchlorate and picrate, which salts are more suitable for the determinations than the nitrate has been measured down to 0.0002 molal solution by Dr. King. He found in experiments not yet published a close approach to the theoretical slope in the dilute solution and thus verified the extrapolated part of the curve in Fig. 7.

In the case of hydroxyl ion catalysis the equations corresponding to (17) and (19) are

$$K_c = K_a \frac{f_B}{f_A f_{OH^-}}$$

and

$$\frac{f_B}{f_A f_{OH^-}} = 10^{z_A \sqrt{\mu}} \quad . \quad . \quad . \quad . \quad . \quad . \quad . \qquad (20)$$

The sign of salt effect in this case is identical with the sign of the corresponding acid molecule.

Measurements of hydroxyl ion catalysis have been made on the decomposition of nitrosotriacetonamine. This reaction has been shown by Francis and his coworkers to proceed proportionally to the hydroxyl ion concentration in dilute solution. Dr. King,[19] who worked in my laboratory on this reaction, studied the salt effect in two buffers: a piperidine-piperidinum and a secondary-normal phosphate solution, in which the following equilibria exist:

$$C_5H_{11}N \rightleftarrows C_5H_{11}NH^+ + OH^-$$

$$PO_4^{---} \rightleftarrows HPO_4^{--} + OH^-$$

in correspondence to the general acid-basic equilibrium in aqueous solution:

$$B \rightleftarrows A + OH^-$$

Since in the first of these reactions $z_A = +1$ and in the second $z_A = -2$ according to Equation (20) a strong positive and a still stronger negative effect in the first and second cases respectively are predicted by the theory. As shown by the graph Figs. 8 and 9 this expectation is completely fulfilled.

In judging the magnitude of a salt effect expected it must be noted that not only the change in the dissociation constant, but also the relative concentrations of the molecules entering the equilibrium scheme are of significance. The smaller the concentration of the catalytically active molecules, in these examples of the hydrogen and hydroxyl ions, in comparison to the concentration of the other participants the more pronounced the kinetic effect will prove to be.

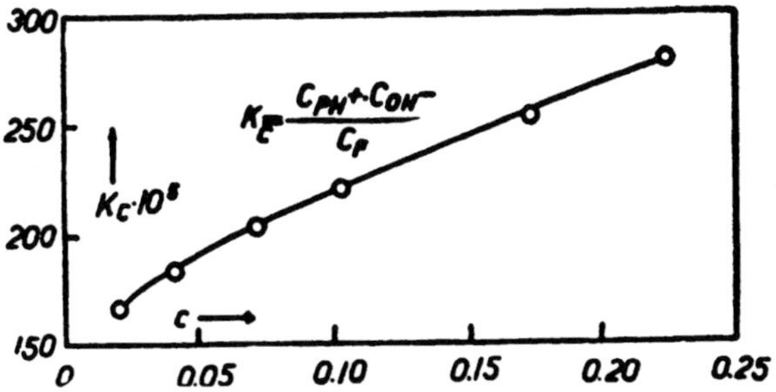

FIG. 8.—Secondary salt effects for the decomposition of nitroso-triacetone-amine at 15° in piperidine buffer solutions.

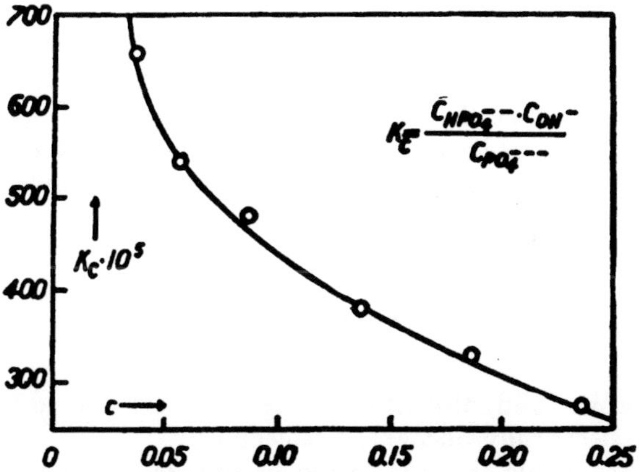

FIG. 9.—Secondary salt effects for the decomposition of nitroso-triacetone-amine in phosphate buffer solutions.

A verification of this conclusion is obtained in different catalytic reactions. Here should be mentioned the decomposition of nitramide,[20] a reaction of extraordinary interest for the theory of acid and basic catalysis on the whole. Nitramide $H_2N_2O_2$ is catalytically decomposed according to the following scheme:

$$H_2N_2O_2 \rightarrow H_2O + N_2O$$

in the presence of any substance of basic nature, for instance acetate and phosphate ions, ammonia, aniline, etc. If the reaction is carried out in an acetic acid-acetate buffer addition of salts has no influence upon the velocity, because in the equilibrium

$$EH \rightleftarrows E^- + H^+$$

with relative great concentration of E^- and EH, only the hydrogen ions can be affected by the equilibrium displacement. On the other hand using an acetic acid-hydrochloric acid buffer, in which the acetate ion concentration is small in comparison with that of EH and H^+ only the acetate concentration is appreciably influenced by salt addition, and a correspondingly great positive salt effect appears. A solution of free acetic acid, finally, with no excess of H^+ neither of E^- holds, as to be expected, an intermediate position as to the effect of salt.

The experimental evidence in favor of the new theory of velocity of ionic reactions covering the phenomena of primary as well as secondary salt effect is doubtless of a sufficiently rigid character to prove the correctness and significance of the theory. Remembering that the expressions for the salt effect were derived on a general basis, without making use of the fact that the ions bear electric charges, one may, however, be induced to believe that the same formulae, for instance:

$$h = kc_A c_B \frac{f_A f_B}{f_X}$$

should prove valid, also when the reaction considered takes place between electrically uncharged molecules.

Neither would this conclusion be untrue, but it must be considered that the theory for two reasons will be of less validity in the case of non-electrolytic reactions. First there are no such general laws of the activity coefficient as exist for the ions. We know that changes in f for non-electrolytes are linear and also that the salt effect is usually so as to increase it. But even this qualitative rule is not obligatory and of a quantitative calculation of the activity coefficient factor of uncharged molecules there exists no general possibility.

Secondly the changes in f are certainly not the only reasons for changing velocity constant on addition of salts or other changes of the solvent. We must expect also a general solvent effect linear in nature and differing from one reaction to another. No doubt, when dealing with ionic reactions this uncertainty is also present and impedes an exact quantitative calculation at ordinary conditions of concentration. The great and unmistakable salt effects here, however, will at all events appear conspicuously. With non-electrolytes the activity effects are small and liable to be masked by solvent effects. Obviously the conditions for applying the new formulae to such substances are much more complicated.

The problem of overcoming such difficulties is actually the problem of extending the present theory into the region of unlimited concentrations. Even if such an attempt undoubtedly is bound to encounter considerable difficulties, there are good reasons to believe that the points of view of the present theory will prove of value also for the development of such a more general theory of reaction velocity in solution.

LITERATURE

1 G. N. Lewis, Proc. Amer. Acad. Sci., **43**, 259 (1907).
2 Harned, J. Am. Chem. Soc., **40**, 1461 (1918).
3 Jones and W. C. M. Lewis, J. Chem. Soc., **117**, 1120 (1920); Moran and Lewis, *Ibid.*, **121**, 1613 (1922).
4 Brönsted, Z. physik. Chem., **102**, 169 (1922).
5 Bjerrum, Z. physik. Chem., **108**, 82 (1924).
6 Price, Z. physik. Chem., **27**, 474 (1898).
7 Holmberg, Z. physik. Chem., **79**, 147 (1912); **80**, 573 (1912); **97**, 134 (1921).
8 Bredig and Riplay, Ber. d. chem. Ges., **40**, 4015 (1907).
9 Harned and Seltz, J. Am. Chem. Soc., **44**, 1475 (1922).
10 Bray and Livingston, J. Am. Chem. Soc., **45**, 1251 (1923); **45**, 2048 (1923).
11 A. A. Noyes, Z. physik. Chem., **16**, 561 (1895).
12 Timofeew, Muchin, and Gurewitsch, Z. physik. Chem., **115**, 161 (1925).
13 Debye and Hückel, Physik. Z., **24**, 185 (1923).
14 Brönsted and La Mer, J. Am. Chem. Soc., **46**, 555 (1924).
15 Brönsted and Delbanco, Z. anorgan. Chem., **144**, 248 (1925).
16 Arrhenius, Z. physik. Chem., **31**, 197 (1899).
17 Brönsted, J. Chem. Soc., **119**, 574 (1921).
18 Brönsted and Teeter, J. Phys. Chem., **28**, 579 (1924).
19 Brönsted and King, J. Am. Chem. Soc., **47**, 2523 (1925).
20 Brönsted and Pedersen, Z. physik. Chem. **108**, 185 (1924).

Physico-Chemical Principles in Electro-Metallurgical Research

BY

COLIN G. FINK

ASSOCIATE PROFESSOR OF CHEMICAL ENGINEERING
COLUMBIA UNIVERSITY

Lecture delivered at Columbia University in the special course on Contemporary Developments in Chemistry given in the Summer Session of 1926 on the occasion of the opening of the Chandler Chemical Laboratories

COLUMBIA UNIVERSITY PRESS

NEW YORK

1927

Printed from type. Published March, 1927

Printed in the United States of America

PHYSICO-CHEMICAL PRINCIPLES IN ELECTRO-METALLURGICAL RESEARCH

COLIN G. FINK

If we were to trace metallurgical research from its earliest days, we would find that the methods of smelting and testing were of the crudest type and it was often very difficult, indeed, to reproduce results and obtain, for example, a steel or a bronze, of the same physical properties as a given standard sample. One of the greatest steps forward in metallurgical research was due to the introduction of chemical analysis. The metallurgist could now correlate to a large extent certain physical properties of his metal with its chemical analysis. He observed, for instance, that a large percentage of carbon in his steel made it very hard or brittle and that very little carbon made it very ductile and malleable.

The second and comparatively recent step in advance has been the introduction of physical chemistry into metallurgical research. Tests were now instituted as to the exact melting point, specific gravity, latent heat of fusion, crystalline structure, electrical conductivity, heat conductivity, thermo-electric and magnetic properties, etc. As these tests were developed, more and more exacting demands were also made upon the analytical chemist. No longer did it suffice to report the percentage of carbon present in the steel but it was necessary to determine whether this carbon was there, either as combined or as graphitic carbon. You all know that two samples of steel may have the same percentage of total carbon and yet their physical properties be entirely different depending upon whether the carbon is present largely as graphite or as combined carbon. A complete steel analysis today may include half a dozen or more of the following elements: carbon, graphite, nickel, chromium, manganese, silicon, sulfur, phosphorus, titanium, vanadium, tungsten, molybdenum, copper, cobalt, zirconium, uranium, tantalum, aluminum, tin, oxygen and nitrogen. This is a goodly array as compared with the report of the first steel analyst who concentrated his attention on the single element, carbon.

Chemical analysis, however important as it may seem, is very often of secondary value in modern metallurgical research. A careful determination of one or more of the physical or physico-chemical constants of the metal or alloy is now of far more crucial importance. For example, in the case of copper for electrical purposes, we are more concerned with its high electrical conductivity than to know whether the impurities present are silver, antimony, bismuth or sulfur, and the absolute percentage of these. The remarkable development in metallurgical research within the last generation is due in a large measure to the perfection of our physical apparatus and instruments, among which the optical instruments are second to none. Without the microscope and camera metallurgical research would today be very badly handicapped.

In the course of our own metallurgical investigations, we have observed one interesting fact and that is this: No matter whether you are investigating the effect of gases on metals or alloys, or the ductility of metals, or the corrosion of metals or any other physical, mechanical or chemical property—the method of attack is very much the same. To put it a little more definitely, it really makes very little difference which particular physical, chemical or mechanical property we were particularly interested in, we have always made it a point to investigate a number of other properties simultaneously and then compare and correlate and draw conclusions and get suggestions for the next experiment. For example in the investigation which finally led to a substitute for platinum [1] as a seal in glass, there were two other properties we studied simultaneously besides the coefficient of expansion, namely, the plasticity of the metal or alloy and the extent to which glass will adhere to or wet the metal or alloy. It is not hard to appreciate that a metal that is very plastic and adheres tenaciously to glass need not necessarily have the same coefficient of expansion as the glass, to make an air-tight seal. The final product we developed was a compromise: All three properties, plasticity, adhesion of metal to glass and expansion and contraction with changes of temperature were important factors and no single one of them was the determining one.

The Insoluble Anode Problem

A research to which we have devoted years of study is the corrosion of alloys, and in particular, the research which had for its object the finding of an insoluble anode which would be resistant to disintegration in a copper solution containing sulfates, nitrates and chlorides. More specifically, the anode sought after must have the following properties: good electrical conductivity;

low electrode potential; mechanical strength; ease of manufacture from cheap raw materials; insolubility in acid copper sulfate solutions containing chlorides and nitrates; and finally freedom from such deleterious constituents, which, though dissolving extremely slowly, would enter the cathode copper and affect its quality.

The problem when brought to us was not a new one. But owing to the rapid development of the leaching and electrolytic precipitation process for copper, the importance of finding a suitable anode became more evident every day.

Here in this country we first tried out *lead* and *carbon*, and providing the electrolyte was a sulfate one and was free from acids such as hydrochloric and nitric, lead was found to be serviceable and is today used in plants (*e.g.*, in that of the Utah Copper Co.) where fairly pure sulfate electrolytes are in order. Carbon anodes had the advantage of a lower electrode potential, but contrary to the experience with carbon anodes in chloride liquors, the life of the carbon anode in sulfate liquors was comparatively short.

Germany had tackled the problem and one of the materials which they suggested and tried out commercially to some extent was *magnetite*. This magnetite anode resisted the action of chlorides and nitrates, besides sulfates, fairly well, but two of the great objections to magnetite anodes were: First—The extreme brittleness; magnetite being more brittle than glass. Second—The poor electrical conductivity.

Although the second objection was more or less set aside when hollow magnetite anodes with inner copper linings were prepared, the first objection, the brittleness, was never removed.

Soon after the vast copper deposits of the northern part of Chile were acquired by the Chile Exploration Company, experiments on leaching and electrolytic extraction of the copper were started here in America at the plant of the American Smelting and Refining Company at Perth Amboy. Lead anodes and magnetite anodes were tried out, but neither proved to be satisfactory. Owing to the high percentage of nitrate and chloride present in the ore, the lead anodes disintegrated very rapidly, losing at the rate of 65 pounds of lead for every hundred pounds of copper deposited. The magnetite anodes were fairly resistant and were tried out at the plant at Chuquicamata, but owing to the excessive breakage and the high cost of manufacture, the magnetite anode had to be abandoned. At about that time a series of acid resistant ferro-silicon alloys were developed in this country, and a number of these were tried out. It was found that ferro-silicon, with about 13 per cent silicon, showed a loss due to corrosion decidedly below that for lead, although not as low as that for magnetite.

The ferro-silicon anode was used in the Chuquicamata plant for a few years and although the loss was only about 5 pounds per 100 pounds of copper deposited, yet there was one serious objection to it and that was the accumulation of iron in the electrolyte. Iron in copper electrolytes has a very depressing effect on the current efficiency. Iron consumes appreciable quantities of electrical energy by being oxidized at the anode, then migrating to the cathode and there being reduced, then migrating back to the anode and there being oxidized again, and so on.

Something better than ferro-silicon was required and our in-

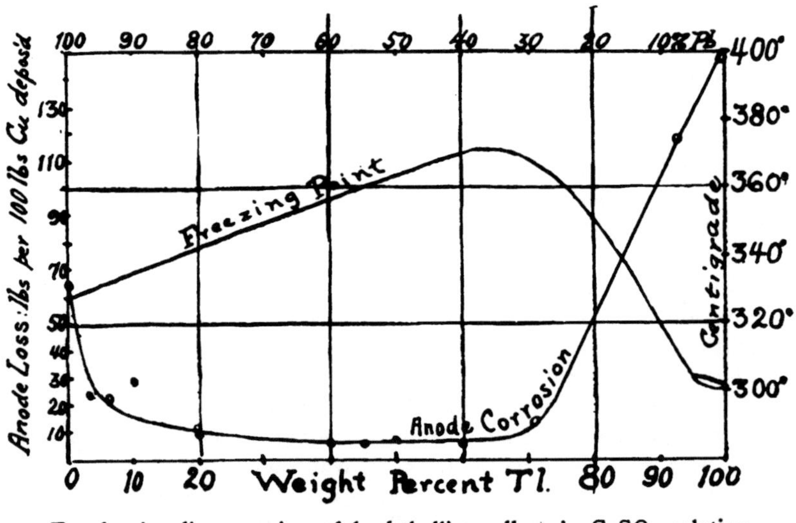

FIG. 1.—Anodic corrosion of lead-thallium alloys in $CuSO_4$ solution.

tensive study started at this stage. We investigated a great number of series of alloys and again our investigations were not confined merely to the one important property, namely resistance to corrosion, but we simultaneously investigated the metallographic structure of the alloys, the hardness, melting point and film forming properties. In this way, for example, we gleaned many a good suggestion by a careful study of the fusion curves and found curiously enough that in a number of cases there was a fairly close coordination between the fusion and corrosion curves. For instance, in the case of the lead-thallium alloys, as the thallium percentage is increased the melting point increases and the resistance to corrosion increases. The alloys with the highest melting points showed lowest corrosion losses [2] (Fig. 1).

The lack of the full appreciation of the value of this coordination of the physico-chemical properties, we have always considered to be the chief reason why previous to our investigations, progress in solving the anode problem proceeded so very slowly. For example, it was generally believed that a metal or alloy that would ordinarily not dissolve in one or the other acid would therefore be likewise resistant to dissolution if used as an anode in that acid. For example, ferro-chromium containing upwards of 60 per cent chromium is ordinarily extremely resistant to sulfuric acid solutions. You can take a sample of the alloy and cover it with sulfuric acid and after months of immersion find little or no loss in weight. Yet this very same sample of alloy made anode in the same acid will disappear in 24 hours. The reason for this great difference is not to be sought in any change in the primary reaction. In either case we are dealing with electrolytic dissolution, but without the application of comparatively high external electromotive forces, the surface of the alloy soon becomes "badly polarized"—that is, a protective film is formed which greatly retards further dissolution. On the other hand, this film is neither strong enough nor of the proper chemical composition to be of any practical value when the alloy is used as an anode.

Catalytic Surface Film on Anodes

Our main line of attack was that dealing with the surface film. We theorized in this fashion: if we could produce on the surface of the anode a film which would have a decided catalytic effect on the discharge of the anions and liberation of oxygen, then the chances of the metal or alloy being attacked and dissolved by the acid would be minimized. Since almost every metal that was tried either alone or in combination was objectionable on account of traces getting into the copper cathode and materially affecting its electrical conductivity, it soon became apparent to us that the only permissible metal was copper. "Would it be possible" we asked ourselves, "to render copper insoluble in an acid electrolyte?" It was not difficult to imagine a cast copper anode which would be insoluble if every individual crystal of copper in the anode were completely covered with a very thin film or glaze of an acid resistant nature.

But we further reasoned this way: There are two reactions at the anode that are of immediate concern:[3] (1) The combination of the anion (acid radical) with the metal of the anode and (2) the discharge of the anions and reaction with water to form oxygen gas. The relative velocity of these two reactions more or less determines the degree of solubility of the anode under the action of the current. It is not difficult to picture the ideal case for a perfect insoluble anode in which the velocity of the second reaction is infinitely greater than that of the first reaction. The more we can catalyze the formation and evolution of gaseous oxygen, the more insoluble will be the anode.

In the hydrometallurgy of copper, an ideal insoluble anode would be one of copper with a surface film that would catalyze the oxygen formation and evolution to such an extent that practically no copper of the anode would go into solution.

In the selection of the film metal the following were tried out: barium, manganese, chromium, uranium, vanadium, tungsten, cobalt, nickel, tin, lead, and silver. Best results to date have been obtained with barium, manganese, cobalt, lead, and silver. Mercury was tried out merely from scientific interest. Two kinds of metals were considered as film metals: first, metals, such as vanadium, that form two or more oxides; and second, metals that form insoluble compounds on the surface of the anode in a very porous or highly dispersed form.

One of the most effective of the catalysts previously tried was the oxide of cobalt. The anode developed based on the behavior of this film was a forerunner of the present copper silicide anode. Here again the cobalt oxide film catalyzed better in the presence of a promoter.[4] Manganese plus cobalt present in the film together showed much greater catalytic activity than either cobalt or manganese alone.

When it came to a final solution of the anode problem, an anode for the copper sulfate electrolyte containing both nitric and hydrochloric acid besides sulfuric, it was again a case of compromise. Instead of consisting of 100 per cent copper with a thin catalytic surface film, the anode as introduced into the large plant at Chuquicamata was essentially a complex cuprosilicon alloy, with about 70 to 80 per cent copper besides small percentages of iron and the catalytic film forming elements lead, tin and manganese. It is well to appreciate that there are only five elements out of the 85 of the periodic table that could be economically used to form the major constituent of the anode.

The standard practice at the plant today is as follows: Before the anodes are put into operation in the regular copper electrolyte, they are hung into a special cell and current is applied until a good catalytic film consisting chiefly of lead dioxide is built up.

A polished section of the anode examined under the microscope shows that the anode consists of primary crystals of free silicon, very resistant to corrosion, and between these a ternary eutectic of copper and the other elements.

The standard anodes are 3 feet 11 inches high by 2 feet 9 inches wide and about 50,000 are in operation 24 hours a day. Due to the catalytic surface film the voltage was reduced to about one half of that for the ferro-silicon anode. In other words, the rapid anodic formation of oxygen not only results in the protection of the metal but—and this is important—the anode potential is very materially reduced.

According to E. L. Jorgensen, Superintendent of the Chuquicamata plant at the time our anode was installed, "the saving in power resulting from the introduction of the new anode made the existing power plant able to take care of increased tank house production, saving several million dollars. The cost of anode maintenance was reduced by about 90 per cent, saving in addition about $1,000,000 per year." [5]

In this copper-silicon or "Chilex" anode research, my chief assistants were Chas. H. Eldridge, who had charge of the metallographic investigation, and Earl M. Anger and Frank L. Stack, who were in charge of the electric furnace work.

The Silver-lead Anode for Alkali-chloride Cells

A research just recently completed at Columbia is that carried out by Dr. L. C. Pan [6] on the insoluble anodes for the electrolysis of brine. The one anode material used throughout the world today in the alkali-chloride cells is graphite. But due to the porosity of the graphite anodes, oxidation of the binder and disintegration takes place—although this may to some extent be offset by proper impregnation. In the past, metal anodes have been tried out and there are two plants today that are using platinum anodes. We started out to find a metallic anode which would be decidedly cheaper than platinum.

The anolyte of the alkali-chloride cell contains besides Cl^- ions, appreciable quantities of OH ions and also small amounts of ClO^- and ClO_3^- ions. Accordingly, an anode material which is to be satisfactory

1. Must resist oxidation or attack by such substances as sodium chloride, hydrochloric acid, hypochlorous acid, chlorates, chlorine gas and oxygen.
2. Must be mechanically strong.
3. Must have a low electrode potential or a low chlorine overvoltage.
4. If it should go into solution or corrode but slightly, must not contaminate nor affect the products.
5. Must be a good electrical conductor.
6. Must not have a high initial cost.

From a purely chemical standpoint, there are two common metals, silver and lead, which form insoluble chlorides. If, under the action of the chloride solution and electric current, these chlorides form on the surface of anodes made of either of these metals, they should remain unchanged. If these solid chlorides should be in the form of a protective film over the anodes, then the corrosion of these anodes might be greatly reduced. Unfortunately, the electric conductivity of these chlorides is extremely low. Therefore, when the chloride forms a perfect protective insoluble chloride film it also cuts off the electric current, and the useful function of the anode ceases. But still more unfortunate is the fact that these chlorides are far from insoluble, when dealing with strong solutions of sodium chloride. Therefore the metals are continuously attacked when fresh brine is supplied from time to time.

Metallic lead, although soluble in neutral sodium chloride solution, is less soluble as the concentration increases beyond a critical point.[7] But in the anolyte of a brine cell, there is the dissolved chlorine which greatly accelerates the formation of $PbCl_2$ and $PbCl_4$.[8] The latter is very soluble, and the $PbCl_2$ is soluble, particularly in strong brine.[9] Therefore metallic lead alone has not been found to be a suitable material for the brine-cell anode.

Metallic silver is easily attacked by sodium chloride solutions,[10] resulting in silver chloride, which is slightly soluble in the sodium chloride solution.[11] It was upon the solubility of silver chloride in sodium chloride solution that the old Augustin's process of extracting silver from its ores was based.[12] In the alkali-chlorine cell, with a constant supply of fresh brine, a silver anode will go into solution slowly but steadily, because the saturation point is never reached. Furthermore, the presence of dissolved and free chlorine will keep silver in solution, and cause more silver to dissolve.[13] We found that within reasonable limits of current density, when the anolyte is made of weak brine, and the anode is of pure silver, AgCl forms much faster than it is dissolved, even at 80° C. Therefore, a film of AgCl is formed over the surface of the anode, which is hard, horny and impervious, cuts off the current, and withstands an impressed voltage of about 110 volts before it breaks down, when a spark discharge takes place. The cell voltage drops down when the AgCl film is punctured, but the punctured spots are healed up within a few minutes, and the cycle repeats itself over and over again. Accordingly, pure silver is by no means a suitable anode material for the brine cell.

As the pure metals, lead and silver, gave little promise we

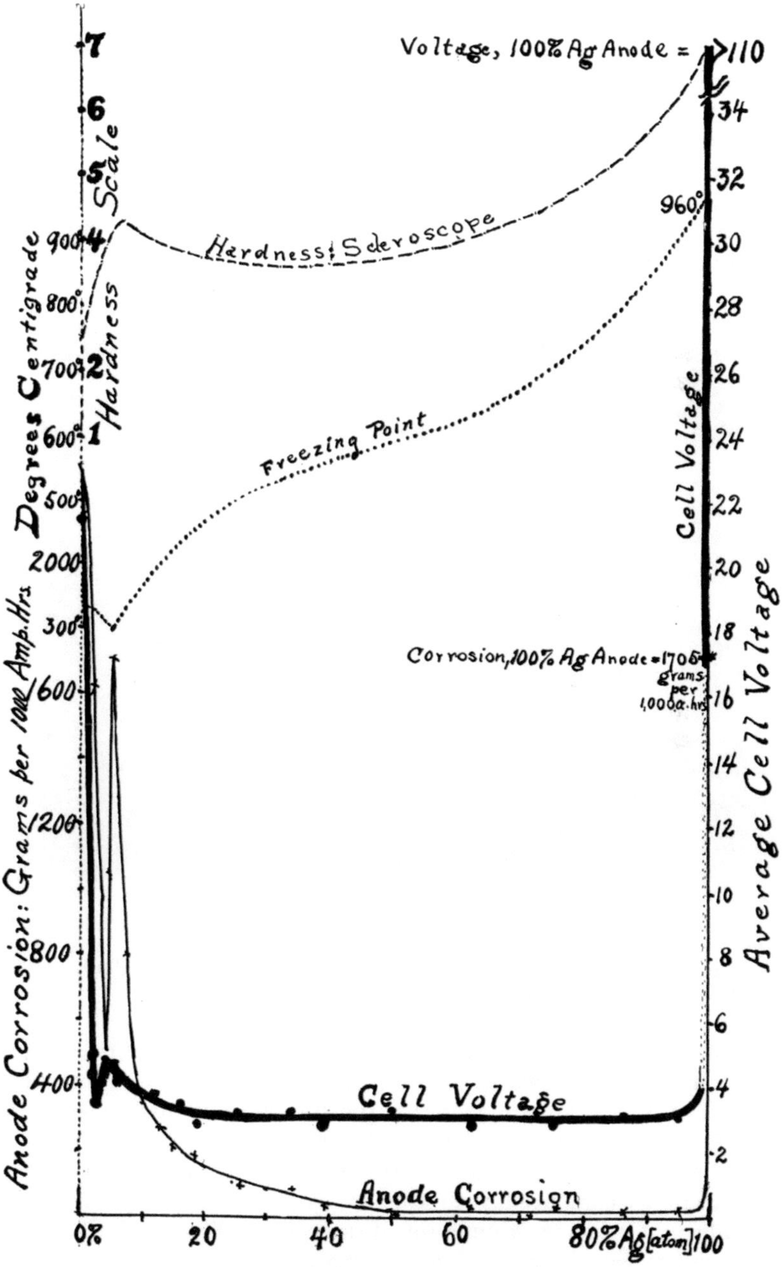

FIG. 2. Cell voltage and net anodic corrosion of Ag.Pb alloys in 5 per cent NaCl solution at 22° C., with diaphragm.

next investigated the alloys of these two elements encouraged by our findings in the case of thallium-lead and other alloys where the properties of the alloys were radically different from that of the constituents. Our results are shown in Fig. 2.

It will be observed that in general the corrodibility of silver or of lead drops exceedingly fast upon the addition of small percentages of the second metal. Thus, for example, on the 510-hour test, pure lead lost 2055 grams, pure silver lost 1709 grams, and the alloy with 61 per cent silver almost nothing. Another interesting and commercially important fact is that the

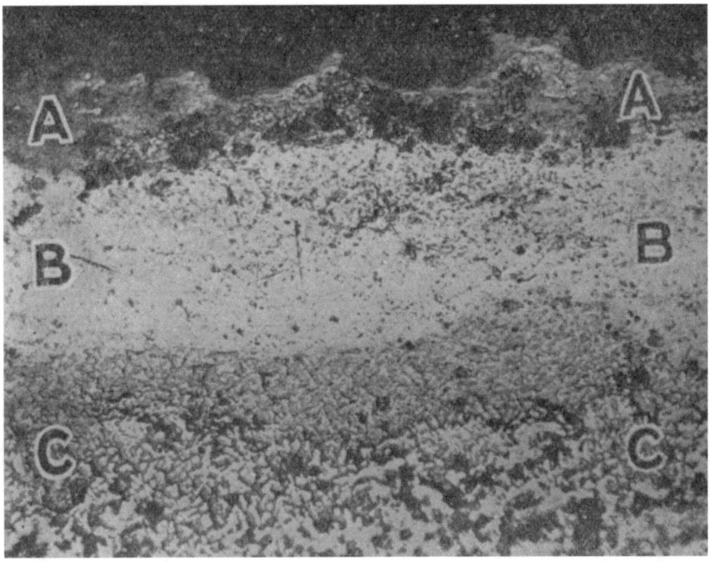

FIG. 3. Cross-section of the anode film, formed over the alloy, 49.3 per cent Ag and 50.7 per cent Pb, in a NaCl solution. Outer layer, A-A composed of PbO_2, Ag_2O; inner layer, B-B composed of PbO_2, AgCl and below this C-C the alloy (Ag-Pb). Magnification: 220.

voltage likewise drops very abruptly. Such behavior of metals and their alloys toward corrosive agents is rather in contradiction to the well-known fact that the solubility of many metals in acids decreases as the purity of the metal increases.

Thus, for eaxmple, zinc 99.99 per cent pure is very insoluble in acids, as compared with zinc of 99 per cent and 98 per cent grades. Ordinary aluminum dissolves readily in HCl, pure aluminum very slowly. Another example is found in the difference between chemical lead, about 99.99 per cent pure, and ordinary common lead, about 99.90 per cent pure. The former resists acid attacks much better than the latter.

However, ordinary contact corrosion on the one hand and forced anodic corrosion on the other often give results directly opposite to each other as we pointed out above in the case of the cupro-silicon anode.

After the completion of the tests, all of the anodes in the lead-silver series ranging from the lowest to the highest silver content were examined under the microscope. The abrupt drop in the voltage indicated to us that here again we were dealing with a surface or film reaction quite distinct from that for pure lead or pure silver. Fig. 3 shows a magnified view of the film of the alloy containing 49.3 per cent silver. The outer layer is composed of PbO_2 and Ag_2O; the inner layer is composed of PbO_2 and AgCl. The function of the horny AgCl is very important. It holds the PbO_2 in place and fills the pores of the inner layer of the film, rendering it more resistant to anodic corrosion. As the silver content increases above 61 per cent by weight, the amount of AgCl produced is more than is required to fill the pores of the PbO_2 layer; consequently, the excess AgCl goes partly into solution and partly coats over the PbO_2 particles, cutting down the electrical conductivity of the film. By direct test, a greater corrosion loss is observed the more silver there is in the alloy in excess of that needed to form AgCl to fill the PbO_2 pores.

In the case of anodic corrosion of lead alloys, insoluble salts are usually formed, the character of which largely determines the rate of corrosion. The presence of impurities has only an indirect effect upon the latter, insofar as it determines the character of the insoluble salts formed on the surface of the metal. Silver has a lower oxygen overvoltage than a chlorine overvoltage; and the oxygen overvoltage of silver is lower than that of lead. The oxygen discharged at the surface of silver is in very active form, and readily attacks lead in juxtaposition, forming lead super oxides.[14] The latter are unstable and are decomposed by chlorine, leaving lead peroxide on the surface.

The more silver in the alloy, the more numerous will be the silver-rich crystals in the metallic surface of the silver-lead anode, and therefore the more uniform will be the layer of lead dioxide on the whole surface, with only two exceptions: (1) With about 2.25 per cent silver by weight, or 4.25 per cent, atomic, the corrosion shows a sharp minimum. (2) With silver content over 61 per cent by weight, or 75 per cent, atomic, the anodic corrosion begins to rise again until we approach pure silver, when the corrosion rate jumps suddenly to an enormous value.

Each of these two exceptional points in the corrosion curves is surprisingly correlated to points on the freezing point curve of the Ag-Pb system,[16] that is, a point about 2.25 per cent is the eutectic, and another point around 60 per cent is a counterflexure of the upward curve. Since, however, there is no compound known corresponding to these latter proportions, explanation from a purely chemical point of view is not so apparent. But a full realization of the composition and behavior of the film easily accounts for this second point.

Aside from the striking correlation between the freezing point and anodic corrosion loss curves, we find similar correlation between the hardness and loss curve. Again at the eutectic point 4.25 atomic per cent of silver there is a break in the curve. This correlation merely further emphasizes our contention that the character and composition of the supporting body has a decided influence on the behavior of the catalytic surface film. To further prove this we made up a film of PbO_2 and AgCl and applied it to the face of a lead battery grid and in another case to the surface of a graphite anode. Corrosion losses were very high.

Conditions under which the anode film is formed on the surface of the silver-lead anode have a direct bearing on the efficiency of the film as a protector against corrosion. Excessively high current densities retard the film formation, because the rapidly discharging gases constitute a disturbing factor. Too concentrated an electrolyte prevents film formation, because it dissolves the anode surface faster than it can be permanently peroxidized. Too high a temperature causes a loose film, because the PbO_2 is too rapidly formed and of a coarse texture. The higher the film-forming temperature, the coarser is the film, and therefore the heavier is the corrosion of the anodes. But if the anode is first treated in a low temperature film-forming bath, and then treated at a moderately high temperature, excellent films can be obtained. A higher "finishing temperature" gives rise to a thicker layer of PbO_2Ag_2O, which makes the film as a whole more protective.

The silver-lead alloy anode (about 61 per cent Ag) is mechanically strong, ductile, malleable, low-melting, easily cast and easily machined. It has a low electrode potential and low chlorine overvoltage, about the same as graphite in these respects. It has a high specific electric conductivity as compared with graphite, and makes easy and good contact joints with copper leads, giving a very low contact resistance as compared with graphite. The silver-lead alloy can be "burned" or welded, like lead, or soldered if desired, so as to eliminate the contact resistance entirely.

Cathode Film Reactions

We have not confined our attention to the behavior of the anode films, but have also investigated the structure and function of cathode films, in particular in the case of nickel, iron, copper, tin, zinc and chromium.

Interesting studies especially from a purely academic point of view, are the cathodic reactions in copper sulfate electrolytes containing an excess of iron ions, and by excess we mean more than 0.25 gram of iron per liter. As pointed out before and as is well known to all analytical chemists, copper will not be electrodeposited in the presence of appreciable quantities of iron in solution due to the so-called "see-saw reaction"; ferrous ions oxidized to ferric ions at the anode and then reduced again to ferrous at the cathode. In one instance, we weighed our cathodes at night, electrolyzed the copper-iron sulfate solution for 15 hours and in the morning found that our cathode had *lost* in weight rather than gained in weight, due very likely to the reaction

$$Fe^{+++} + Cu = Fe^{++} + Cu^{+}.$$

The commercial significance of this deleterious effect of iron on the deposition of copper will at once be appreciated if we consider that at one copper plant alone electrolyte, which has built up in iron is run to waste into the ocean carrying with it over 100,000 tons of copper per annum.[16]

Now, it ought not to be difficult to separate the copper from the iron. Copper is a noble metal as compared with iron, its cathode potential in normal sulfate solutions is almost 0.7 volt lower than that of iron. We can keep the ferrous ions from migrating to the anode by interposing a diaphragm. But diaphragms are out of the question when it comes to handling large volumes of solution amounting to as much as 300,000 cubic meters per annum. Furthermore, diaphragms always greatly increase the voltage and therefore the power consumption of your cell.

Without going into a detailed description of our experiments we found that over 80 per cent of the copper could be recovered by merely rotating the cathode at a speed of about 500 R.P.M. The copper is hard and bright. No diaphragm is used. If the noxious ferric ions prevent the deposition of copper, violent stirring of the electrolyte due to the rotating cathode ought to intensify the solvent action. But the opposite result is obtained. Our further investigation lead us to believe that the rotation of the cathode reduces the hydrogen film on the cathode to a minimum, thick enough to favor the deposition of copper but not thick enough to bring about the reduction of ferric iron.

Another case showing the importance of the hydrogen film and the effect of catalyzers is the electrodeposition of chromium.

Electrodeposition of Chromium

We have been interested in this subject for a good many years, but the introduction of our process into commercial production occurred but a few years ago. From a physico-chemical point of view, the term "passivity" has been linked with chromium from the very first investigation. Our researches had indicated that chromium does form an oxide film but the important feature was that the coefficient of expansion of this oxide film is practically the same as that of the underlying metal thus serving as an ideal protective coating. The lower oxide once formed is very stable and is reduced back to metal with difficulty. This is quite contrary to the behavior of the oxides of the two metals, tungsten and molybdenum, metals of the same periodic group as chromium. Whereas hydrogen gas will readily reduce oxides of W and Mo at comparatively low temperatures and at low pressures, no appreciable reduction of chromium oxide is obtained below 5 atmosphere hydrogen pressure, and the hydrogen must be very free from oxygen or oxygen compounds.

Accordingly, it was not entirely surprising to find that investigators in the past met with considerable difficulty in the electrolytic reduction and precipitation of chromium metal. However, we have in the case of electrolytic hydrogen, in particular at the very instance of its formation, a comparatively strong reducing agent. Without going into details of our research, which was pursued very much on the same principle as other researches here recorded, the ultimate results were these. Chromium metal can be electrodeposited from solutions of chromic acid providing there is always present on the cathode surface, a hydrogen film of well defined thickness. It must not be too thin otherwise chromium is not deposited nor must it be too thick as otherwise the deposit is dull and comparatively rough. Now we further found that the proper film is readily produced in the presence of certain stable ions in the bath. Curiously enough these ions ordinarily have negative charges. But that these acid radicals are important primarily on account of their reaction in the catholyte and not on account of any reaction in the anolyte was easily demonstrated by splitting the cell and confining the additions to the cathode compartment.

Increasing the temperature of the electrolyte—and thereby

increasing the temperature of the cathode surface—decreases the hydrogen film and ultimately no chromium metal is deposited. However, a rise in temperature can be partly offset by an increase in current density.

Electrodeposition of Nickel

Hydrogen also plays a very important part in the deposition of thick plates of malleable, ductile, nickel. Working with C. P. Madsen we found that the brittleness of electrodeposited nickel, so familiar to everybody who has ever studied the subject, can be entirely eliminated by the reduction of the hydrogen gas evolution or hydrogen concentration to a very low minimum. We followed our work very closely with potential measurements and of the various "depolarizers" tried, hydrogen peroxide was found most satisfactory and is today regularly employed in plants all over the world.

Reduction of Stannic Oxide

Closely related to these cathodic hydrogen cases cited above is the research still in progress on the reduction of SnO_2. Mr. C. L. Mantell of our laboratory has tried out every conceivable aqueous solvent on the natural stannic oxide, so called tin stone or cassiterite. Not one of the solvents or combination of solvents would bring into solution appreciable quantities of tin within a commercially reasonable time.

Now, the old field test for cassiterite is to bring the sample in contact with a small piece of zinc and cover it with a few drops of sulfuric acid. Immediately metallic tin appears. Ordinary hydrogen gas passed into a suspension of stannic oxide of almost colloidal fineness does not reduce it. Apparently it is the so-called nascent hydrogen in the case of the field test for tin stone that brings about the reduction. But our experiments are not yet completed.

Ductility of Metals and Other Researches

Throughout our various investigations, we have always appreciated the advantage of the application of physical and physicochemical methods in place of the purely mechanical or chemical analytical methods so common in most industrial laboratories. In the production of ductile tungsten, the microscope, the resistance thermometer and the Wheatstone bridge were the most important tools. Chemical analytical methods are of little avail when determining the presence of very small amounts of impurities in metals. On the other hand the temperature coefficient of resistance will indicate the presence of the smallest traces.

In the chlorination of zinc blende, a research carried out by E. W. Hale of our laboratory, the success of the process was based on the full realization of the catalytic effect of small amounts of bromine added to the chlorine.

In the research on the restoration of ancient bronzes carried out by C. H. Eldridge of our laboratory, it was the substitution of physico-chemical methods for the old mechanical and chemical methods that brought about results so completely novel and surprising. Objects of art have been restored and the original details of design revealed that could never have been restored by any of the older crude methods.

Conclusion

We have tried to emphasize the great importance in the application of physico-chemical methods to metallurgical research. The importance of the surface film reactions has been evident in many of our researches. The commercial success of the insoluble anode was due to a surface film with high oxidizing powers. In the case of the copper-clad leading-in wire, used in place of platinum as a seal in glass, the immediate solution of the problem was due to the valuable properties of the copper surface film in contact with the glass. In the electrodeposition of chromium and chromium alloys, it was the control of the cathode surface film of hydrogen that transformed a laboratory stunt into an industrial process. In the ductilization of tungsten, it was the study and control of the surface film of the individual tungsten crystals that indicated the proper path of solution.

We have also tried to stress the importance of studying those properties and characteristics of the material under investigation that seem to have no direct bearing on the solution. Many valuable suggestions can be traced right back to so-called "irrelevant data."

REFERENCES

1 Fink. Brit. Patent No. 23,775 (1912); U. S. Patent No. 1,498,908 (1924).
2 Fink and Eldridge, Trans. Am. Electrochem. Soc., 40, 55 (1921).
3 Fink, Industrial Eng. Chem., 16, 566 (1924).
4 Fink. U. S. Patent 1,437,507 (December 5, 1922).
5 Eichrodt, Trans. Am. Electrochem. Soc. 45, 403 (1924).
6 Fink and Pan, Trans. Am. Electrochem. Soc., 49, (1926).
7 J. N. Friend and J. S. Tidmus, J. Inst. Minerals, 31, 183 (1924).
8 A. Nathansohn, Brit. Pat. 187, 195 (1921); D. R. Pat. 366, 483-4-5 (1922).
9 M. Demassieux, Comt. Rendus, 158, 702 (1914).
10 G. Wetzlar, Schweigger's J., 52, 473 (1828); J. L. Proust, J. Phys., 62, 211 (1806); H. St. C. Deville, L'Aluminum, ses propriétés, sa fabrication, et ses applications, Paris, 33 (1859); J. Percy, Silver and Gold, London, I, 67 (1880).
11 G. S. Forbes, J. Am. Chem. Soc., 33, 1973 (1911).

12 A. Grutzner, Die Augustin'sche Silberextraktion, in ihrer Anwendung auf Hüttenprodukte und Erze, Braunschweig (1851).

13 O. C. Ralston, Tainton Process, Trans. Am. Inst. Min. Met. Eng., **70**, 447 (1924); G. Tamann and W. Köster, Zeitsch. Anorg. u. Allgemeine Chem., **123**, 196 (1922).

14 S. Glasstone, J. Chem. Soc., **121**, 2091–8 (1922).

15 Heycock and Neville; Friedrich; G. J. Petrenko; loc. cit.

16 The world's production of copper in 1925 amounted to 1,600,000 tons.

REACTIONS IN LIQUID AMMONIA

BY

EDWARD C. FRANKLIN

Professor of Organic Chemistry

Stanford University

Lecture delivered at Columbia University in the special course on Contemporary Developments in Chemistry given in the Summer Session of 1926 on the occasion of the opening of the Chandler Chemical Laboratories

COLUMBIA UNIVERSITY PRESS

NEW YORK

1927

Printed from type. Published March, 1927

Printed in the United States of America

REACTIONS IN LIQUID AMMONIA

An Experimental Lecture

By Edward C. Franklin

Everyone is familiar with the extremely important part played by water in the processes of nature. Without enumerating in detail the properties which give to this substance its commanding position among chemical compounds, it will be sufficient for present purposes to recall its exceptional power as an electrolytic solvent, to note the fact that by far the larger number of the more familiar and important reactions take place in water solutions, and especially to emphasize the relation existing between water, on the one hand, and the important classes of substances known as bases, acids, salts, on the other. In a very real sense the ordinary oxygen bases, acids, salts, alcohols, ethers, esters, aldehydes, etc., are derivatives of water and as such may be said to constitute a water system of compounds.

Now of all known substances ammonia, in the liquid form, most closely approaches water in all those properties which make for the latter its important place among solvents. Liquid ammonia is an excellent solvent for many salts and a large variety of organic compounds, the former in general, as well as many of the latter, forming solutions which conduct electricity with remarkable facility. Especially important is the existence of compounds which are to be looked upon as constituting an ammonia system of compounds in the same sense that the ordinary oxygen acids, bases, salts, alcohols, aldehydes, etc., may be said to make up a water system.

Liquid Ammonia as a Solvent

A cylinder of commercial liquid ammonia is connected with the vacuum jacketed container, *C*, as shown in Figure 1. *D* is a bottle or carboy half filled with water. Stopcock, *E*, is sealed to *C* in a plane at right angles to the plane of the figure. Liquid ammonia is withdrawn from the cylinder, *A*, in the following manner. When valve, *B*, is opened liquid ammonia runs from the cylinder in which it exists under a pressure of 9 to 10 atmospheres, and collects in the container, *C*, at atmospheric pressure. The ammonia gas formed in cooling the liquid collecting in *C* to its boiling point at atmospheric pressure passes into water in bottle, *D*, where it is absorbed. Any convenient quantity of liquid may thus be transferred from the stock cylinder to the container, *C*. Closing stopcock, *F*, brings about the accumulation of pressure in container, *C*. On opening stopcock, *E*, liquid is forced out and is collected in desirable quantities in four small vacuum jacketed beakers and in a plain test tube. That the liquid boiling in the test tube is cold is shown by the rapid deposition of hoar frost on the outside of the tube and by the downward current of fog-laden air.

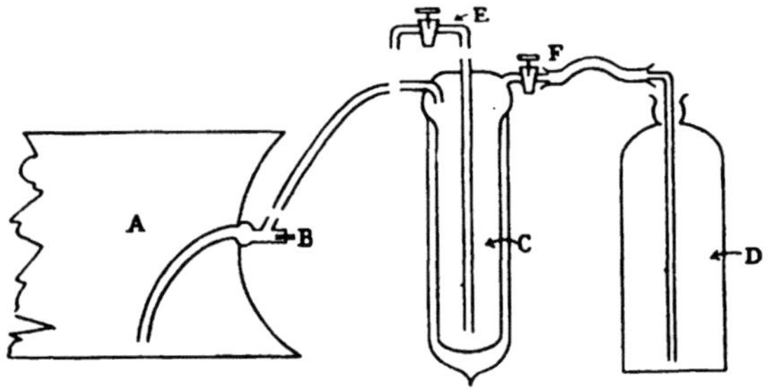

FIG. 1

The following experiments show the remarkable solvent power of liquid ammonia.

Experiment 1. To the four beakers respectively—each half filled with liquid ammonia—cupric nitrate, potassium permanganate, mercuric iodide and fluorescein are added. All three salts as well as the dyestuff dissolve abundantly. The solution in the first beaker is intensely blue, the potassium permanganate solution, in the second beaker, differs not at all in appearance from a water solution of the salt, the solution of the red iodide of mercury is colorless, while the contents of the fourth beaker show the familiar fluorescence of a water solution of potassium fluorescein. Allowing the solvent to evaporate away from the solution of mercuric iodide—the solution being transferred to a plain beaker to accelerate evaporation—the colorless ammonated salt left behind is seen to effloresce to the red anammonous * salt.

Experiment 2. For showing further the properties of liquid ammonia as a solvent sealed glass tubes containing solutions of a number of salts and organic compounds are successively immersed in a refrigerating bath of liquid ammonia as shown in Figure 2.

From solutions of sodium chloride, barium nitrate and silver nitrate beautiful needle-like crystals are seen growing with a speed that is most striking. Silver chloride is slightly soluble, silver bromide is moderately soluble, while silver iodide is extremely soluble at laboratory temperature. On cooling the three solutions we observe the separation of a moderate amount of silver chloride from the one solution, of a much more abundant crop of crystals from the second, while the silver iodide solution is converted into a solid mass of crystals. All of these salts separate from solution with ammonia of crystallization. Specimen tubes containing solutions of benzene and urea respectively show an abundant separation of solute when similarly cooled. A concentrated solution of cane sugar has the consistence of a syrup.

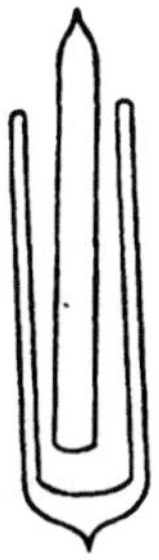

FIG. 2

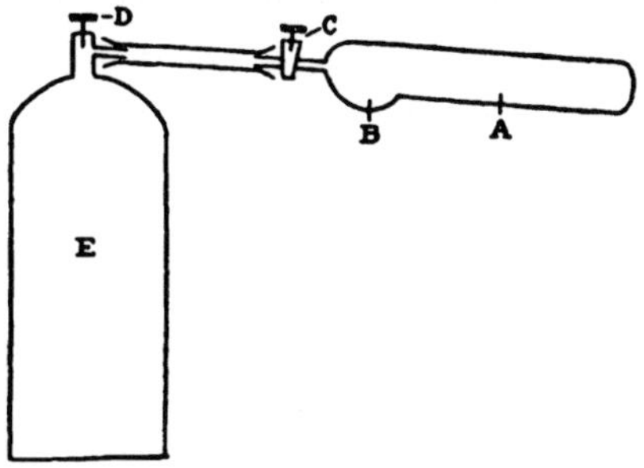

FIG. 3

When cooled in the refrigerating bath the syrup becomes so viscous as scarcely to flow at all when the tube is removed from the bath and inverted. Ethylamine and diethylamine, which are ethyl alcohols of the ammonia system, and triethylamine, an ammono ethyl ether, are miscible with ammonia at laboratory temperature. At —33° the mixture of triethylamine and ammonia separates into two layers.

Experiment 3. Liquid ammonia possesses the remarkable power of dissolving the alkali and alkaline earth metals. From these solutions the metals separate in beautiful crystals, when the solvent is carefully evaporated.

A glass receptacle, *A,* containing a small piece of metallic lithium, *B,* is attached to a small cylinder of liquid ammonia, *E,* as shown in Figure 3. With stopcock, *C,* open, ammonia gas enters the tube, *A,* when valve, *D,* is opened and reacts with the metal to form a fiery yellow liquid. This brilliantly colored liquid is a concentrated solution of metallic lithium in liquid ammonia. By properly tipping the container tube a small quantity of the yellow solution is allowed to run down into the lower end of the tube. Cooling this posterior end of the tube by means of a swab wet with liquid ammonia brings about the dilution of the metal solution and with the dilution a change of color from the brilliant metallic yellow to an intense blue. Disconnecting now the tube, *A*, from the cylinder, *E*, and allowing the ammonia to escape from the former through the stopcock, *C*, the free metal will be left behind. It is interesting to note that these metal solutions conduct the electric current with remarkable facility and that in such solutions, according to Kraus, the metal is ionized as represented by the equation, $M \rightleftharpoons M^+ + e^-$, into the metallic cation and the electron.

The Electrical Conductance of Liquid Ammonia Solutions

It might well be expected that a substance which resembles water so closely in its physical properties and which shows such conspicuous power as a solvent for salts, would form solutions with marked capacity for carrying the electric current. That liquid ammonia is in fact an excellent electrolytic solvent is shown by the following experiment.

Experiment 4. The piece of apparatus sketched in Figure 4 is used for showing first, the high resistance of pure liquid ammonia and second, the very striking effect on the conductance of the liquid produced by the introduction therein of a minute quantity of a salt solute.

Liquid ammonia runs from the steel cylinder, *B*, of highly purified ammonia into the conductivity cell, *A*, when the valve, *D*, is partly opened. The gaseous ammonia formed by the volatilization of a portion of the liquid during the process of cooling is taken care of by absorption in water contained in flask, *D*. The two platinum electrodes, close together [1] and of large area, are connected in series with an incandescent lamp to the terminals of an ordinary 110-volt lighting circuit. The high resistance of the pure liquid is shown by the failure of the relatively high difference of potential to force the passage of sufficient current to heat the filament of the lamp even to the faintest luminosity. As a matter of fact the specific conductivity of pure liquid ammonia is far below that of the purest water.

When a minute quantity * of a soluble salt is introduced into the cell through the tubulure, *C*,—the stopper being removed momentarily for the purpose—the very dilute solution thus formed shows its conducting power by permitting sufficient cur-

[1] Much closer than shown in the figure.
* In our experiment 0.12 mg. NH_4NO_3 was dissolved in 20 cc. of solvent.

rent to pass to render the lamp filament distinctly luminous. When a much larger quantity of solute is now added to the solution the resistance of the cell becomes so low as to constitute but a negligible element in the resistance of the circuit as shown by the fact that short-circuiting the cell produces a scarcely perceptible change in the luminosity of the lamp.

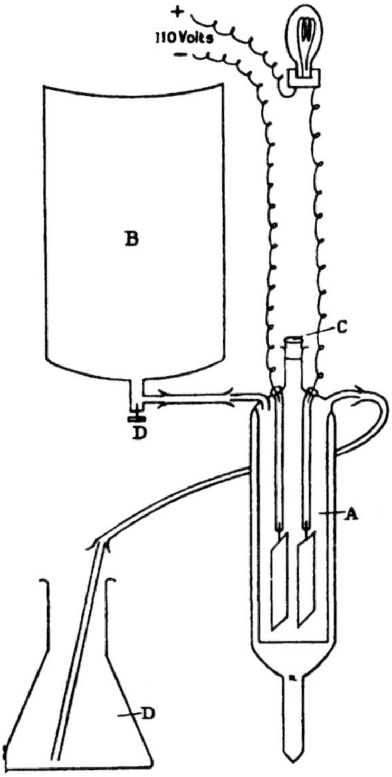

FIG. 4

The molecular conductivity values for potassium nitrate in water solution at 18° and in ammonia solution at —33° respectively are summarized in Figure 5 in which the ordinates represent molecular conductivities in Kohlrausch units and the abscissas the logarithms of the dilutions.

Metathetic Reactions in Liquid Ammonia Solutions

If the facility with which metathetic reactions take place in water solutions is dependent upon the ionized state of the reacting substances then—since the preceding experiment shows that salts are likewise ionized in liquid ammonia solutions—we should expect reactions to take place in this solvent with all the ease which characterizes reactions in aqueous solutions. That such in fact is the case is shown by the following experiments.

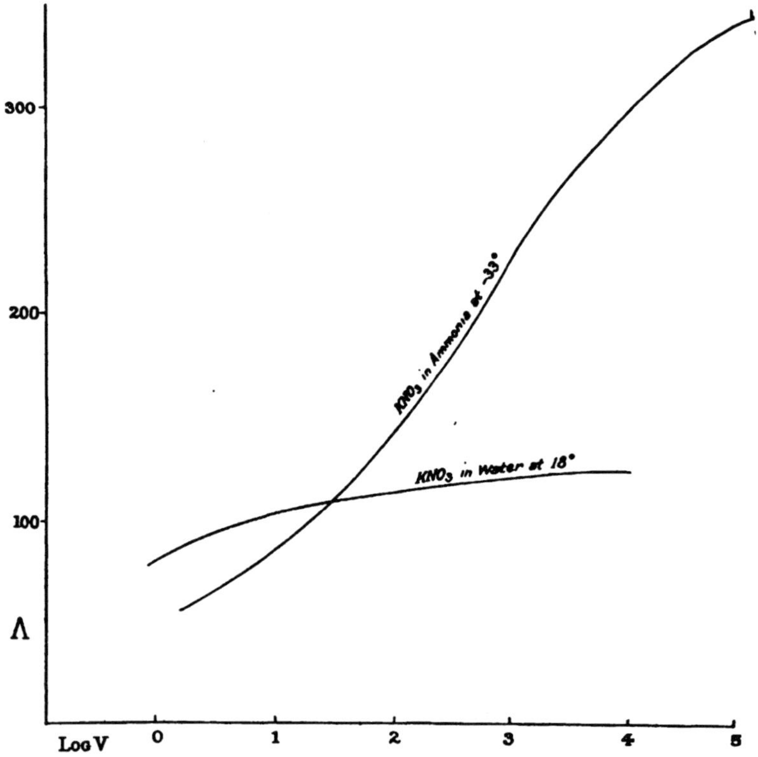

FIG. 5

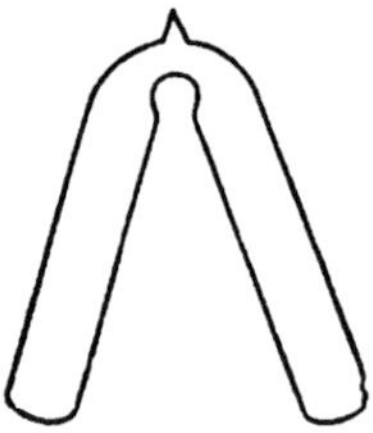

FIG. 6

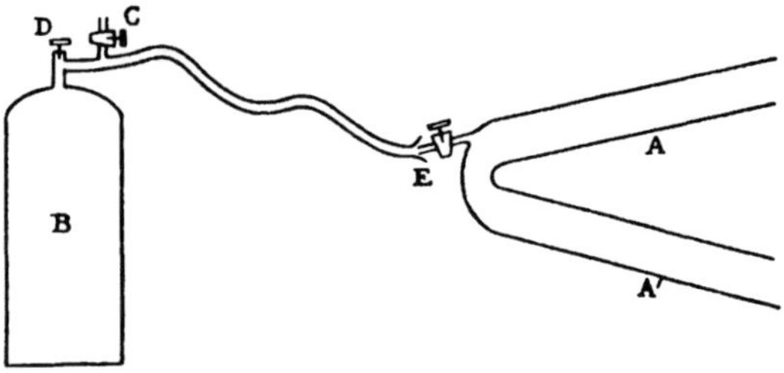

FIG. 7

Reactions Between Salts

Experiment 5. In the one leg of the reaction tube shown in Figure 6, is contained a liquid ammonia solution of barium nitrate, in the other a similar solution of sodium chloride.* Upon pouring a portion of the sodium chloride solution into the solution of barium nitrate—or vice versa—a precipitate of barium chloride, $BaCl_2 \cdot 8NH_3$, is formed. The reaction is represented by the equation, $2NaCl + Ba(NO_3)_2 = BaCl_2 + 2NaNO_3$. Similarly when a saturated solution of silver chloride is added to a solution of barium nitrate a precipitate of barium chloride is formed.

The Action of Acids on Bases and Metals

At first thought it may seem absurd to speak of acids existing in solution in liquid ammonia and reacting therein with bases. Ammonia is frequently spoken of as a base and everyone knows that acids unite directly with ammonia to form the familiar ammonium salts. The truth is, however, that ammonia is not a base. It is only in the presence of water that ammonia exhibits basic properties. The water-free liquid behaves quite as indifferently as water itself toward indicators. Now it is an interesting fact that many ammonium salts dissolve freely in liquid ammonia to form solutions which exhibit certain truly acid properties.

Experiment 6. A two-legged reaction tube, Figure 6, contains in the one leg a solution of ammonium iodide (any other soluble ammonium salt may be used) in the other a few small lumps of sodium hydroxide covered with a very dilute solution of phenolphthalein. The red color of the phenolphthalein solution is due to the alkaline action of the all but insoluble sodium hydroxide. When a few drops of the ammonium iodide solution is poured into the alkaline solution, the red color of the latter is discharged instantly. Shaking the reaction tube gradually

* Such reaction tubes were prepared before the lecture in the following manner. The reaction tube, *AA′*, is connected to a steel cylinder, *B*, Figure 7, containing liquid ammonia, by means of a small lead tube soldered to the nipple at *C* and attached to the reaction tube at *E* by means of sealing wax. With the exit of the one leg, say *A*, closed by means of a stopper, the other is heated with a bunsen burner while a slow current of ammonia from the cylinder, *B*, is passing through it. Then transferring the stopper to the second leg, the first is similarly heated. The drying of the tube is thus thoroughly accomplished. With a current of ammonia passing through *A*, a proper quantity of solute is introduced by means of a long-stemmed platinum spoon. By means of a hand blowpipe a glass rod is attached to the end of leg, *A*, at a point sufficiently removed from the open end to avoid the entrance of water, following which the tube is closed. The necessary gentle pressure required for blowing a symmetrical end being obtained by closing momentarily the exit *A′*—the stopper having been removed from *A′* just prior to the final closing of *A*—by means of the finger. Solute is then placed in leg, *A′*, which is closed after the manner just described. The pressure required for "blowing" the closed end being brought into existence by placing the finger over the exit at *C*, stopcock, *C*, having been opened to the atmosphere just before finally closing the exit at *A′*. Closing now stopcock, *C*, both legs of the reaction tube are immersed in a bath of ice and water. With stopcock, *E*, and valve, *D*, open, ammonia distils from the stock cylinder into the reaction tube. When a sufficient quantity of ammonia has distilled over the two legs of the reaction tube are cooled until the pressure within the apparatus is reduced to that of the atmosphere following which stopcock *E* is sealed off. The final result is a piece of apparatus, in essence two test tubes, shown in Figure 5, containing the two solutions which are to be mixed during the lecture.

restores the red color to the solution. The explanation of the phenomena which have thus been observed is fairly obvious. Sodium hydroxide is sufficiently soluble in liquid ammonia to give an alkaline reaction with the indicator. Addition of ammonium iodide (hydriodic acid) neutralizes the sodium hydroxide in solution in accordance with the equation,

$$NaOH + NH_3{\cdot}HI = NaI + H_2O + NH_3,$$

thus discharging the red color of the indicator. On shaking the tube the excess of acid in contact with solid sodium hydroxide is rapidly neutralized, the products of the interaction of acid and base, sodium iodide and water, go into solution, and soon sufficient sodium hydroxide dissolves to restore the red color characteristic of alkaline solutions of phenolphthalein.

Just as many metals react with acids in water solutions so certain metals are attacked by liquid ammonia solutions of ammonium salts with the evolution of hydrogen and the formation of the corresponding metallic salts.

Experiment 7. In the one branch of a reaction tube is contained a spiral of clean magnesium ribbon, in the other a liquid ammonia solution of ammonium nitrate. When the acid solution is poured upon the metal the evolution of gas begins immediately and after a time the metal has gone completely into solution. The gas evolved is easily shown to be hydrogen while the other product of the reaction is magnesium nitrate which is formed in accordance with the equation,

$$Mg + 2NH_4NO_3 = Mg(NO_3)_2 + 2NH_3 + H_2$$

The salt crystallizes from solution with six molecules of ammonia of crystallization. $Mg(NO_3)_2{\cdot}6NH_3$. Other metals such as sodium, calcium, beryllium, aluminum and iron react in a similar manner with liquid ammonia solutions of ammonium salts.

Bases, Acids and Salts of the Ammonia System

Compounds which are related to ammonia as the familiar bases are related to water are the metallic amides and imides, while the metallic nitrides are basic nitrides corresponding to the basic oxides of the water system. Two ammono bases are potassium amide, KNH_2, and lead imide, $Pb{=}NH$. An example of a basic nitride is cuprous nitride, Cu_3N. Amides of the electro negative or acid-forming elements, together with their partially deammonated products, are acids of the ammonia system while the final deammonation products, the non-metallic or acid nitrides, as the analogs of the acid anhydrides, are acid anammonides. Examples of acids belonging to the ammonia system are cyanamide,

H_2NCN, dicyanimide, $HN(CN)_2$, hydrocyanic acid, HNC, hydrazoic acid, $H-N{=}N{\equiv}N$, acetamidine, $CH_3C(NH)NH_2$, and monochloramine, H_2NCl. The first two compounds are ammono carbonic acids, the third is an ammono carbonous acid, the fourth an ammono nitric acid, the fifth, an ammono acetic acid and the last, an ammono hypochlorous acid.

Several mixed acids derived simultaneously from water and ammonia, are cyanic acid and urea, acetamide and benzenesulfonamide. The first two compounds are aquo ammono carbonic acids, the third is an aquo ammono acetic acid, the last an aquo ammono benzenesulfonic acid.

Proceeding now on the assumption that the relations thus indicated are real then, in liquid ammonia solutions, ammono acids must react with ammono bases and with the more electropositive metals to form ammono salts. Similarly aquo ammono salts should be formed by the action of aquo ammono acids on ammono bases and on metals.

Four experimentally established reactions are represented by the equations,

$$(1)\quad Mg + 2H_2NCN = Mg(HNCN)_2 + H_2$$

$$(2)\quad C_6H_5C\begin{matrix}\nearrow NH \\ \searrow NH_2\end{matrix} + KNH_2 = C_6H_5C\begin{matrix}\nearrow NH \\ \searrow NHK\end{matrix} + NH_3$$

$$(3)\quad C_6H_5C\begin{matrix}\nearrow O \\ \searrow NH_2\end{matrix} + KNH_2 = C_6H_5C\begin{matrix}\nearrow O \\ \searrow NHK\end{matrix} + NH_3$$

$$(4)\quad 2C_6H_5SO_2NH_2 + Zn = (C_6H_5SO_2NH)_2Zn + H_2,$$

which may be read as representing, (1) the action of an ammono carbonic acid on magnesium to form an acid magnesium ammono carbonate, (2 and 3) the action of potassium amide, an ammono base, on an ammono benzoic acid and an aquo ammono benzoic acid to form respectively a potassium ammono benzoate and a potassium aquo ammono benzoate, and (4) the action of aquo ammono benzene sulfonic acid on zinc to form a zinc salt.

Experiment 8. A solution of cyanamide contained in the one leg of a reaction tube is poured upon a spiral of metallic magnesium in the other. Evolution of hydrogen begins immediately. After a time crystals of ammonated magnesium cyanamide begin to separate from the solution.

Experiment 9. In the two legs of a reaction tube are contained respectively solutions of benzamidine and potassium amide. On bringing together the fairly concentrated solutions of the acid and the base a crop of well formed crystals of the salt separates.

Experiment 10. In another reaction tube a crystalline precipitate of monopotassium benzamide is produced when a potassium amide solution is poured into a solution of benzamide. When reciprocally the latter solution is poured into an excess of the former a precipitate of the dipotassium salt is formed.

Bases and acids of the ammonia system, again of course in liquid ammonia solution, must bring about the same reciprocal color changes of indicators which are so familiar a characteristic of the action of bases and acids in aqueous solutions. The following experiment illustrates this statement in a striking manner.

Experiment 11. A liquid ammonia solution of triphenylmethane contained in the middle branch of the three-legged reaction tube shown in Figure 8 is entirely without color. A single drop

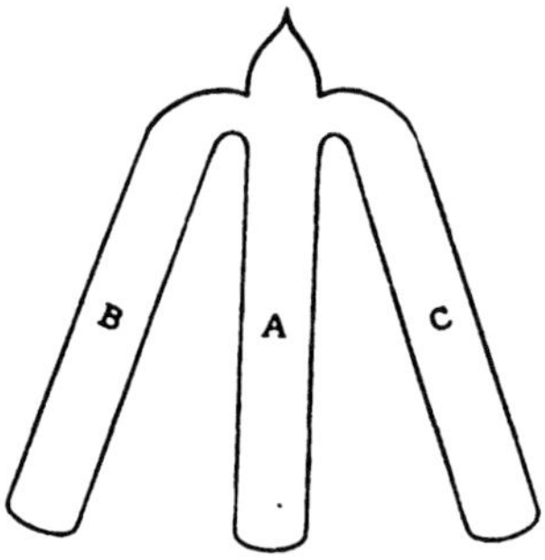

FIG. 8

of a dilute solution of potassium amide poured from the one of the side tubes into the triphenylmethane solution develops instantly a characteristic orange red color. Addition of ammonium azide * solution from the second side tube neutralizes the base present and discharges the red color quite as sharply as an ordinary acid in water solution destroys the color of an alkaline aqueous solution of phenolphthalein.

Bases of the Ammonia System

Unlike the great majority of ammono acids the ammono bases are incapable of existence in the presence of water for which reason but few representatives of this class of compounds have been known in the past. Potassium amide and sodium amide, which were first prepared by Gay Lussac and Thénard by the action of ammonia gas on the metals at elevated temperatures, and zinc amide, prepared by Frankland by the action of ammonia on zinc ethyl in ethereal solutions, are practically the only compounds of their kind recorded in the older literature.

* Hydrazoic acid, H—N=N≡N, is the ammonia analog of ordinary nitric acid. It is an ammono nitric acid.

The amides of potassium, sodium, lithium and calcium are conveniently made by the action of liquid ammonia on the respective metals in the presence of platinum black or metallic iron which act as catalysts.

Experiment 12. In the two legs of a reaction tube are contained respectively a solution of metallic potassium and pure ammonia along with a small quantity of platinum black. On pouring a portion of the blue solution of the metal into the second leg of the reaction tube the color is observed to disappear rapidly. The reaction involved, which is represented by the equation, $K + NH_3 = KNH_2 + H$, is strictly analogous to that which takes place when potassium is brought into contact with water—excepting that the speed of the latter reaction is incomparably greater.

Experiment 13. A tube containing a concentrated solution of potassium amide was placed in an ice bath at the beginning of the lecture. Removing the tube from the bath it is seen to be filled with beautiful crystals of the ammono base.

Other ammono bases may be prepared by reactions in liquid ammonia which are analogous to those whereby the insoluble metallic hydroxides are formed in water solution. Moreover just as in aqueous solutions potassium hydroxide precipitates sometimes the hydroxide and at other times the oxide of the heavy metals, so when potassium amide is used as the precipitant in liquid ammonia solutions one may expect in some instances the precipitation of a metallic amide, in others of an imide, while in still other instances a metallic nitride will be formed.

In the following experiments the formation of representatives of these three classes of compounds is demonstrated.

Experiment 14. In the two legs of a reaction tube are contained respectively liquid ammonia solutions of silver nitrate and potassium amide. On pouring a portion of the latter into the former a white precipitate of silver amide is formed in accordance with the equation,

$$AgNO_3 + KNH_2 = AgNH_2 + KNO_3.$$

Experiment 15. When in a similar tube a solution of potassium amide is run into a solution of lead iodide a precipitate of lead imide is formed.

$$PbI_2 + 2KNH_2 = Pb = NH + 2KI + NH_3.$$

Experiment 16. Potassium amide solution and thallous nitrate contained in a reaction tube are brought together. A black precipitate of thallous nitride is formed as represented by the equation,

$$3TlNO_3 + 3KNH_2 = Tl_3N + 3KNO_3 + 2NH_3.$$

The reactions which have thus been demonstrated are entirely analogous to those which take place when potassium hydroxide and metallic salts are brought together in water solution. In the latter class of reactions a metallic hydroxide or its dehydration product, a metallic oxide is formed, in the other a metallic amide or the one or the other of its deammonation products, a metallic imide or nitride.

Ammonolysis and Ammonobasic Salts

It is interesting to find that the resemblances between water and ammonia extend to reactions in the latter solvent which are strictly analogous to hydrolytic reactions in the more familiar solvent. When mercuric chloride, for example, is dissolved in water the solution formed shows an acid reaction which is due to slight hydrolysis of the salt. Without perpetrating serious error the reaction involved may be represented by the equation,

$$HgCl_2 + H_2O \rightleftharpoons HOHgCl + HCl.$$

Experiment 17. Similarly when mercuric chloride is treated with liquid ammonia, a white precipitate of ammonobasic mercuric chloride is formed in accordance with the equation,

$$HgCl_2 + NH_3 \rightleftharpoons NH_2HgCl\,^{*} + HCl.$$

Experiment 18. In the branches of a three-legged reaction tube, such as is shown in Figure 8, are contained in legs, *B, C,* and *A,* respectively, a solution of ammonium iodide, a solution of potassium amide and dry mercuric iodide. At the pressure prevailing within the apparatus ammonated mercuric iodide is stable. One consequently observes the white color of the addition product, $HgI_2{\cdot}2NH_3$, instead of the red of the anammonous (ammonia free) salt. When now ammonia is distilled upon the salt from leg, *B*, or *C*, it is seen to dissolve rapidly, but not completely. An insoluble, yellow, ammonobasic salt remains which is formed in accordance with the equation,

$$2HgI_2 + 4NH_3 = Hg = N\text{–}HgI + 3NH_4I.$$

Addition of potassium amide from leg, *C*, increases the amount of the basic precipitate. That is to say, equilibrium is shifted toward the right as the result of the neutralization of acid by potassium amide, $NH_4I + KNH_2 = KI + 2NH_3$. Addition of potassium amide in amount over and above that necessary

* It is interesting to note in passing that this ammonobasic salt is identical with the long known infusible white precipitate.

for the precipitation of all the mercury as basic salt gradually converts the latter into mercuric nitride,

$$3Hg = NH\text{-}gI + 3KNH_2 = 2Hg_3N_2 + 3KI + 2NH_3.$$

The transition is accompanied by a change of color from yellow to brown and can therefore be readily followed ocularly. Addition of ammonium iodide solution, from leg, *B*, causes the reaction represented by the equation above to proceed from right to left with the ultimate result that the precipitate dissolves to a clear, colorless solution of mercuric iodide.

Experiment 19. In the one leg of a reaction tube, such as is shown in Figure 6, is contained a concentrated solution of stannic iodide made slightly acid by the presence of a small quantity of ammonium iodide in order to prevent the formation of an ammonobasic salt. On pouring pure ammonia, from the second leg of the tube, carefully upon the surface of the solution of stannic iodide, a white precipitate of an ammonobasic iodide of tin is formed occupying a zone between the concentrated stannic iodide solution and the layer of pure solvent. On shaking the apparatus the precipitate goes into solution. Again pouring pure solvent upon the surface of the solution the precipitate reappears to disappear again when mixed with the concentrated solution. The precipitate may be produced and caused to disappear several times in the manner just described. After the solution reaches a certain degree of dilution the precipitate becomes permanent.

Amphoteric Metallic Amides and Imides

If now we return to the reaction tubes in which the precipitation of silver amide, lead imide, and thallium nitride was shown we find that the precipitates dissolve when treated with an excess of potassium amide. This behavior recalls the familiar solubility of the hydroxides of lead, aluminum and zinc in aqueous potassium hydroxide and suggests the question as to whether ammonia analogs of potassium plumbite, potassium aluminate and potassium zincate may be prepared. As a matter of fact a very considerable list of such compounds has been prepared.

Experiment 20. In the two legs of a reaction tube are contained respectively, one equivalent of silver iodide and two equivalents of potassium amide. Gradually pouring the potassium amide solution into the solution of silver iodide brings about the formation of a precipitate of silver amide which is observed to increase in quantity until all the silver is removed from the solution and then to dissolve in excess of potassium amide. The resulting clear, colorless solution is then cooled to a low temperature when a crop of beautifully formed crystals of potassium ammono argentate is deposited from the solution. The reactions involved are represented by the equations,

$$AgI + KNH_2 = AgNH_2 + KI$$
$$AgNH_2 + KNH_2 = AgNHK + NH_3.$$

Experiment 21. Similarly a crop of yellow crystals of potassium ammono thallite is obtained when the black thallium nitride, formed by the addition of potassium amide to a solution of thallous nitrate, is dissolved in excess of potassium amide and the solution thus formed is cooled to a low temperature. The reactions involved in this experiment are represented by the equations,

$$3TlNO_3 + 3KNH_2 = Tl_3N + 3KNO_3 + 2NH_3$$
$$Tl_3N + 6KNH_2 = 3TlNK_2 + 4NH_3.$$

While the many analogies which have been shown to exist between water and ammonia and many of their respective derivatives have prepared us for the existence of such compounds as potassium ammono zincate, potassium ammono aluminate and potassium ammono plumbite they would hardly lead us to predict a potassium ammono argentate or a potassium ammono thallite. Nor does the formation of these compounds exhaust the possibilities in this direction for it has been shown that the metallic amides in general, including the amides of the strong base forming metals, barium, strontium, calcium and even lithium and sodium, react with potassium amide in solution in liquid ammonia to form compounds of the same type as that to which the plumbites, aluminates and zincates belong.

Experiment 22. The two legs of a reaction tube contain liquid ammonia solutions of barium nitrate and potassium amide respectively. When the barium nitrate solution is run into the solution of potassium amide the precipitate which appears is formed in accordance with the equation,

$$Ba(NO_3)_2 + 3KNH_2 = BaNK + 2KNO_3 + 2NH_3.$$

The compound has the composition represented by the formula, $BaNK \cdot 2NH_3$ (or $Ba(NH_2)_2 \cdot 2KNH_2$) and has been named potassium ammono bariate.

Experiment 23. In a similar manner a solution of sodium iodide contained in the one leg of a reaction tube when run into a solution of potassium amide gives a crystalline precipitate of potassium ammono sodiate in accordance with a reaction represented by the equation,

$$NaI + 3KNH_2 = NaNK_2 \cdot 2NH_3 \text{ (or } NaNH_2 \cdot 2KNH_2) + KI.$$

Experiment 24. An experiment in which is shown the formation, successively, of a basic lead iodide, lead imide and potassium ammono plumbite by the action of potassium amide on lead iodide and the inverse conversion of potassium ammono plumbite through lead imide and ammonobasic lead iodide to normal lead iodide by the action of ammonium iodide is conducted as follows:

In the two side branches of a three-legged reaction tube, such as is shown in Figure 8, are contained solutions respectively of potassium amide and ammonium iodide. The middle leg contains a solution of lead iodide. The contents of the middle leg are seen to be milky in appearance, a condition which is due to the presence of a small quantity of an ammonobasic iodide of lead formed by the ammonolytic action of the solvent on normal lead iodide. The addition of a few drops of ammonium iodide solution clears up the cloudy solution instantly. When now the slight excess of acid is neutralized by the addition of a small quantity of potassium amide solution the white ammonobasic salt reappears. Further addition of the base brings about the precipitation of additional quantities of the ammonobasic salt. After the lead iodide is completely removed from solution as ammonobasic iodide then additional quantities of potassium amide gradually change the white basic salt into red lead imide, the change from the pure white of the basic salt through varying shades of red to the pure red of the ammono base being plainly visible. Further addition of potassium amide now gradually dissolves the lead imide to a clear, colorless solution of potassium ammono plumbite. The order of the phenomena just described may now be reversed by the gradual addition of ammonium iodide solution from leg three to the contents of the middle leg. The precipitation of lead imide is first observed, then this precipitate is changed into the white basic salt and finally the latter dissolves to a clear, colorless solution of lead iodide. In addition to lead iodide the solution in the middle leg now contains potassium iodide in an amount equivalent to the potassium amide and ammonium iodide used. The reactions involved in the above described transformations carried on in the one direction may be represented by the equations,

$$2PbI_2 + 2NH_3 \rightarrow NH_2\text{–}Pb\text{–}NH\text{–}PbI + 3HI$$

$$H_2N\text{–}Pb\text{–}NH\text{–}Pb\text{–}I + KNH_2 \rightarrow 2Pb{=}NH + KI + NH_3$$

$$Pb{=}NH + KNH_2 \rightarrow Pb{=}NK + NH_3,$$

in the reverse direction by the equations,

$$Pb{=}NK + HI \rightarrow Pb{=}NH + KI$$

$$2Pb{=}NH + HI \rightarrow NH_2\text{–}Pb\text{–}NH\text{–}PbI$$

$$H_2N\text{–}Pb\text{–}NH\text{–}Pb\text{–}I + 3HI \rightarrow 2PbI_2 + 2NH_3.$$

Alcohols, Aldehydes, Carboxylic Acids of the Ammonia System

Having acquainted ourselves with the existence of derivatives of ammonia which show all the properties respectively of bases, acids and salts we come to enquire whether among the many organic nitrogen compounds there may be such as are to be looked upon as ammonia analogs of the alcohols, aldehydes, ketones, the carboxylic acids, etc.

Methylamine and dimethylamine, which are related to ammonia as methyl alcohol is related to water, are formerly methyl alcohols of the ammonia system, while trimethyl amine is clearly an ammono ether. The methods of preparation and the general and specific properties of these compounds are in harmony with this view.

If methylamine is in reality an ammono methyl alcohol then just as ordinary methyl alcohol may be oxidized successively to formaldehyde, formic acid and carbon dioxide, as represented by the familiar scheme,

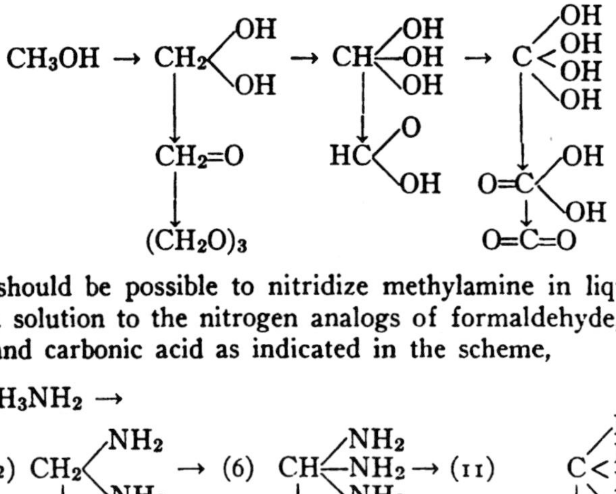

so it should be possible to nitridize methylamine in liquid ammonia solution to the nitrogen analogs of formaldehyde, formic acid and carbonic acid as indicated in the scheme,

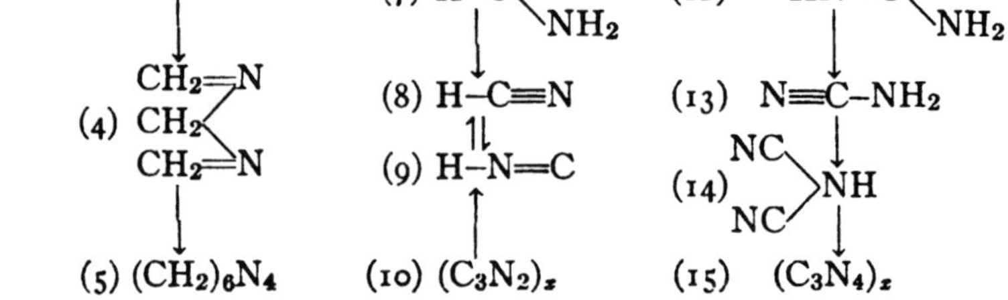

The compounds thus formulated are (1) an ammono methyl alcohol, (2) the ammonia analog of methylene glycol, (3) an aldehyde-alcohol, (4) in so far as the two $CH_2=N^{\cdot -}$ groups are concerned, a formaldehyde, (5) hexamethylenetetramine a dimer of ammono formaldehyde, (6) ortho ammono formic acid, (7) the ammonia analog of ordinary formic acid, (8) formic anammonide, (9) ammono carbonous acid, (10) carbonous anammonide,* (11) ortho ammono carbonic acid, (12, 13, 14) ammono carbonic acids and (15) carbonic anammonide.

In view of the limited time at our disposal for a discussion of the organic chemistry of nitrogen compounds, we may first say that observations in wide variety are on record in the chemical literature giving convincing support to the general statement that the formal relations involved in the above table are real and then proceed to the consideration of a limited number of significant reactions most of which have been accomplished in the Stanford laboratory.

Ammono Alcohols

Methylamine, dimethylamine and benzylamine, are respectively two ammono methyl alcohols and an ammono benzyl alcohol. As such they should be susceptible to the action of nitridizing agents just as their oxygen analogs yield to the action of oxidizing agents. As a matter of fact guanidine is formed when a liquid ammonia solution containing methylamine and hydrazoic acid (ammono nitric acid) is heated for a time at 100°, and an abundant yield of cyanamide is obtained when, at laboratory temperature, iodine and methylamine are brought together in liquid ammonia solution. Similarly benzamidine and phenylcyanide are formed by the action of iodine on benzylamine, again of course in liquid ammonia solution. The reactions involved are represented by the equations,

$$(1)\quad CH_3NH_2 + 3HNNN = HNC(NH_2)_2 + NH_3 + 3N_2$$

$$(2)\quad CH_3NH_2 + 3NH_2I = NCNH_2 + 2NH_3 + 3HI$$

$$(3)\quad C_6H_5CH_2NH_2 + 2NH_2I = C_6H_5C(NH)NH_2 + NH_3 + 2HI$$

$$(4)\quad C_6H_5CH_2NH_2 + 2NH_2I = C_6H_5CN + 2NH_3 + 2HI,$$

and may be said to consist in the nitridation, (1, 2) of an ammono methyl alcohol into the one or the other of the two ammono carbonic acids and (3, 4) of an ammono benzyl alcohol into a mixture of an ammono benzoic acid and benzoic anammonide.

* The nitrogen analog of the acid anhydrides have been named acid anammonides.

Ammono Aldehydes

Methylene nitride, which is unknown, and hydrobenzamide, $(C_6H_5CH)_3N_2$, the nitrogen analogs of methylene oxide, $CH_2{=}O$, and benzal (or benzylidene) oxide, $C_6H_5CH{=}O$, are respectively an ammono formaldehyde and an ammono benzaldehyde. The only known compound among those represented by the formulas, (2), (3), (4) and (5) given in the table above is hexamethylenetetramine which is formed by the dimerization of the ammono formaldehyde, (4), just as trioxymethylene is formed by the trimerization of aquo formaldehyde. The closely parallel polymerization reactions are represented by the schemes,

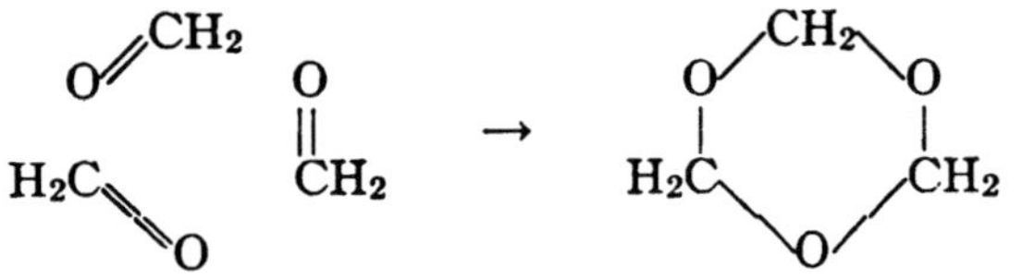

and

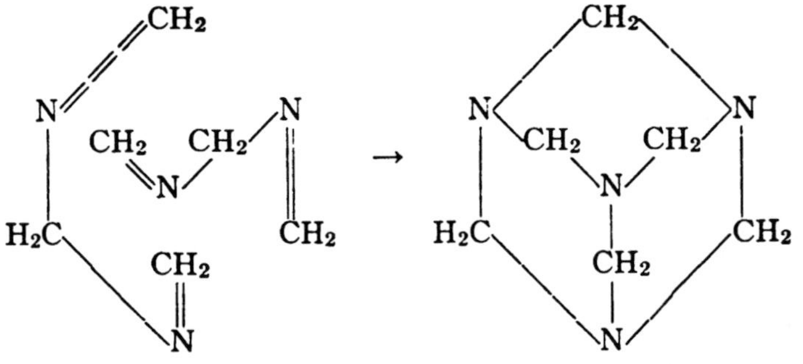

In the one case three molecules of ordinary formaldehyde coalesce to the familiar trioxymethylene ring, in the other two molecules of ammono formaldehyde react with each other in a strictly analogous manner giving Duden and Scharff's stereo-formula of which the above is a plane projection.

By the action of hydrazoic acid in liquid ammonia solution at 100° this dimer of ammono formaldehyde has been nitridized to guanidine.

As is well known hydrobenzamide, $\begin{matrix} C_6H_5CH{=}N \\ C_6H_5CH{=}N \end{matrix}\!\!>CHC_6H_5$,*

* It is to be noted that true aldehydic properties are carried by the two $C_6H_5CH{=}N{-}$ groups. The group, $C_6H_5CH{<}^{N=}_{N=}$, is, approximately at least, the nitrogen analog of $C_6H_5CH{<}^{O-}_{O-}$.

is formed by the action of ammonia on benzaldehyde, which is to say that an ammono benzaldehyde is formed by the ammonolysis of aquo benzaldehyde.

Among the large number of observations on record which show the aldehydic properties of hydrobenzamide, the following may be here recalled. Reducing agents convert hydrobenzamide into a mixture of benzylamine and dibenzylamine both of which are primary ammono alcohols. In liquid ammonia solution it has been nitridized to ammono benzoic acid. In its reactions with alcohols, amines, hydrocyanic acid, hydroxylamine, sulfurous acid, malonic acid and Grignard's reagent it follows, as closely as the trivalency of nitrogen as compared with the divalency of oxygen permits, the well known behavior of ordinary benzaldehyde toward the same reagents. Like aquo benzaldehyde it also undergoes the benzoin condensation. Simply on heating hydrobenzamide is converted into amarine in accordance with the scheme,

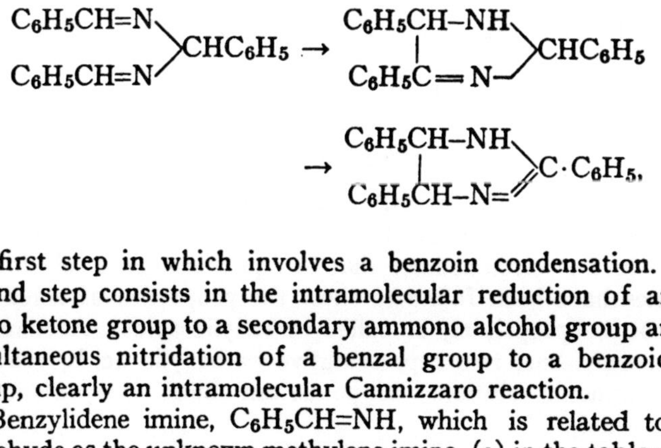

the first step in which involves a benzoin condensation. The second step consists in the intramolecular reduction of an ammono ketone group to a secondary ammono alcohol group and the simultaneous nitridation of a benzal group to a benzoic acid group, clearly an intramolecular Cannizzaro reaction.

Benzylidene imine, $C_6H_5CH{=}NH$, which is related to benzaldehyde as the unknown methylene imine, (3) in the table above, is related to formaldehyde, has been obtained by the action of liquid ammonia on hydrobenzamide in accordance with the equation, $(C_6H_5CH)_3N_2 + NH_3 = 3C_6H_5CH{=}NH$. Working with liquid ammonia as solvent it has been found that this compound may be nitridized to an ammono benzoic acid, namely to benzamidine, that it reacts with hydroxylamine in accordance with the equation,

$$C_6H_5CH{=}NH + H_2NOH = C_6H_5CH{=}NOH + NH_3,$$

to form benzaldoxime and that it undergoes the Cannizzaro reaction, forming equimolecular quantities of the ammono benzyl alcohol, $C_6H_5CH_2NH_2$, and the ammono benzoic acid, $C_6H_5C\begin{smallmatrix}\nearrow NH \\ \searrow NH_2\end{smallmatrix}$ The nitridation of the one molecule of benzylidene imine and the simultaneous reduction of the other is brought about, it may be assumed, by the interchange of a hydrogen atom and an amino group as represented by the equation,

$$C_6H_5CH\begin{matrix} NH_2 \\ NH_2 \end{matrix} + C_6H_5CH{=}NH = C_6H_5CH_2NH_2 + C_6H_5C\begin{matrix} {=}NH \\ NH_2 \end{matrix},$$

it being fair to assume that, potentially at least, the molecular species $C_6H_5CH(NH_2)_2$ is present in the solution.

Clearly benzylidene imine has the properties of an aldehyde. At the same time it behaves as an acid (or an alcohol) to the extent that it reacts with potassium amide in liquid ammonia solution in accordance with the equation,

$$C_6H_5CH{=}NH + KNH_2 = C_6H_5CH{-}NK + NH_3,$$

to form a potassium salt.

Carbazylic Acids

The amidines are formally the nitrogen analogs of the carboxylic acids. In view of this relationship and in consideration of the further circumstance that in liquid ammonia solution the amidines have been found to show the properties of acids they may not inappropriately be called carbazylic acids. Acetamidine and benzamidine, two representative carbazylic acids, react, in liquid ammonia solution, with potassium amide as represented by the equation,

$$RC\begin{matrix} {=}NH \\ NH_2 \end{matrix} + KNH_2 = RC\begin{matrix} {=}NH \\ NHK \end{matrix} + NH_3,$$

to form respectively a potassium ammono acetate and a potassium ammono benzoate. Both salts separate beautifully crystallized from cold, concentrated solutions.

Salts of carbazylic acids are also formed by the direct addition of metallic amides to nitriles. For example, ethyl cyanide, which is the anammonide of ammono propionic acid, reacts with potassium amide, an ammono base, to form potassium ammono propionate just as propionic anhydride reacts with potassium hydroxide to form potassium aquo propionate. The parallel reactions are represented by the equations,

$$C_2H_5CN + KNH_2 = C_2H_5C\begin{matrix}\diagup NH \\ \diagdown NHK\end{matrix}$$

$$(C_2H_5CO)_2O + 2KOH = 2C_2H_5COOK + H_2O.$$

The one reaction takes place in liquid ammonia solution, the other in water.

The close relationship existing between the carboxylic acids and the carbazylic acids is further shown by such reactions as are represented by the equations,

$$CH_3C\begin{matrix}\diagup O \\ \diagdown OK\end{matrix} + KOH = CH_4 + K_2CO_3$$

$$CH_3C\begin{matrix}\diagup NH \\ \diagdown NHK\end{matrix} + KNH_2 = CH_4 + K_2CN_2 + NH_3.$$

Just as potassium aquo acetate when heated with potassium hydroxide yields methane and potassium aquo carbonate so when a mixture of potassium ammono acetate and potassium amide is heated methane and potassium ammono carbonate are formed.

Potassium acetamide, CH_3CONHK or $CH_3C(OK)NH$, which is the potassium salt of a mixed aquo ammono acetic acid, on heating breaks down into methane and the potassium salt of a mixed aquo ammono carbonic acid, namely potassium cyanate. The acid amides may be conveniently referred to as carboxazylic acids.

Hydrocyanic acid, which possesses simultaneously the properties of an ammono aldehyde, of a formic anammonide and of an ammono carbonous acid, has been nitridized, by the action of iodine in liquid ammonia solution, to an ammono carbonic acid, namely to cyanamide, in accordance with the equation,

$$HNC + H_2NI = HN{=}C{=}NH\ (H_2NCN) + HI,$$

while potassium cyanide, which is to be looked upon as potassium ammono carbonite, has been nitridized to a potassium ammono carbonate of the formula, KC_2N_3, and also to an ammono carbonate of the formula, K_2CN_2.

Ammono Carbonic Acids

Guanidine, biguanide, cyanamide, dicyandiamide, melamine, dicyanimide, tricyanmelamine and Liebig's hydromelonic acid are carbonic acids of the ammonia system. Guanidine and biguanide are extremely weak acids, melamine, dicyandiamide and cyanamide are stronger in the order named, while tricyanmelamine, hydromelonic acid and dicyanimide are not much inferior in strength to the ordinary mineral acids. Metallic salts of all these ammono carbonic acids have been prepared, the salts of the extremely weak acids by reactions carried on in liquid ammonia solution. Alkyl and aryl derivatives of these compounds are of course to be looked upon as esters. It has accordingly been found that dimethyl cyanamide and triphenyl guanidine, for example, may be ammonolyzed in liquid ammonia as represented by the equations,

$$(CH_3)_2NCN + NH_3 = (CH_3)_2NH + H_2NCN$$

$$C_6H_5N{=}C(NHC_6H_5)_2 + 3NH_3 = 3C_6H_5NH_2 + HNC(NH_2)_2$$

AGRICULTURAL CHEMISTRY

BY

CHARLES A. BROWNE
Chemist, and Chief of the Bureau of Chemistry
U. S. Department of Agriculture

Lecture delivered at Columbia University in the special course on Contemporary Developments in Chemistry given in the Summer Session of 1926 on the occasion of the opening of the Chandler Chemical Laboratories

COLUMBIA UNIVERSITY PRESS
NEW YORK
1927

Printed from type. Published March, 1927

Printed in the United States of America

AGRICULTURAL CHEMISTRY

By C. A. Browne

In beginning my lecture to you in this course it will be necessary first of all to inquire what agricultural chemistry is and in what respects it differs from other forms of chemistry. It has been defined in general terms as the chemistry of farm operations, yet when we analyze this definition it is found to include too much. The farmer employs concrete for the foundations of his barn, paint for his sheds, gasoline for his traction motor and dynamite for clearing his field of stumps, yet the chemistry of concrete, paint, gasoline or dynamite is not a part of agricultural chemistry. We come a little closer to the truth by defining it as the chemistry of the commodities which are produced upon the farm such as crops, fruit, meat, milk, and the like. Yet here again we must limit ourselves. A part of the speaker's boyhood experience upon a New England farm was the making of soap from two home products—the potash of wood ashes and the waste fat of slaughtered animals. Yet the making of soap is not regarded at present as coming within the purview of agricultural chemistry, although in early times soap, potash, saltpeter, leather, distilled spirits, indigo and many other manufactured products were all made upon the farm; and should these rural industries through necessity ever be revived they would all very properly come within the scope of agricultural chemistry in so far as they played a part in the economics of farm operations. The burning of lime and the making of brick from the mineral resources of the farm have also been part of the business of some landholders but such chemical operations as these are so foreign to the central operation, about which all other agricultural activities revolve, namely the production of crops, that they can be dismissed without question from the province of agricultural chemistry.

In its generally accepted meaning, agricultural chemistry may be defined as that branch of chemistry which treats of the chemical composition and mutual chemical relations of soils, crops and farm animals in so far as they concern the production of the means of human subsistence and welfare. It will be seen at once that the field comprised by this definition is almost limitless. In the formation of soils by the weathering and decomposition of rocks, questions of geology, mineralogy, meteorology and physics are involved as well as of chemistry. In the production of crops and farm animals we enter into the complex field of organic and physiological chemistry to any single phase of which a chemist can devote the entire period of his life. While it is necessary and important for the agricultural chemist to specialize in some branch of his subject, whether it be the chemistry of some particular crop, or the action of fertilizers or the feeding of animals, the agricultural chemist must be more than a specialist. He must have a broad outlook upon all the chemical aspects of his subject and upon their social, industrial and economic relationships.

It is held by many that agricultural chemistry relates solely to the production of raw materials, such as grain, vegetables, fruit, livestock, milk, hides, wool, etc., and not to the utilization of these. Yet in many cases the manufacture of butter, cheese, vinegar, sirup, sugar, canned goods, etc., from raw materials is performed directly on the place of production and constitutes an integral part of the farm operations. The opinion has also been expressed that human nutrition and the chemical control of human foods lie outside the province of agricultural chemistry. Yet the nutritive needs of the farmer and his family must be considered as essential factors in the successful management of the farm; the farmer, moreover, cannot ignore the regulatory requirements which govern the sale of his milk and other produce. Agricultural chemistry is therefore closely concerned with the utilization and disposition of manufactured products of farm origin although it is less actively engaged in these phases of the subject than in the production of raw materials.

A compilation of projects of American Agricultural Experiment Stations, prepared in 1923 by E. R. Flint of the United States Department of Agriculture, shows that of the investigations which involve chemistry 188 relate to soils; 150 to fertilizers; 111 to animal nutrition; 73 to insecticides, fungicides, etc.; 57 to dairy products; 49 to plant nutrition; and 45 to various miscellaneous studies of foods, etc. These figures show in a general way the relative stress which is now being placed by the experiment states of the United States upon certain applications of chemistry to agriculture. It is to some of the subjects indicated in this compilation that I wish to invite your attention.

Soils

The chemistry of soils is of fundamental importance to agriculture, for without good soil the operations of the farm are futile. Yet there is no factor which is so much ignored in agricultural activities as soil chemistry. Only a few years ago a large $25,000,000 agricultural enterprise was launched in the island of San Domingo, but it was doomed to financial failure from the very beginning owing to the neglect of the promoters to make a chemical survey of the new lands which they proposed to bring under cultivation. After spending millions of dollars in expensive preparations, the directors of this enterprise discovered to their consternation that the subsoil of their fields was so heavily impregnated with salt that the crops were almost valueless. They ascertained when it was too late that an agricultural chemist should have been consulted before making such lavish expenditures of money. False ideas of soil fertility have also led promoters astray with respect to the agricultural value of certain lands in the Everglades of Florida. The abundance of black decaying vegetable matter in some of these lands seemed an indication of vast agricultural wealth. But what seemed to be good soil was in reality only peat, which is worthless for the raising of crops.

A knowledge of geological origins will often assist the agricultural chemist in diagnosing the agricultural value of soils and in removing some of the false conceptions which so frequently exist in the popular mind. In British Guiana the wonderful fertility of the alluvial low lands, which were the first to meet the attention of the settlers, seemed to many only a presage of even greater fertility in the extensive plateaus farther back. Those uplands came thus to be regarded as of almost inexhaustible fertility. The late Sir John Harrison, who was Director of Science and Agriculture in British Guiana, showed, however, that this popular idea was a false one, for the potassium and other elements essential to fertility had already been largely removed from the uplands by various meteorological, chemical and physical forces. The sand, kaolin, bauxite, limonite and other residues left behind gave rise to soils of low agricultural value, while the deposition of the materials leached away contributed to the richness of the alluvial lowlands.

Through the ignorance of our ancestors the soils in many parts of the United States have been robbed of much of their native fertility. The exportation of potash, prepared by burning trees and leaching the resultant ashes, removed many millions of dollars' worth of a valuable fertilizing material from the virgin lands of North America. This agricultural blunder of our forefathers was not realized until after agricultural chemistry one hundred years ago had demonstrated that potash was necessary for the growth of crops. The value of the potash and other fertilizing elements contained in grain, tobacco and other agricultural exports was also lost sight of for over two centuries, and the excessive drain upon the fertility of the land during this long period of time had its final inevitable result in abandoned farms. The millions of dollars spent during the last half century for importations of potash salts from Germany and of sodium nitrate from Chile is the penalty which American agriculture has had to pay for previous neglect.

When the agricultural lands of the Eastern States began to show a falling off in productivity, as a result of soil depletion, there was a general movement of the agricultural population towards the virgin lands of the West. It has only been within recent decades, when all the best unoccupied agricultural areas of the United States have been preempted, that the questions of maintaining and improving soil fertility have begun to receive serious consideration.

The degree of attention which a nation gives to its soil problems usually varies with the density of population. In overcrowded countries, such as China, where the feeding of immense numbers of people is dependent upon the highest productivity of the land, we find the greatest development of the multiple crop system and the greatest efforts to maintain soil fertility by means of manures, composts, mulches, thorough cultivation, rotation of crops, and all the other expedients of an intensive agricultural system. In those densely populated regions where land is dear and labor is cheap, we find the highest production of crops per unit of land. The conditions are opposite in thinly settled countries where land is cheap and labor is dear. Under this situation the effort is to secure the highest production of crops, not per unit of land but per unit of individual man-power, and for attaining this objective the utmost use is made of farm machinery and the other labor-saving devices of an extensive system of agriculture. Unless these differences of the intensive and extensive systems of agriculture are fully understood we are apt to be misled by statistics which compare the agricultural efficiency of different countries exclusively upon yield of crops per acre.

The United States by its development of the extensive system of agriculture has led the nations of the world in the yield of crops per man power, although it is far down upon the list in the yield of crops per acre. The time is fast approaching when with the rapid increase in the population of our country, more and more attention must be given to intensive methods of agriculture and to increasing the productivity of our fields.

When unproductiveness is due to a lack of available potassium, nitrogen, phosphorus, or other element in a soil, the damage can be corrected by applications of certain amounts of the necessary lacking element in an easily assimilable form. In many cases, however, the unproductiveness is due not to an absence of any particular element but to an injurious excess of acid, alkali, salt, water, or some other soil ingredient. Researches at the experiment stations of Rhode Island and other States have shown that many fields, abandoned as worthless, were simply suffering from an excess of acid and that they could be restored to high productivity by neutralizing this acid with lime. Excess of water can be removed by drainage. Excess of alkali can be eliminated by irrigation and drainage, or by means of certain corrective agents such as gypsum, or by suitable methods of cultivation, which prevent the transportation by capillarity of alkaline salts from the subsoil to the surface.

In many cases the deficiencies of a soil as regards tilth, moisture retention and other desirable physical characteristics are due to a lack of organic matter or humus, a certain amount of which is almost indispensable for the successful production of crops. Eighty and more years ago it was held by many agricultural chemists that crops derived a considerable amount of their organic matter from the humus of the soil by absorption through the roots. This theory of the direct absorption of humus is now no longer held although some agricultural chemists believe that assimilation of carbon dioxide, produced by the decay of humus in the soil, may take place through the roots and thus contribute to the nourishment of plants. The acid decomposition products of humus no doubt also play an important role in rendering the insoluble phosphoric acid and potassium of the soil more readily available. The process, however, is a slow one although it can be hastened by a well regulated system of composting.

The art of composting has reached its highest perfection in China, Japan and other countries of Asia, where density of population has brought about methods of husbandry, which are intensive as well as highly economic. The agricultural areas of these countries are intersected by a network of canals which serve not only for purposes of drainage and transportation but also as overflow reservoirs for rivers during periods of floods and freshets. A large amount of fine alluvial mud is deposited in the canals during such overflows and this mud, when mixed with crop residues, excreta of animals and other refuse, is soon converted into most valuable fertilizer, which the oriental farmer adds to his fields according to the requirements of each particular crop. For further information as to the elaborate processes of oriental agriculture (processes which the growing density of our population will force us also some day to consider), I will refer you to a most interesting book by the late Professor King entitled "Farmers of Forty Centuries." The bacteriology and chemistry of composting, by which the organic matter of straw, leaves and similar plant residues can be quickly converted into humus for the soil, is a subject of research which is now attracting considerable attention both in foreign countries and the United States. The specific organisms which are best suited for this purpose, the specific conditions which are most favorable for the growth of such organisms and the specific organic constituents of the plant residues, whether cellulose, pentosans or lignin, which give the maximum yield of humus, are phases of some very complex yet highly attractive agricultural chemical problems.

If none of the produce of a farm is removed and all of the plant and animal residues are restored to the fields it is possible to maintain a self-supporting or permanent system of agriculture. The farmer, however, is always selling grain, vegetables, tobacco, fruit, poultry, milk, livestock and other produce of his fields, so that there is a continuous loss of nitrogen, phosphoric acid, potassium, calcium and the other elements essential to soil fertility. The economic restoration of these losses is the secret of successful farming and it is the duty of the agricultural chemist, by keeping a check upon the income and outgo of the farm, to inform the owner as to whether his fertility account is properly balanced or overdrawn.

The atmosphere above and the subsoil beneath are two almost inexhaustible reserves upon which the farmer can always draw to restore a part of the losses occasioned by the sale of produce. The gradual turning up of new depths of soil is a method followed by the European farmer to increase the fertility of his fields. The occasional planting of crops with deep root systems is another means of bringing the elements of fertility from the subsoil to the surface. The introduction into the rotation of legumes, such as clover, alfalfa, beans, peas, etc., which can assimilate the free nitrogen of the air, is a third method which the farmer has at his disposal for adding richness to his fields.

In methods for preserving the elements of fertility contained in the excreta of horses, cows, sheep, swine, and other livestock. European farmers are far more advanced than their American colleagues. As a result of leaching by rain and of fermentation by denitrifying bacteria, the neglected manure piles of many American farms lose a large amount of valuable plant food.

Fertilizers

The addition to the soil of phosphates, nitrates, potash salts, ammonium compounds and other commercial fertilizers is the readiest method which the farmer has at his disposal for increasing the depleted fertility of his fields. In the old days when transportation was much cheaper than it is now, the purchase of mixed fertilizers of comparatively low nitrogen, phosphorus and potassium content was a far more economical proposition than it is at present when the rates of transportation have increased. The farmer at present is realizing more and more the advantage of buying the unmixed concentrates and of doing his own diluting at the place of application, for in this way he avoids the cost of transporting sand, clay and other inert materials that were used by the manufacturer in the mixing.

The amount of fertilizing elements which are necessary to be applied to a field to produce the most economical results is an agricultural chemical problem to which more time and study have been devoted than to all other agricultural chemical problems combined. The supplying of crops with the necessary amounts of raw materials, from which they can synthesize the maximum quantity of starch, cellulose, sugar, fat, protein, alkaloid, essential oil, dye stuff, or other desired ingredient, has been, and will always remain, the great basic problem of agricultural chemistry. The determination of the nature and quantity of this supply for each individual crop is the purpose of fertilizer plot experimentation, a method of research which in its present form was originated by Lawes and Gilbert at Rothamsted in 1843 and is now pursued in every agricultural experiment station of the world. The method is to test the particular crop with applications of varying known amounts of ammonium salts, nitrates, phosphates, potassium salts, lime, etc., singly and in combination until the best mixture is found for securing a maximum yield on a given soil under a given method of cultivation. The possible combinations of mixtures and the possible effects of rainfall, temperature, cultivation, rotation, drainage, etc., introduce so many variables into plot experimentation that the interpretation of the results is rendered extremely difficult.

Two much discussed laws in fertilizer experimentation are the "law of the minimum," first announced by Liebig, and the "law of diminishing returns." According to the "law of the minimum" the yield of a crop is limited by the amount of the minimum essential constituent which is present in the soil whether it be nitrogen, potash, phosphoric acid, lime, or other ingredient. If potash, for example, is the deficient constituent, then no increase of phosphoric acid, nitrogen, or other fertilizing substance will be of much benefit to the crop. According to the "law of diminishing returns" additional increases of fertilizer, beyond a certain amount, although perhaps at first financially remunerative, give proportionately smaller increases of crop until finally a point is reached beyond which further additions give no extra return.

The "law of diminishing returns" has given rise to a vast amount of discussion among economists, for it is an important factor in determining the profitable limit to which crops can be fertilized for increased production. Obviously the profitable limit will vary according to the relation between the costs for labor, seed, fertilizer, transportation, etc., and the selling price of the farm product. The "law of the minimum" and "the law of diminishing returns" are both fundamentally important since they apply not only to crop fertilization but to cattle feeding and other agricultural operations.

The possible future shortage of food supplies, with the increase of population, the decrease of soil fertility and the depletion of available fertilizing materials, has given rise to frequent dismal predictions. Sir William Crookes in his presidential address before the British Association for the Advancement of Science, in 1898, issued a warning with respect to the rapid exhaustion of the world's supplies of combined nitrogen for agricultural uses. While this particular danger has been largely averted by the recent discovery of commercial methods of fixing atmospheric nitrogen, there is cause for distant future apprehension in the case of potash and phosphoric acid, which have no such unlimited storehouse of reserves.

Sir Daniel Hall in his presidential address before the British Association for 1926 issued a similar warning with respect to the future shortage of food supplies.

The chemistry of soils and fertilizers is of such magnitude and complexity that a whole series of lectures could be devoted to this subject alone. For further information I can best refer you for collateral reading to the excellent treatise "Soil Fertility and Permanent Agriculture" by the late Professor Cyril G. Hopkins of the Illinois Agricultural Experiment Station.

Crops

The chemistry of crops is the third great subdivision of Agricultural Chemistry, and as an introduction to this phase of the subject I would recommend that you read two excellent little works of the late Professor S. W. Johnson of Yale University, entitled "How Crops Feed" and "How Crops Grow." Although written many years ago, these books are still among the best for introducing the beginner to the broad subject of plant chemistry.

The chemist may study crops from three points of approach. He may study them first from the standpoint of chemical composition; second, from the standpoint of functional operation, such as germination, respiration, assimilation, and the like; and third, from the standpoint of utilization. The first of these approaches interests the analyst, the second the plant physiologist, and the third the technologist.

The opportunities which are open to the agricultural chemist in the field of crop chemistry are unlimited. A complete chemical history of an individual species of plants has never as yet been written and we are probably safe in saying that it never will be written. Such a history involves a knowledge of every chemical compound in the roots, stalks, leaves, fruits and other organs of the plant; of the manner in which these various compounds are produced; of the chemical processes which underlie all the various physiological activities such as germination, assimilation and growth; of the chemical reactions involved in the relations between the plant and its surrounding, such as soil, water, air, sunlight, insects, bacteria and other externals; and finally of the chemical utilization of the various parts of the plant for the specific needs of commerce and industry. Even the crops which have been most studied, such as corn or wheat, appear to the chemist only in the resemblance of an unsolved, very complicated cross-word puzzle; with here and there a section filled in, with a few suggestive key words perhaps inserted, but with many blank spaces still remaining to be worked out.

I can indicate only very briefly a few phases in this field of my subject. Crops at one time were grown largely for net weight of production without much regard to the question of composition. But during the last half century the capacity of a crop to produce some specific ingredient, such as sugar, oil, starch, cellulose, or protein, has attracted more and more attention. Breeding experiments have been conducted to develop varieties of corn which excel in protein or oil or starch, according to the specific purposes of utilization. The classic example of what can be done in the improvement of a crop by chemical selection is the sugar beet which, by the simple expedient of selecting for seed production only the roots of highest sucrose content, has had its sugar producing capacity more than doubled. The maintenance of a quality, which has been developed in this way, requires, however, constant chemical supervision, otherwise the crop would quickly degenerate into its former state of inferiority.

The complete chemical utilization of our crops can be said to be only in its infancy. Dr. Arno Behr, some forty years ago, was the first to show what possibilities could be realized in the chemical utilization of corn, in the manufacture of starch, dextrin, glucose syrup, dextrose, gluten, oil, cattle feeds and other products too numerous to mention. No attempt to utilize the valuable storehouse of oil, protein and other products contained in cottonseed was made until after the Civil War. In 1873 only 2 per cent of the cottonseed crop was utilized for oil and feed production; the remainder was cast into streams or otherwise wasted, the accumulation of this most valuable agricultural product about the cotton gins constituting at that time a great nuisance. The utilization of the oil of cottonseed in countless ways as a human food and industrially is one of the greatest benefits which chemistry has conferred upon American agriculture, millions of dollars being thus added each year to the value of our cotton crop.

The utilization of unmarketable and surplus agricultural products for new industrial uses is one of the most important problems which American agriculture at present is depending upon chemistry to solve. Much has already been done in this connection. Some years ago the disposal of the inferior culls of the California lemon crop was a problem which the Bureau of Chemistry was called upon to solve. Such unmarketable lemons are now all profitably utilized in the manufacture of citric acid, lemon oil, marmalade, pectin and cattle feed. Similar researches are now being carried on upon the orange and other crops. A perfect system of utilizing surplus fruits and vegetables by dehydration still remains to be worked out by the agricultural chemist. The old process of sulfuring and sun-drying must soon give way to a process which is more sanitary and less objectionable. There are great difficulties, such as the discoloration of the cut plant tissues by oxidizing enzymes, which have to be overcome. We can be confident, however, that they will be overcome and the consumer will then no longer be suffocated by the odor of sulfurous acid when he opens an occasional package of evaporated fruit. The chemical problems which await solution in the field of dehydration, canning and preserving are innumerable.

The reduction of the enormous losses which occur each year from the spoilage of fruits, vegetables, foods, cattle feeds, farm fabrics, timber and other products is another field in which the chemist can render inestimable services to agriculture. The destruction of perishable material by means of bacteria, moulds, yeasts and fungi, by enzymes, by insects, by oxidation and other destructive agencies, occasions the American farmer a loss of several billion of dollars each year. The chemist is the farmer's chief reliance for aid in the elimination of these losses. New insecticides, new fungicides, and new preserving agents must be discovered; we must have better knowledge of the conditions which promote decay and of the means for controlling these conditions. In this field I can only cite one or two examples which must serve as general illustrations of the whole.

Insecticides

It has been estimated that the losses to American agriculture from the depredations of the cotton boll weevil exceed $300,000,000 a year. One of the most successful means of combating this pest has been by means of calcium arsenate, of which the proper conditions for manufacture and application had to be worked out by the chemist. Cotton fields now are dusted with calcium arsenate from airplanes and thus a new system of chemical warfare has been inaugurated. Yet the present promiscuous system of applying insecticides is a most wasteful one since for every one part of chemical which kills an insect many hundred parts are scattered without effect. The tremendous excess of arsenicals which are employed in our present system constitutes in fact a grave danger in some cases, the spray residues upon certain shipments of celery, fruit and other products having been found to occur in quantities which threatened the health of the consumer.

Agricultural chemists are at present endeavoring to devise methods of applying insecticides which shall not only avoid the present extravagant waste of material and danger to health, but shall make the poison more effective in its action. It is assumed that the attracting power of a special family of plants for a certain insect is due to some definite volatile substance. If this substance can be determined and prepared synthetically, its incorporation with the poison as a bait offers a means of luring the insect to its own destruction. Amyl acetate has been used in this way as a lure for attracting grasshoppers and geraniol for attracting the Japanese beetle. Out of some sixteen volatile organic substances obtained by Power and Chesnut of the Bureau of Chemistry from the cotton plant, one of them, trimethylamine, was found to have an attracting influence upon the cotton boll weevil. Other possibilities of employing such attractants for the destruction of insects upon an economic scale have not, however, been sufficiently worked out. If the principle is sound an immense field of service will be opened up for the present generation of agricultural chemists.

The use of volatile poisons, such as hydrocyanic acid, within confined spaces has been employed successfully for destroying scale insects upon fruit trees and for killing weevils in stored cereal products. After the insects are killed, aeration will remove the adhering traces of volatile poison. The perfection of newer insecticides, which are more effective in action, cheaper in price, and harmless not only to the plant but also as regards toxicity of food products to the consumer, are problems that demand the exercise of the highest chemical knowledge.

Microbiology

The spoilage of agricultural products by the attacks of microorganisms offers an almost endless number of problems to the chemist. The souring of milk, the moulding of bread, the swelling of canned goods, the heating of cattle feeds, the rotting of fruits, the fermenting of syrups and the decay of timber are all phases of the same general destructive process called deterioration. This process may proceed so slowly as to be barely noticeable from one day to the next or it may proceed so rapidly as to cause a haystack to catch fire spontaneously and burn to the ground. The chemical transformations which take place in these different processes of deterioration are very imperfectly understood and there is a wealth of opportunities in this field for valuable chemical research upon the nature of the destructive changes produced and upon the means for their prevention.

It must not be inferred, however, that the relationship of microorganisms to agriculture is wholly injurious in character. The moulds, bacteria and yeasts when properly controlled are exceedingly useful to the farmer and without their assistance the operations of agriculture and of all animate nature would speedily come to a standstill. The resolution of the dead material of plants and animals into their original elements, which can again enter the cycle of Nature's operations, is largely the work of microorganisms. The conversion of organic nitrogen compounds in the soil into nitrates and the fixation of atmospheric nitrogen by the root nodules of leguminous plants are accomplished by means of bacteria. The ripening of cheese, the fermenting of vinegar, the preparation of sauer kraut, ensilage and many other agricultural products are the result of bacterial action. Lactic acid, citric acid, and other organic substances are now being made on a commercial scale by the carefully controlled fermentation of sugar and other agricultural products. There is apparently no limit to the opportunities which await the investigator in this field of microbiological chemistry, and they are bound to attract an increasingly large number of workers.

Animal Nutrition

Experimental work in animal nutrition ranks third in the amount of attention which American agricultural stations are giving to chemical projects. But if the total amount of work in the colleges, hospitals and private laboratories of the whole country be considered, the researches in animal nutrition would easily rank first in number among agricultural chemical investigations. To speak upon this subject before an audience of Columbia University students is very much like bringing coals to Newcastle, for so much excellent work has been done at this institution in the field of nutrition under the direction of Professors Sherman, Rose, Eddy, MacLeod and others. This is a subject in which American chemists during the past fifty years have taken a leading part. The late Professor Atwater of Middletown, Conn., was the pioneer in this field and his experiments with the respiration calorimeter with human subjects excelled all previous work in accuracy and importance.

The methods of Atwater upon human subjects were applied to farm animals by the late Henry P. Armsby, Founder and First Director of the Pennsylvania State College Institute of Animal Nutrition. By means of his respiration apparatus Armsby was able to place the science of stock feeding for the first time upon a strictly scientific basis. The chief result of this method of investigation has been the introduction of net energy as a basis for measuring the nutritive value of food materials. According to this conception the difference between the gross energy of the food before consumption and that expended and lost in the process of its utilization represents the net energy value of the food for purposes of maintenance and production. Armsby's determinations of the net energy values of the principal American feeding stuffs and his use of these in place of the old digestion coefficients for computing the rations of farm animals form one of the most important contributions to agricultural chemistry which have been made in recent years.

The horse and ox are being rapidly superseded upon farms by the tractor and the feeding of animals for work production is attracting less attention than was formerly the case. The feeding of animals for meat and milk production is, however, being given an increasing amount of attention. In the old days it was considered that if a sufficient amount of the basal requirements of carbohydrates, fat, protein and mineral matter were given to the animal, all else was provided for. But as the result of investigations conducted in this country since the beginning of the present century we know that this is no longer the case. The classic work of Professor Osborne of Yale University upon the vegetable proteins shows that all proteins are not alike in tissue-building capacity. Some of them contain more and others less of the essential amino acids and feeding experiments by Professors Osborne and Mendel upon rats show that some kinds of vegetable protein are inadequate to promote growth. In consequence of this discovery the investigation of feeding materials as to the best growth-producing kinds of protein has become a subject of increasing importance.

In the old days it was also supposed that if the animal were given a sufficient amount of salt and calcium phosphate in its ration that nothing more in the nature of mineral ingredients was required. We know at present that this is very far from being the case. The necessity of balancing the acid-forming with sufficient base-forming elements in the ration was largely overlooked in nutritional work until Dr. Sherman of Columbia called attention to this very important factor. The necessity of iodine in the diet, with special reference to the prevention of goiter in animals, has only recently begun to attract serious attention. The deficiency of iodine in the soils and waters of certain regions of the United States has caused a prevalence of goiter among the farm animals, as well as among the human residents, which subsist upon the grain and other farm products of such sections. The assertion has been made that the iodine taken up from the soil by our cereal crops is largely localized in the germ of the grain, and, for this as well as for the organic phosphorus, vitamins and other necessary food adjuncts which the germ contains, a greater use in the diet of this rejected portion of our cereals has been advocated by many authorities upon nutrition.

The subject of vitamins is another subject which is engaging the attention of many nutritional chemists but as this topic has been assigned to Professors McCollum and Steenbock in this course, it is omitted from the present discussion.

Agricultural Chemical Analysis

The agricultural chemist is constantly being called upon to improve his analytical methods for the purpose of determining the amounts of some of the less abundant and hitherto neglected constituents of our food. The needs in our diet of traces of fluorine for teeth formation, of iodine for the prevention of goiter, of iron for the production of haemoglobin, and of other less common elements for specific physiological uses, call for a greater improvement in many of our analytical processes. Indeed, the agricultural products of certain sections of our country are sometimes advertised as superior because of their higher iodine content and thus, as so often happens, a subject of scientific interest assumes forthwith a regulatory aspect. The reputed beneficial effect of minute quantities of certain substances, such as boron, zinc, manganese, and arsenic, as stimulators of development and the surmised injurious effects of traces of mineral and organic poisons as promoters of malignant growths, in plants and animals, are additional reasons for our devoting more attention to analytical refinements. The reported poisonous action of soluble aluminum compounds upon the plant and animal organism has recently been brought to the front again in connection with the reopening of the old alum baking powder controversy and as one wades through the voluminous mass of testimony presented at the hearings upon this subject he is impressed with the need of more accurate methods for determining traces of aluminum in animal and vegetable tissues.

Agricultural Chemical Technology

Another field of agricultural chemistry which is attracting a vast amount of attention is the technical utilization of the crops, fruits, milk and other commodities which are produced upon the farm. The manufacture of sugar, syrup, vinegar, butter, cheese, dried apples, and many other farm products involves a thorough knowledge of chemistry at every step of each process. The curing of tobacco, the tanning of leather, the canning of vegetables, the preserving of fruits, the pressing of oil, the distilling of turpentine, and the utilization of lemons for citric acid are other illustrations of which almost countless examples could be given. A knowledge of engineering is very desirable for those who specialize in the field of agricultural chemical technology. The director of such processes must have not only an excellent knowledge of the chemical properties of plant and animal substances, but he must understand the construction and operation of the various mills, extractors, digesters, clarifiers, evaporators, filters, crystallizers, driers, rectifiers, and other appliances which are used in manufacturing plants. The opportunities in this field are almost unlimited for much remains to be done in the chemical utilization of agricultural residues. Thousands of tons of straw, corn stalks and similar refuse are wastefully burned each year for the purpose of disposal. The destructive distillation of such refuse for the production of gas, carbon, pyroligneous acid, tar and other products opens up vast possibilities. The chemical utilization of the cellulose in agricultural wastes, such as straw, is already carried out in certain paper mills. The conversion of the pentosans of such materials as corn cobs and oat hulls into the valuable chemical furfural is being performed upon a small scale. The important lignin complex in waste wood, straw, corn stalks and similar agricultural refuse probably offers as many methods of utilization for the manufacture of tanning materials, dye-stuffs, and other industrial products as was offered eighty years ago by that other trade waste coal tar which through the ingenuity of the chemist has proved to be an almost inexhaustible source of wealth.

Law Enforcement

Another important phase of agricultural chemistry, which I must mention before concluding, is that referring to law enforcement. Many years ago gigantic frauds were perpetrated upon the farmer through the adulteration of fertilizers, cattle-feeds, insecticides, fungicides, and other supplies which he purchased. But now efficient State or Federal laws, which require the accurate labeling of such products as to their chemical composition, protect the farmer against such impositions. The farmer is also required to do his part and the milk, butter, cheese, grain, fruit, syrup, etc., which he sells must come up to a certain standard of excellence if he is not to be penalized by some one of the Federal or State laws against adulteration. This field is almost unlimited in its possibilities, for with the constant introduction of new agricultural processes and products, new standards of excellence and new methods of examination must be devised. It seems almost inevitable that every new valuable discovery in agricultural chemistry should immediately be put to an illegal use. The valuable properties of commercial pectin were no sooner discovered than hosts of fraudulent manufacturers began to use it for adulterating fruit jellies. The use of ethylene gas for imparting a yellow color to mature citrus fruits of a green color was immediately misapplied by fraudulent producers in order to give immature oranges, lemons and grapefruit a fictitious appearance of ripeness. Hundreds of examples of this nature could be cited and the agricultural chemist must be constantly alert in order to prevent the consumer from being deceived.

Conclusion

This lecture has given you only a very incomplete picture of the vastness and extent of the field of agricultural chemistry. In the short time at my disposal it has been necessary to omit many branches of my subject. To those of you who may be interested to pursue the subject further, I would refer you to the recent volume "Chemistry in Agriculture," published by the Chemical Foundation, which gives in non-technical language a series of chapters upon various phases of the subject, written by experts in their respective fields. With this popular low-priced book as an introduction and with the assistance of the bibliographies and references, which it contains, many of you may perhaps be induced to make agricultural chemistry your life work.

Statistics contained in "American Men of Science" show that one-fifth of the chemists listed in this volume are devoting themselves to some line of agricultural chemistry. We might conclude from this that one-fifth of the students contained in this audience will take up some phase of the teaching or practical application of agricultural chemistry. It is an ancient and most honorable chemical calling, filled with opportunities for service and one which such men as Priestley, Davy, Liebig, Pasteur, and many others celebrated in the annals of chemistry, have not only zealously cultivated but which they have also greatly enriched by the value of their contributions.

COMPLETING THE PERIODIC TABLE

BY

B. S. HOPKINS
PROFESSOR OF INORGANIC CHEMISTRY
UNIVERSITY OF ILLINOIS

Lecture delivered at Columbia University in the special course on Contemporary Developments in Chemistry given in the Summer Session of 1926 on the occasion of the opening of the Chandler Chemical Laboratories

COLUMBIA UNIVERSITY PRESS
NEW YORK
1927

Printed from type. Published March, 1927

Printed in the United States of America

COMPLETING THE PERIODIC TABLE

B. S. Hopkins

One of the most outstanding scientific accomplishments of all time is the grouping together of the entire list of chemical elements in the chart which is commonly known as the Periodic Table. Before its publication chemistry was a bewildering jumble of isolated facts, becoming more and more confusing as knowledge increased. Gradually there grew up the notion that there must be some systematic order in the relation of these simple forms of matter which could serve as a compass to guide the adventurer through the maze of chemical explorations. Chemists of many times and many countries have studied the interrelationship of the elements and as each made his contribution to the problem there gradually grew up a table which makes the attempt to classify all the known elements, revealing Nature's plan of orderly, systematic arrangement. To Mendeléeff belongs the major honor for this useful table because he first dared to use the table for predicting new discoveries. Since the days of Mendeléeff, chemists have continued to work upon this table and many useful and extremely helpful suggestions have been made. It is probable that the chemists of the future will continue this work for it is no simple plan that Nature has followed in building the universe.

When Mendeléeff published [1] his Periodic Table (Fig. 1) there were recognized 63 distinct elements. In order to locate each of these elements in related groups it was found necessary to leave many blank spaces for elements at that time unknown. A study of Mendeléeff's table shows the presence of all the 63 elements together with 31 blank spaces, most of which were left for elements with atomic weights greater than that of cerium. From these facts it might be concluded that Mendeléeff definitely predicted the existence of 94 elements, a number which corresponds remarkably with our present belief. On examining Mendeléeff's Table more closely we observe that he made no provision whatever for the Zero Group and certainly his method of distributing the Rare Earth Group leaves much to be desired in the light of present day information. Hence it is safe to as-

Mendeléeff's Table[1] *(1871)*

	I	II	III	IV	V	VI	VII	VIII
	R_2O	RO	R_2O_3	RH_4 RO_2	RH_3 R_2O_5	RH_2 RO_3	RH R_2O_7	RO_4
1	H = 1							
2	Li = 7	Be = 9.4	B = 11	C = 12	N = 14	O = 16	F = 19	
3	Na = 23	Mg = 24	Al = 27.3	Si = 28	P = 31	S = 32	Cl = 35.5	
4	K = 39	Ca = 40	— = 44	Ti = 48	V = 51	Cr = 52	Mn = 55	Fe = 56, Co = 59 Ni = 59, Cu = 63
5	(Cu = 63)	Zn = 65	— = 68	— = 72	As = 75	Se = 78	Br = 80	
6	Rb = 85	Sr = 87	Y = 88	Zr = 90	Nb = 94	Mo = 96	— = 100	Ru = 104, Rh = 104 Pd = 106, Ag = 108
7	(Ag = 108)	Cd = 112	In = 113	Sn = 118	Sb = 122	Te = 125	I = 127	
8	Cs = 133	Ba = 137	Di = 138	Ce = 140	—	—	—	— — — —
9	(—)	—	—	—	—	—	—	
10	—	—	Er = 178	La = 180	Ta = 182	W = 184	—	Os = 195, Ir = 197 Pt = 198, Au = 199
11	(Au = 199)	Hg = 200	Tl = 204	Pb = 207	Bi = 208	—	—	
12	—	—	—	Th = 231	—	U = 240	—	— — — —

[1] *Annalen Suppl.* 8 151 (1872).

FIG. 1.

Periodic Table (1914)

0	I	II	III	IV	V	VI	VII	VIII
	H 1 1.008							
He 2 4.00	Li 3 6.94	Be 4 9.1	B 5 10.85	C 6 12.005	7 N 14.008	8 O 16.	9 F 19.0	
Ne 10 20.2	Na 11 23.00	Mg 12 24.32	Al 13 26.97	Si 14 28.1	15 P 31.04	16 S 32.06	17 Cl 35.46	
A 18 39.9	K 19 39.10	Ca 20 40.07	Sc 21 45.1	Ti 22 48.1	V 23 51.0	Cr 24 52.0	Mn 25 54.93	Fe 26 55.84 Co 27 58.97 Ni 28 58.68
	29 Cu 63.57	30 Zn 65.37	31 Ga 70.1	32 Ge 72.5	33 As 74.96	34 Se 79.2	35 Br 79.92	
Kr 36 82.92	Rb 37 85.45	Sr 38 87.63	Y 39 89.	Zr 40 90.6	Cb 41 93.1	Mo 42 96.0	43	Ru 44 101.7 Rh 45 102.9 Pd 46 106.7
	47 Ag 107.88	48 Cd 112.40	49 In 114.8	50 Sn 118.7	51 Sb 121.8	52 Te 127.5	53 I 126.92	
Xe 54 130.2	Cs 55 132.81	Ba 56 137.37	La 57 139.0	Ce 58 140.25				
			59–72 Rare Earths		Ta 73 181.5	W 74 184.	75	Os 76 190.9 Ir 77 193.1 Pt 78 195.2
	79 Au 197.2	80 Hg 200.6	81 Tl 204.0	82 Pb 207.2	83 Bi 209.0	84 Po 210.	85	
Rn 86 222.	87	Ra 88 226	Ac 89	Th 90 232.12	Pa 91	U 92 238.2		

FIG. 2.

sume that when this table was prepared little attention was devoted to the total number of elements.

Mendeléeff's blank spaces represented such definite realities in his mind that he ventured to predict the discovery of elements to fill these spaces and in certain cases he foretold the properties of the elements and their compounds. Within four years de Boisbaudran discovered gallium which corresponds with astonishing accuracy to the predictions made for eka-aluminium. Scandium, which had been predicted as eka-boron, was discovered in 1879 by Nilson while Winkler in 1886 located eka-silicon and named it germanium. These three confirmations of Mendeléeff's predictions have done much to establish our faith in the Periodic Table and in the resultant study of the interrelationship of the elements.

In spite of our admiration for this remarkably clear conception concerning the periodicity shown by the elementary forms of matter, efforts to fill in the remaining blank spaces in the original Mendeléeff Table are decidedly disappointing. In no case does a discovered element fulfill the predictions with the same startling accuracy as we find in the case of gallium, scandium and germanium. The table runs with perfect smoothness until the rare earth group is encountered and then difficulties are experienced due to inaccurate atomic weights as well as the puzzling problem presented by a group of 15 elements of like valence and strikingly similar properties. Most chemists of our own day who have tried to solve the enigma of the rare earth group are led to a profound sympathy for Mendeléeff in his efforts to find a simple and satisfactory means of locating this group in any simple tabulation.

The period from 1871 to 1914 saw the discovery of three distinct groups of elements—the noble gases, radioactive elements and many members of the rare earth group. In accordance with the information gained in this period the Mendeléeff table was modified, but still the rare earth group was a source of trouble. It was insisted that cerium, which shows a definite valence of four, must be placed in Group IV (see Fig. 2). This made necessary an awkward break in the sequence in order to get the other members of the rare earth group back into Group III where they very obviously belonged. Following this group came tantalum, in Group V, which necessitated another puzzling irregularity in the omission of Group IV.

The year 1914 marks a distinct advance in the development of the Periodic Table, because in that year the work of Moseley made it possible to define with accuracy the position of any element in the Table. This outstanding accomplishment also made it necessary to assume that from hydrogen to uranium there must be a total of 92 elements. Not only was the number of blank spaces definitely determined but the position of each was clearly shown. Accordingly the investigator who was interested in completing the Periodic Table was led to search for a missing element in material containing a closely related element. In addition, the method of X-ray analysis has been of the greatest service because it permits the definite identification of an element by the determination of its atomic number through the wave length of its X-ray lines. From such considerations the chemist of 1914 who accepted the common belief in the existence of 87 known elements was convinced that there must be 5 missing elements within the usual limits of the Table. Three of the five unknown elements were in Group VII, No. 43, eka-manganese, No. 75 dwi-manganese, and No. 85 eka-iodine. There still remained one undiscovered member of the rare earth group, No. 61, eka-neodymium, while the existence of No. 87, eka-caesium, has been suspected for many years.

The next important step in the completion of the Periodic Table came in the form of a simplification in the region of the rare earth group. In 1911 Urbain, working with the extreme soluble end of an extensive series from gadolinite, had isolated a few oily drops of mother liquor which refused to crystallize. Examination of this material by the aid of arc spectrum analysis revealed the presence of several lines of what appeared to be a new element. Evidence of the existence of this new element was also found in the fact that something gave this new substance a much lower magnetic susceptibility and a weaker basicity than either lutecium or ytterbium with which it was associated. Accordingly Urbain assumed that these phenomena were due to the presence of a new rare earth to which the name celtium was assigned. In 1922, Dauvillier found in this material two extremely faint X-ray lines which correspond closely to the positions assigned to $L\beta_2$ and $L\alpha_1$ of element No. 72. Accordingly Urbian argues [2] that the identification of the element celtium, with the atomic number 72, is complete.

On the other hand Coster and Hevesy [3] studying the probable atomic structure of element No. 72 in the light of the Bohr theory reached the conclusion that this element cannot be a member of the rare earth group. According to the Bohr theory the presence in the Periodic Table of a group of similar elements with a uniform valence is to be accounted for by the addition of successive electrons to an inner shell while the number of electrons in the outer shell, which determines the valence, remains unchanged (see Fig. 3). When electrons are added in

Arrangement of Electrons in Atomic Shells

38	Sr	2	8	18	8	2	
39	Y	2	8	18	8	3	
40	Zr	2	8	18	8	4	
47	Ag	2	8	18	18	1	
48	Cd	2	8	18	18	2	
49	In	2	8	18	18	3	
55	Cs	2	8	18	18	8	1
56	Ba	2	8	18	18	8	2
57	La	2	8	18	18	8	3
58	Ce	2	8	18	18+1	8	3
59	Pr	2	8	18	18+2	8	3
71	Lu	2	8	18	32	8	3
72	Ct or Hf	2	8	18	32	8	4
73	Ta		8	18	32	8	5

FIG. 3.

the rare earth group all the inner shells become complete when element No. 71, lutecium, is reached. Consequently the only possible way to form element No. 72 is to add the extra electron to the outer shell, which gives it a structure similar to zirconium and makes it a quadrivalent element. When these investigators examined samples of Norwegian zircon by the methods of X-ray analysis, they were able to discover distinct lines in the exact positions predicted for α_1, α_2, β_1, β_2, β_3 and γ_1 of the L series of No. 72. Accordingly they also announced the discovery of element No. 72 for which they have proposed the name Hafnium. Immediately a controversy arose concerning the possible presence of this element in Urbain's material. It is to be hoped that the International Committee will decide the question upon its merits so that we shall not be compelled to add to our list one other case of an element which is known by two different names. Before the decision is made in favor of Urbain, it seems necessary for some one to prove beyond the shadow of a doubt that element No. 72 is sometimes found in rare earth minerals, especially gadolinite of the type used by Urbain. If the decision is made in favor of Coster and Hevesy it will be necessary for them to explain why it is that No. 72 is never found in rare earth minerals while its two closest relatives, zirconium and thorium, are very common stray components of rare earth minerals.

However this controversy is settled, certain it is that the Periodic Table is simplified by a recognition of the fact that No. 72 is a member of Group IV and closely related to zir-

Periodic Table (1926)

0	I	II	III	IV	V	VI	VII	VIII		
	H 1 1.008									
He 2 4.00	Li 3 6.94	Be 4 9.1	B 5 10.85	C 6 12.005	7 N 14.008	8 O 16.	9 F 19.0			
Ne 10 20.2	Na 11 23.00	Mg 12 24.32	Al 13 26.97	Si 14 28.1	15 P 31.04	16 S 32.06	17 Cl 35.46			
A 18 39.9	K 19 39.10	Ca 20 40.07	Sc 21 45.1	Ti 22 48.1	V 23 51.0	Cr 24 52.0	Mn 25 54.93	Fe 26 55.84	Co 27 58.97	Ni 28 58.68
	29 Cu 63.57	30 Zn 65.37	31 Ga 70.1	32 Ge 72.5	33 As 74.96	34 Se 79.2	35 Br 79.92			
Kr 36 82.92	Rb 37 85.45	Sr 38 87.63	Y 39 89.	Zr 40 90.6	Cb 41 93.1	Mo 42 96.0	Ma 43	Ru 44 101.7	Rh 45 102.9	Pd 46 106.7
	47 Ag 107.88	48 Cd 112.40	49 In 114.8	50 Sn 118.7	51 Sb 121.8	52 Te 127.5	53 I 126.92			
Xe 54 130.2	Cs 55 132.81	Ba 56 137.37	57-71 Rare Earths	Ct or Hf 72 178.6	Ta 73 181.5	W 74 184.0	Re 75	Os 76 190.9	Ir 77 193.1	Pt 78 195.2
	79 Au 197.2	80 Hg 200.6	81 Tl 204.0	82 Pb 207.2	83 Bi 209.0	84 Po 210.	85			
Rn 86 222.	87	Ra 88 226	Ac 89	Th 90 232.12	Pa 91	U 92 238.2				

57–71 Rare Earths	La 57 139.0	Ce 58 140.25	Pr 59 140.9	Nd 60 144.3	Il 61 146. (?)	Sm 62 150.4	Eu 63 152.0
Gd 64 157.3	Tb 65 159.2	Dy 66 162.5	Ho 67 163.5	Er 68 167.7	Tm 69 168.5	Yb 70 173.5	Lu 71 175.0

FIG. 4.

conium and thorium (Fig. 4). By putting all members of the rare earth group from lanthanum to lutecium in one space, an arrangement is obtained which is as simple as could be expected.

Numerous efforts have been made to reveal the presence of eka-manganese, in fact the discovery of this element, in confirmation of the predictions of Mendeléeff, was announced at least as early as 1908. This announcement has failed of confirmation and element No. 43 has continued to baffle the scientists of the world, until its discovery in Berlin early in 1925, by Noddack and Tacke.

In studying the probable occurrence of elements Nos. 43 and 75, Noddack and Tacke reasoned[4] that these two elements should be found together since nowhere in the Periodic Table do we have closer resemblances than between the corresponding members of Series 5 and 6. As a consequence elements 43 and 75 should be found in the same minerals. Moreover these minerals need not necessarily be manganese ores since the elements in Series 4 are somewhat distinct from their relatives in Series 5. It was also argued that we should expect elements 43 and 75 to be found both in oxygen-containing ores like those of molybdenum, tungsten, columbium and tantalum and also in the free state like the platinum metals. Noddack and Tacke reasoned extensively concerning the chemical properties which should distinguish elements 43 and 75 from their neighbors. In accordance with the conclusions reached in this way, several likely minerals were brought into solution by fusion, then subjected to hydrogen sulfide treatment in such a manner as to separate 43 and 75 if they had the properties predicted for them. In this manner a small quantity of material was obtained from columbite which upon examination by X-ray spectrum analysis revealed the α_1, α_2 and β_1 lines of the K series of element 43, and α_1, α_2 and β_1 of the L series of element 75. In similar manner various platinum ores were heated alternately in hydrogen and in oxygen when a small sublimate was obtained whose behavior agreed with that predicted for elements 43 and 75. Although the amounts obtained are exceedingly small, the evidence seems good and thus we are permited to add * to the table the element masurium with atomic number 43 and the element rhenium, No. 75.

Loring and Druce,[5] working in their London laboratory, have examined critically some manganese salts of commerce, and samples of commercial pyrolusite. From solutions prepared from

* In Nature, 118, 262 (1926), Professor Zvjaginstsev reports that work at the Russian Platinum Institute with large amounts of platinum residues fails to confirm the conclusions of Noddack and Tack so far as the presence with platinum is concerned. The effect of this announcement is to cast doubt upon the statement that Nos. 43 and 75 have been located.

such material, manganese was removed by hydrogen sulfide and calcium was precipitated as the oxalate. From the filtrate there was precipitated by means of ammonium hydroxide a white flocculent substance which darkened on standing. When this material was subjected to X-ray examination in the Hilger laboratory, lines were obtained which correspond nicely with the positions calculated for $L\alpha_1$ and $L\beta_1$ of element No. 75. The amount of this element varies in the different sources, but it is estimated that "pure" manganese sulfate contains about 1 per cent of dwi-manganese, some of whose chemical properties have been determined. These results have been confirmed by workers in the chemical laboratory of Charles' University, Prague, who have suggested that the name dwi-manganese be retained. Thus element No. 75 seems to be definitely established, since its presence has been shown by investigators in three laboratories.

By similar methods Loring and Druce have examined exhaustively the residues obtained from pyrolusite and found some evidence for belief that they contain traces of eka-iodine (No. 85), eka-caesium (No. 87) and the element next beyond uranium which would have an atomic number 93. The evidence is not completely convincing and the authors suggest that it must be interpreted as foreshadowing the existence of these three strangers in such material. Additional work is promised to assist in clearing up the questions which naturally arise concerning these elements.

Much more than usual interest has attached to the possible existence of element No. 87 which is known conveniently as eka-caesium. As long ago as 1902 Professor T. W. Richards and Dr. E. H. Archibald, at Harvard, searched [6] diligently for eka-caesium in the mineral pollucite. In 1915 Professor G. P. Baxter continued [7] this search in the same institution and in 1919 Professor L. M. Dennis and Dr. R. W. G. Wyckoff, of Cornell University, attempted [8] to find evidence of the presence of an alkali element whose atomic weight is greater than that of caesium. All of these efforts failed to locate the missing element but this group of investigations stands out as a brilliant achievement, planned with unusual scientific skill and carried out with rare devotion and accuracy. Dr. Archibald is continuing the search with heroic devotion in spite of discouragements and it is to be sincerely hoped that his work will be rewarded by the discovery of this elusive stranger.

If eka-caesium exists it must be radioactive and the recent investigations of the peculiar radioactive properties of the alkali family only serve to revive our interest in this element. Professor Harkins, of Chicago, reminds [9] us that the elements beyond bismuth which have odd atomic numbers are short-lived. It seems reasonable to conclude that both 85 and 87 may be elements with a very brief half-life period, or they may not exist at all. But to those who pin their faith in the Periodic Table there remains a hope that this last and most trying test for Mendeléeff's extraordinary generalization may result in a completed table of 92 or more elementary forms of matter.

The most recent announcement of the isolation of a chemical element came in March 1926 when the University of Illinois announced[10] the discovery of the element illinium as No. 61, the missing member of the rare earth group. The discovery resulted from the work on this puzzling group of elements which had been in progress in the laboratory for about 15 years, during five of which there was a definite search being made for the missing element.

Early in 1919 there was arranged a cooperative plan between the Department of Chemistry of the University of Illinois and the U. S. Bureau of Standards for the mapping of the arc spectrum lines of certain rare elements, especially the members of the rare earth group. The University furnished samples from its collection and the Bureau made precision measurements of the lines in the red and infra-red regions of the spectrum. Under date of December 14, 1921, Dr. C. C. Kiess, of the Bureau reported as follows:

"I have about completed the work on neodymium and samarium. Both samples submitted by you seem to be very pure, with one exception. In the spectrum of each appear numerous faint lines of identical wave-length. So far I have been unable to identify them with the lines of known elements. It would be a hasty conclusion to announce that they belong to a new element but this idea should not be laid aside in view of the fact that the present accepted scheme of the periodic table calls for an element (as yet unknown) of the order number 61 between neodymium and samarium."

Following this suggestion, work was begun upon the concentration and identification of the substance which produced the unknown lines. Dr. L. F. Yntema, working as Fellow of the National Research Council, carried out X-ray examination in the Sloane Laboratory of Physics at Yale, while fractionations were being continued in Urbana. The result of two years' work was the conclusion[11] that if No. 61 were present at all, it must be in quantities of less than one part in a thousand.

It was evident that some method of obtaining a greater concentration must be found before the identity of the element could be established. Previous attempts at concentration had been made by the fractional crystallization of the double magnesium nitrate. In such a series the elements arrange themselves in the order of increasing atomic numbers, which would lead us to expect No. 61 to crystallize between neodymium and samarium. Since both of these elements are relatively abundant, it would not be surprising to find these more common elements completely obscuring an intervening element if it were present in very small proportions. Consequently it seemed advisable to run a series using some other salt. Professor James has shown [12] that the solubility of the bromates decreases (Fig. 5) from neodymium to samarium, reaches a minimum in europium and then increases

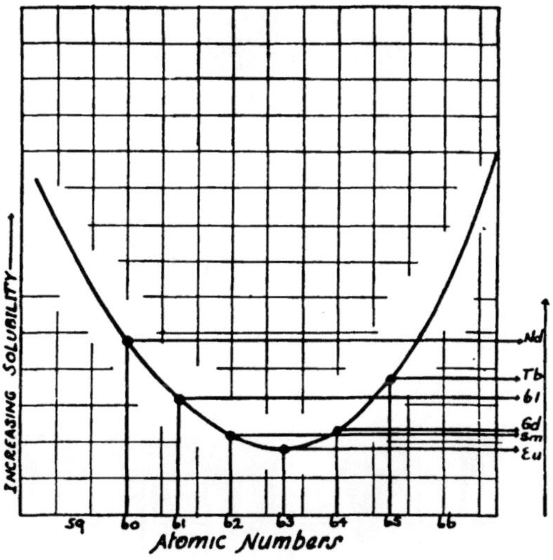

FIG. 5.—Solubility curve for some rare earth bromates. The order of increasing solubility is shown at the right.

again in gadolinium and terbium. As a result these elements when fractionated in a bromate series do not arrange themselves in the order of atomic numbers, but europium appears at the least soluble end while neodymium is well out toward the more soluble end with terbium, gadolinium and samarium distributed between.

When Dr. J. A. Harris fractionated some of our material as the bromate it was found that samarium was crystallizing out at the less soluble end, while neodymium was concentrating in the more soluble fractions. As the fractionation continued the characteristic absorption bands of both neodymium and samarium began to grow less evident in the middle fractions. Soon there was noticed at wave length 5816 A° a faint line which was growing stronger as the fractionation proceeded while the characteristic absorption lines of both neodymium and samarium were growing weaker. Later another line was detected at 5123 A° whose intensity increased as the familiar absorption lines of neodymium and samarium were becoming more faint. These lines evidently belonged to a new element whose presence had been concealed by the broad and persistent absorption bands of neodymium and samarium (Fig. 6). This unknown element we believe to be element No. 61 since our theories lead us to expect that such an element should exist at this point. When a bromate series is examined europium and samarium appear in the least

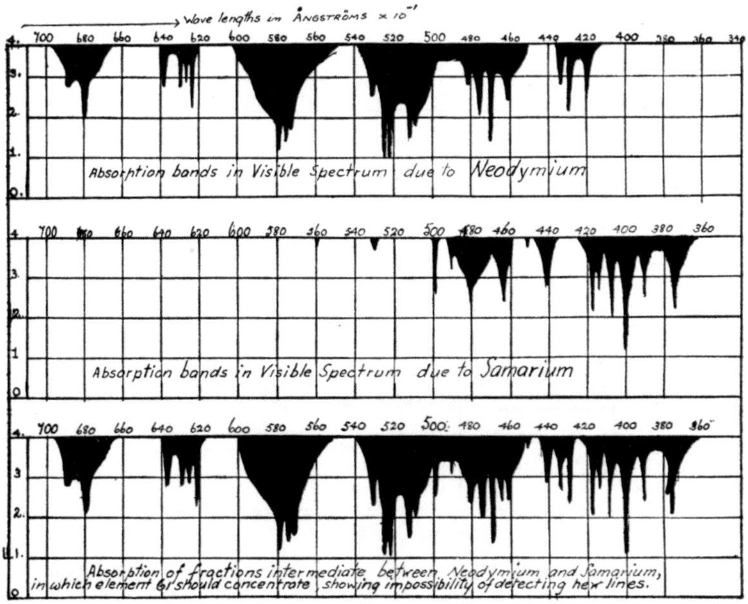

Fig. 6.—Absorption bands of neodymium and samarium, showing how the bands broaden as concentration of these elements increases.

soluble fractions followed in order by gadolinium, No. 61, terbium and neodymium. This order is especially fortunate for the detection of element No. 61 by means of absorption spectra because it is separated from samarium by gadolinium and from neodymium by terbium. Since the absorption bands of gadolinium fall in the invisible portion of the spectrum and terbium has only a single narrow line in the visible spectrum this leaves almost an uninterrupted field for the detection of any new absorption lines which might reveal the presence of a new element. As long as No. 61 remained associated with neodymium and samarium, its detection by absorption spectra was impossible.

But as soon as it was sufficiently separated from these elements, the new absorption lines were quickly made out.

When the most prominent absorption lines of the elements of the rare earth group are plotted according to their wave lengths [13] there is observed a drift (Fig. 7) in their positions which must be significant in dealing with atomic structures.

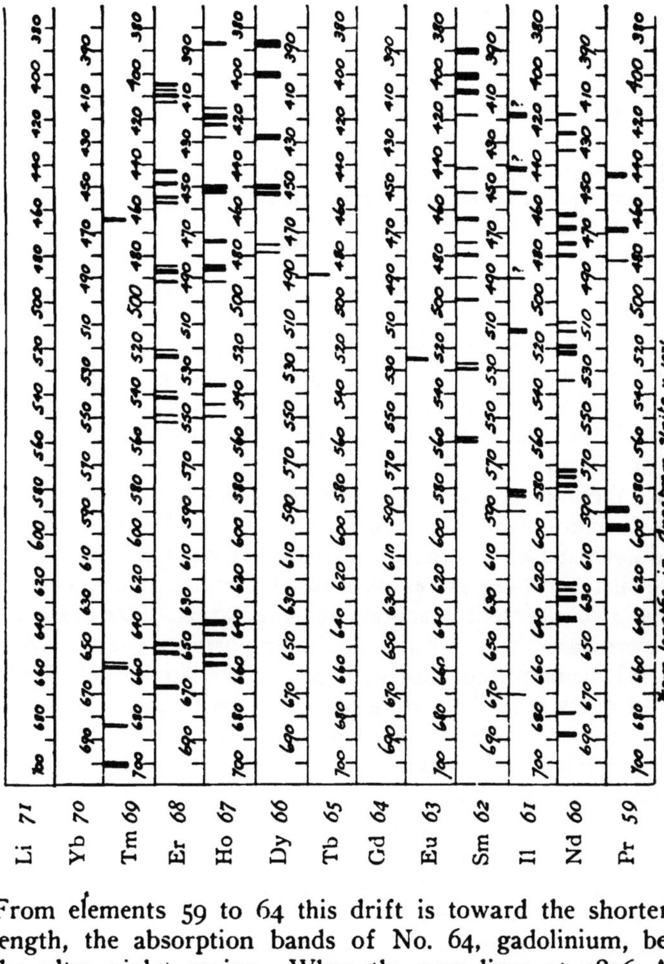

FIG. 7.—Most prominent bands in the absorption spectra of the rare earth group.

From elements 59 to 64 this drift is toward the shorter wave length, the absorption bands of No. 64, gadolinium, being in the ultra violet region. When the new lines at 5816 A° and 5123 A° are assigned to element No. 61, these lines fall into perfect accord with the drift in the lines of the neighboring elements.

In these days the identification of a new element is never complete without adequate measurements of its X-ray spectral lines and the determination of its atomic number. Accordingly there was designed and built in our laboratory a glass X-ray tube equipment with water-cooled electrodes. The cathode was made of pure aluminium and the target of copper, the face being a plate with a depression in the center for holding the fused rare earth oxide. The usual voltage was about 30,000 with an average current of 8–10 milliamperes. The equipment could not be used with safety above 40,000 volts, consequently we were unable to study the K emission spectra but were limited to the L series. We found it necessary to use exposures of 8–16 hours and considerable difficulty was experienced in keeping our rare earth target in place during such a long exposure. By fusing the oxide into the depression on the plate which formed the face of the target, this difficulty was effectively overcome. Because of the skillful cooperation between our mechanician and our glassblower, we were able to obtain a perfect focus upon the rare earth button on the target, so that it was the only source of X-radiation. We were unable to get the copper lines on our plate for reference.

The crystal used in our work was calcite, three different crystals being used in order to avoid the effect of invisible defects. The rotation of the crystal was limited to 30 minutes because of the low energy input which we were compelled to use. The dispersion between $L\alpha_1$ and $L\alpha_2$ of samarium was 0.5 mm., the former line being our principal line of reference, but the deviation from the theoretical value of $L\alpha_1$ of 61 being checked against $L\alpha_2$ of samarium or $L\alpha_1$ of neodymium.

Five successful exposures were made for $L\alpha_1$ of element No. 61, using two different fractions for the target. The lines obtained were faint, but in each case their positions were verified by several disinterested observers before the final reading was made. In every case the setting was made with the utmost care and after the reading had once been taken no re-reading of the line was permitted. The maximum deviation from the theoretical wave length was less than 0.005 A°, the average value for the five determniations differing by 0.0004 A° from the Siegbahn calculated value. These deviations are well within the limits of error of measurement as shown by checking against the lines of samarium and neodymium.

A single determination of $L\beta_1$ of element 61 varies from the calculated value by 0.004A°. On our last plate there was a very faint line corresponding to $L\beta_3$ of No. 61, but since this was at the edge of the area of general radiation covered by the oscillation, it is not safe to accept it without confirmation. Other measurements of the lines of both the L series and of the K series will be made as soon as possible.

In order to make sure that the lines which we have measured are due to element No. 61 and not to some related element from our rare earth material, we have studied carefully all lines of the first, second, and third orders which are so situated that they might be mistaken for the lines of the missing element. Table I shows these lines, with the wave lengths, the elements and the series to which they are due. In addition a study of the relative intensities of these lines shows that about the only possibilities of mistaken identity would be in the case of chromium, cerium,

TABLE I

LINES POSSIBLE IN THIS REGION OTHER THAN ELEMENT 61

Wave Length	Element	Series		Wave Length	Element	Series
2.270	La	$L\beta_7$				
2.273	Cr	$K\alpha^3$				
2.277	Ce	$L\beta_6$	Lines of	2.067	Cr	$K\beta_2$
2.277	La	$L\beta_9$	first order	2.072	Ba	$L\gamma_4$
2.280	Va	$K\beta_1$	in	2.080	Cr	$K\beta_1$
2.285	Cr	$K\alpha_1$	this region	2.082	Ho	$L\rho$
2.285	La	$L\beta_{10}$				
2.285	Va	$K\beta'$				
				1.030	Hg	$L\beta_2$
1.129	W	$L\gamma_5$		1.037	Tl	$L\beta_4$
1.133	Ir	$L\beta$	Second order	1.038	Hg	$L\beta_2$
1.135	Ta	$L\gamma_1$	lines	1.038	As	$K\beta_2$
1.138	Ir	$L\beta_3$	possible	1.038	Br	$K\alpha_1$
1.140	Pt	$L\beta_4$		1.038	Au	$L\beta_5$
				1.042	Br	$K\alpha_2$
0.745	U	$L\beta_4$		0.664	Cb	$K\beta_1$
0.749	Cb	$K\alpha^2$	Third order	0.688	Zr	$K\beta_2$
0.752	Th	$L\beta_3$		0.700	Zr	$K\beta_1$
2.277	61	$L\alpha_1$		2.073	61	$L\beta_1$

lanthanum, mercury and bromine. Of these possibilities chromium could scarcely be present because no compounds of this element were used in our work and chromium is not a common constituent of rare earth minerals. Both cerium and lanthanum were present in the minerals from which our materials were extracted but neither of these elements could have been present in our final samples because these were subjected to hundreds of fractionations after the last trace of the prominent praseodymium lines had disappeared and both cerium and lanthanum are farther removed from our materials than is praseodymium. The presence of mercury in our X-ray tube is scarcely possible because of the thorough evacuation and the presence at all times of a liquid air trap between the pump and the tube. The presence of silver bromide on our photographic plate makes the presence of bromine absorption line possible, and on our last plate we had a rather sharp edge of radiation close to the $L\beta_1$ line of 61 which might possibly be due to the second order of the K absorption limit for bromine. The fact that both lines were distinguishable makes it highly improbable that our line was confused with the bromine line.

The evidence for the existence of a new element is based on three independent facts:

1. The presence in our neodymium-samarium fractions of arc spectrum lines of unknown origin: of these the wave lengths of 135 have already been measured.

2. The development of new absorption bands in the fractions intermediate between neodymium and samarium; these bands become stronger as the familiar characteristic bands of neodymium and samarium are reduced in intensity through fractionation. Bands at 5816 A° and 5123 A° are evidently due to an intermediate element and the position of these bands confirms the conclusion that there is a regular shift in the absorption bands of the adjacent members of the rare earth group.

3. The identification in the X-ray emission spectra of lines corresponding closely with the theoretical positions for $L\alpha_1$ and $L\beta_1$ of element No. 61.

For the new element we propose the name Illinium, with symbol Il and an estimated atomic weight of approximately 146.

REFERENCES

1 Mendeléeff, Annalen Suppl., **8**, 151 (1872).
2 G. Urbain, Chem. and Ind. Rev., **42**, 764 (1923).
3 G. Hevesy, Chem. Rev., **2**, 1 (1925).
4 Walter Noddack and Ida Tacke, Naturwiss., **26**, 567 (1925).
5 Loring and Druce, Nature, **116**, 782 (1925); Jan. 30, 1926, p. 159, Chem. News, **131**, 273, 337 (1925).
6 Richards and Archibald, Proc. Am. Acad., **38**, 449 (1903).
7 Gregory P. Baxter, J. Am. Chem. Soc., **37**, 286 (1915).
8 Dennis and Wycoff, J. Am. Chem. Soc., **42**, 985 (1920).
9 W. D. Harkins, J. Am. Chem. Soc., **42**, 985 (1920).
10 Harris, Yntema and Hopkins, J. Am. Chem. Soc., **48**, 1585, 1594 (1926).
11 L. F. Yntema, J. Am. Chem. Soc., **46**, 37 (1924).
12 C. James, J. Am. Chem. Soc., **36**, 2060 (1914).
13 L. F. Yntema, J. Am. Chem. Soc., **48**, 1598 (1926); Harris and Hopkins, ibid., **48**, 1593 (1926).

Bei Fragen zur Produktsicherheit wenden Sie sich bitte an:
If you have any questions regarding product safety,
please contact:

Walter de Gruyter GmbH
Genthiner Straße 13
10785 Berlin
productsafety@degruyterbrill.com